2003 International Residential Code®
Study Companion

ICC
INTERNATIONAL
CODE COUNCIL®

Craftsman

2003 International Residential Code
Study Companion

Publication Date: January 2004
First printing
Second printing: July 2004
ISBN 1-892395-98-3

Project Editor:	Roger Mensink
Illustrator:	Mike Tamai
Layout Design:	Alberto Herrera
Cover Design:	Lisa Jachymiak
Publications Manager:	Mary Lou Luif

INTERNATIONAL
CODE COUNCIL®

TABLE OF CONTENTS

STUDY SESSION   **PAGE**

Study Session 1: Chapters 1 and 43 — Administration and Referenced Standards 1
  Quiz 15

Study Session 2: Sections R301 and R302 — Building Planning I 21
  Quiz 36

Study Session 3: Sections R303 through R310 — Building Planning II 41
  Quiz 55

Study Session 4: Sections R311 through R323 — Building Planning III 61
  Quiz 78

Study Session 5: Chapter 4 — Foundations 83
  Quiz 98

Study Session 6: Chapter 5 — Floors 103
  Quiz 118

Study Session 7: Chapters 6 and 7 — Wall Construction and Wall Covering 123
  Quiz 138

**Study Session 8: Chapters 8 and 9 — Roof/Ceiling Construction and
Roof Assemblies** 143
  Quiz 159

Study Session 9: Chapters 10 and 11 — Chimneys/Fireplaces and Energy Efficiency . .165
  Quiz 180

**Study Session 10: Chapters 13 and 14 — General Mechanical System Requirements
and Heating and Cooling Equipment** 185
  Quiz 198

**Study Session 11: Chapters 15, 16 and 19 — Exhaust Systems, Duct
Systems and Special Fuel-Burning Equipment** 205
  Quiz 218

**Study Session 12: Chapters 17, 18, 20, 21, 22 and 23 — Combustion Air, Chimneys
and Vents, Boilers/Water Heaters, Hydronic Piping, Special Piping and Storage
Systems, and Solar Systems** 225
  Quiz 238

Study Session 13: Chapter 24 — Fuel Gas 245
  Quiz 258

**Study Session 14: Chapters 25, 26 and 27 — General Plumbing Requirements and
Plumbing Fixtures** 265
  Quiz 279

**Study Session 15: Chapters 28, 29 and 30 — Water Heaters, Water Supply and
Distribution, and Sanitary Drainage** 285

Quiz 298

Study Session 16: Chapters 31 and 32 — Vents and Traps 303
Quiz 317

**Study Session 17: Chapters 33, 34, 35 and 36 — General Electrical Requirements,
Definitions, Services, and Branch Circuits and Feeder Requirements** 323
Quiz 336

**Study Session 18: Chapters 37, 38, 39, 40 and 41 — Wiring Methods, Power and
Lighting Distribution, Devices and Lighting Fixtures, Appliance Installation and
Swimming Pools** 341
Quiz 357

Answer Keys 363

INTRODUCTION

This study companion provides practical learning assignments for independent study of the provisions of the 2003 *International Residential Code*® (IRC®). The independent study format affords a method for the student to complete the study program in an unlimited amount of time. Progressing through the workbook, the learner can measure his or her level of knowledge by using the exercises and quizzes provided for each study session.

The workbook is also valuable for instructor led programs. In jurisdictional training sessions, community college classes, vocational training programs and other structured educational offerings, the study guide and the IRC can be the basis for code instruction.

All study sessions begin with a general learning objective, the specific sections or chapters of the code under consideration and a list of questions summarizing the key points of study. Each session addresses selected topics from the IRC and includes code text, a commentary on the code provisions and illustrations representing the provisions under discussion. Exercises and quizzes are provided at the end of each study session. Before beginning the exercises and quizzes, the student should thoroughly review the referenced IRC provisions, particularly the key points.

The workbook is structured so that after every question, the student has an opportunity to record his or her response and the corresponding code references. The correct answers are indicated in the back of the book in the answer key.

This study companion was developed by Douglas W. Thornburg, AIA, C.B.O., for the 2000 *International Residential Code*. In this publication, he has updated and revised the material based on the 2003 *International Residential Code*. Mr. Thornburg is currently a code consultant and educator on both the IBC and IRC. Formerly Vice-President of Education for the International Conference of Building Officials, Mr. Thornburg presents building code seminars nationally and has developed numerous educational texts and resource materials. He was presented with ICBO's prestigious A. J. (Jack) Lund Award in 1996.

Questions or comments concerning this workbook are encouraged. Please direct your comments to ICC Publications Department, 4051 W. Flossmoor Rd, Country Club Hills, IL 60478-5795 (Phone 708-799-2300).

INTERNATIONAL RESIDENTIAL CODE
Study Session 1
Chapters 1 and 43—Administration and Referenced Standards

OBJECTIVE: To obtain an understanding of the administrative provisions of the *International Residential Code*, including the scope and purpose of the code, duties of the building official, issuance of permits, submission of construction documents, inspection procedures, certificate of occupancy and referenced standards.

CODE REFERENCE: Chapters 1 and 43, 2003 *International Residential Code*

KEY POINTS:
- What is the purpose and scope of the *International Residential Code*?
- How does the *International Existing Building Code*® relate to the IRC?
- When materials, methods of construction or other requirements are specified differently in separate provisions, which requirement shall govern?
- When there is a conflict between a general requirement and a specific requirement, which provision shall be applicable?
- When do the provisions of the appendix apply?
- How are existing buildings to be addressed?
- What term is applied to the building department? How are deputies appointed?
- What are the powers and duties of the building official in regard to the application and interpretation of the code? In regard to right of entry?
- What degree of liability does the building official have in regard to performance of their duties?
- Under what conditions may the building official grant modifications to the code?
- How may alternative materials, designs and methods of construction be approved?
- When is a permit required? What types of work are exempted from permits?
- Is work exempted from a permit required to comply with the provisions of the code?
- What is the process outlined for obtaining a permit?
- What information is required on the permit application?
- What conditions or circumstances would bring the validity of a permit into question?
- When does a permit expire? What must occur when a permit expires prior to completion of a building?
- When are construction documents required? When must plans be prepared by a registered design professional?
- How long must approved construction documents be retained by the building department?
- How are temporary buildings addressed? To what levels of conformance must they comply?
- What fees are set forth in the code? How are fees to be determined?
- What types of inspections are specifically required by the code? When are inspections required?
- When is a certificate of occupancy required? For what reasons is revocation permitted?
- When may a temporary certificate of occupancy be issued?
- What is the purpose of a board of appeals? Who shall serve on the board?
- What limitations are placed on the authority of the board of appeals?
- When should a stop-work order be issued?
- What are referenced standards? How are they applied to the IRC?

Topic: Scope

Reference: IRC R101.2

Category: Administration

Subject: General Requirements

Code Text: *The provisions of the* International Residential Code for One- and Two-Family Dwellings *shall apply to the construction, alteration, movement, enlargement, replacement, repair, equipment, use and occupancy, location, removal and demolition of detached one- and two-family dwellings and multiple single-family dwellings (townhouses) not more than three stories in height with a separate means of egress and their accessory structures.*

Discussion and Commentary: The *International Residential Code* is intended to regulate the broad spectrum of construction activities associated with residential buildings and structures. The provisions address building planning and construction aspects, as well as energy efficiency, mechanical, plumbing and electrical aspects.

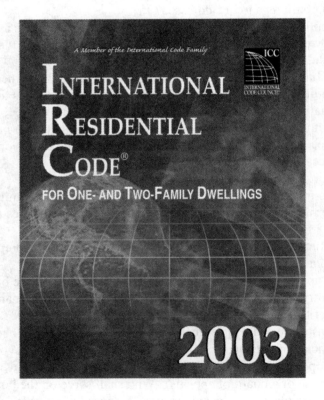

Townhouses regulated under the IRC are not only limited to three stories and provided with individual egress facilities, they must also be designed with open space to the exterior on at least two sides. Each dwelling unit must extend from the foundation to the roof as a single unit.

Topic: Dwelling Units
Reference: IRC R101.2, Exception

Category: Administration
Subject: General Requirements

Code Text: *Detached one- and two-family dwellings and multiple single family dwellings (townhouses) not more than three stories high with a separate means of egress and their accessory structures shall comply with the* International Residential Code.

Discussion and Commentary: Many residential structures are exempt from the requirements of the *International Building Code* and are regulated instead by the *International Residential Code*, a separate and distinct document. The IRC contains prescriptive requirements for the construction of detached single-family dwellings, detached duplexes, townhouses and all structures accessory to such buildings. Limited to three stories with individual egress facilities, these residential buildings are fully regulated by the IRC for building, plumbing, mechanical, electrical and energy provisions.

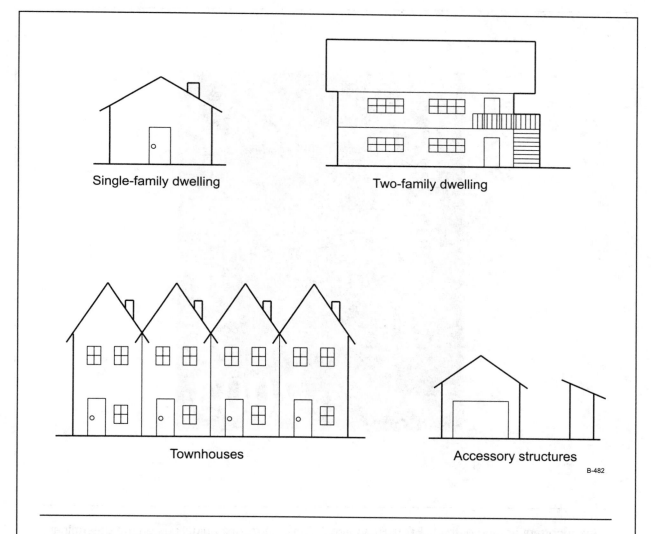

Single-family dwelling

Two-family dwelling

Townhouses

Accessory structures

B-482

In addition to the general requirements, a townhouse is also limited by definition in the IRC. To fall under the scope of the *International Residential Code*, each townhouse must be a single unit from the foundation to the roof, with at least two sides having open space to the exterior.

Topic: Existing buildings

Reference: IRC 101.2, Exception

Category: Administration

Subject: General Requirements

Code Text: *Existing buildings undergoing repair, alterations or additions, and change of occupancy shall be permitted to comply with the* International Existing Building Code®.

Discussion and Commentary: As an alternative to IRC Section R102.7 for the regulation of work or an occupancy change in an existing building, the provisions of the *International Existing Building Code* (IEBC)® may be utilized. The IEBC establishes minimum regulations for existing buildings using prescriptive and performance-related provisions. It is founded on broad-based principles intended to encourage the use and reuse of existing buildings while requiring reasonable upgrades and improvements. The IEBC is fully compatible with all of the *International Codes*® published by the International Code Council®. The scope of the code addresses various classifications of work including repair, three levels of alterations, change of occupancy, additions, historic buildings and relocated structures.

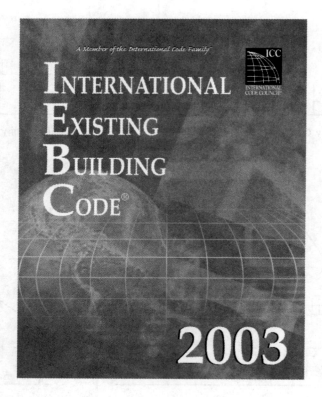

An important portion of the IEBC is Resource A, which provides guidelines on the determination of fire ratings for archaic materials and assemblies. In addition, IRC Appendix Chapter J, when adopted, addresses a more comprehensive approach to existing buildings than Section R102.7.

Topic: Application   **Category:** Administration
Reference: IRC R102.5   **Subject:** Appendices

Code Text: *Provisions in the appendices shall not apply unless specifically referenced in the adopting ordinance.*

Discussion and Commentary: The appendix chapters of the IRC address subjects that are inappropriate as a mandatory portion of the code. Rather, the appendices are optional, with each jurisdiction adopting all, some or none of the appendix chapters, depending on its needs for enforcement in any given area. There are various reasons why certain issues are placed in the appendix. Often, the provisions are limited in application or interest. Some appendix chapters are merely extensions of requirements set forth in the body of the code. Others address issues that are often thought of as outside of the scope of a traditional building code. For whatever reason, no appendix chapter is applicable unless specifically adopted.

IRC Appendix Chapters

Appendix A	Sizing and Capacities of Gas Piping
Appendix B	Sizing of Venting Systems Serving Appliances Equipped with Draft Hoods, Category I Appliances, and Appliances Listed for Use and Type B Vents
Appendix C	Exit Terminals of Mechanical Draft and Direct-Vent Venting Systems
Appendix D	Recommended Procedure for Safety Inspection of an Existing Appliance Installation
Appendix E	Manufactured Housing Used as Dwellings
Appendix F	Radon Control Methods
Appendix G	Swimming Pools, Spas and Hot Tubs
Appendix H	Patio Covers
Appendix I	Private Sewage Disposal
Appendix J	Existing Buildings and Structures
Appendix K	Sound Transmission
Appendix L	International Residential Code Electrical Provisions/National Electrical Code Cross Reference

---

Appendix chapters not adopted as a portion of a jurisdiction's building code may still be of value in application of the code. Provisions in the appendices might provide some degree of assistance in evaluating proposed alternative designs, methods or materials of construction.

Topic: General Requirements
Reference: IRC R104.1

Category: Administration
Subject: Duties and Powers of Building Official

Code Text: *The building official is hereby authorized and directed to enforce the provisions of the IRC. The building official shall have the authority to render interpretations of the IRC and to adopt policies and procedures in order to clarify the application of its provisions. Such interpretations, policies and procedures shall be in conformance with the intent and purpose of the IRC. Such policies and procedures shall not have the effect of waiving requirements specifically provided for in the IRC.*

Discussion and Commentary: The building official must be knowledgeable to the extent that he or she can rule on those issues that are not directly addressed or are unclear in the code. The basis for such a determination is the intent and purpose of the *International Residential Code*, which often takes some research to discover.

City of (Jurisdiction)

Department of Building Safety

_____
Name of individual

_____
Job function

The individual identified on the badge is a duly authorized employee
of the (Jurisdiction) and is a designated representative of the Department of Building Safety.

Valid
through   _____   _____
                   Date   Building Official

Proper Identification Mandated by Section R104.5 when Inspecting Structures in Premises

---

Although the IRC gives broad authority to the building official in interpreting the code, this authority also comes with great responsibility. The building official must restrict all decisions to the intent and purpose of the code; the waiving of any requirements is strictly prohibited.

Topic: Alternative Materials   **Category:** Administration
Reference: IRC R104.11   **Subject:** Duties and Powers of Building Official

Code Text: *The provisions of the IRC are not intended to prevent the installation of any material or to prohibit any design or method of construction not specifically prescribed by the IRC, provided than any such alternative has been approved.*

Discussion and Commentary: The building official is granted broad authority in the acceptance of alternative materials, designs and methods of construction. Arguably the most important provision of the IRC, the intent is to implement the adoption of new technologies. Furthermore, it gives the code even more of the character of a performance code. The provisions encourage state-of-the-art concepts in construction, design and materials as long as they meet the performance level intended by the IRC. In a portion of the provision not quoted above, reference is also made to the use of any performance-based provisions of the other *International Codes* as acceptable alternates.

Building official may approve alternative materials, design and methods of construction, if:

- Proposed alternative is satisfactory
- Proposed alternative complies with intent of code
- Material, method or work is equivalent in:
  (1) Quality
  (2) Strength
  (3) Effectiveness
  (4) Fire Resistance
  (5) Durability
  (6) Safety

Advisable for building official to:

- Require sufficient evidence or proof
- Record any action granting approval
- Enter information into files

The building official should ensure that any necessary substantiating data or other evidence be submitted to show that the alternative is in fact equivalent in performance. Moreover, where tests are performed, reports of such tests must be retained by the building official.

Topic: Submittal Documents

Reference: IRC R106

Category: Administration

Subject: Construction Documents

Code Text: *Construction documents, special inspection and structural observation programs, and other data shall be submitted in one or more sets with each application for a permit.* See exception that authorizes building official to waive the submission of construction documents. *Manufacturer's installation instructions, as required by* the IRC, *shall be available on the job site at the time of inspection. The construction documents submitted with the application for permit shall be accompanied by a site plan showing the size and location of new construction and existing structures on the site and distances from lot lines.*

Discussion and Commentary: After the construction documents have been submitted, they are to be reviewed by the building department for compliance with the adopted codes and other jurisdictional requirements. Once the construction documents have been approved, a permit is to be issued.

As a part of the necessary recordkeeping process, the retention of at least one set of approved construction documents is required. Such documents must be retained for a minimum of 180 days after the completion date of the work, or as otherwise mandated by state or local laws.

Topic: Permits Required and Exempted

Reference: IRC R105.1, R105.2

Category: Administration

Subject: Permits

Code Text: *Any owner or authorized agent who intends to construct, enlarge, alter, repair, move, demolish, or change the occupancy of a building or structure . . . shall first make application to the building official and obtain the required permit.* See multiple exceptions where a permit is not required. *Exemption from permit requirements of the IRC shall not be deemed to grant authorization for any work to be done in any manner in violation of the provisions of the IRC or any other laws or ordinances of the jurisdiction.*

Discussion and Commentary: Except in those few instances specifically listed, such as minor repairs, maintenance, finish work and limited equipment replacement, all construction-related work requires a permit and is subject to subsequent inspections.

Work exempt from building permit:

- One-story detached accessory buildings limited to 200 square feet
- Fences not over 6-feet high
- Retaining walls not over 4-feet high
- Water tanks supported directly on grade, if the capacity does not exceed 5,000 gallons and the ratio of height to diameter does not exceed 2 to 1
- Sidewalks and driveways not over 30 inches above grade
- Painting, papering, tiling, carpeting, cabinets, counter tops and similar finish work
- Prefabricated swimming pools less than 24-inches deep
- Swings and other playground equipment
- Window awnings supported by an exterior wall that do not project more than 54 inches from the exterior wall

---

Whether or not a building permit is required by the code, all work must be done in accordance with the code requirements. It is important that the owner be responsible for proper and safe construction.

Topic: Required Inspections

Reference: IRC R109.1

Category: Administration

Subject: Inspections

Code Text: *For onsite construction, from time to time the building official, upon notification from the permit holder or his agent, shall make or cause to be made any necessary inspections and shall either approve that portion of the construction as completed or shall notify the permit holder or his or her agent wherein the same fails to comply with* the IRC.

Discussion and Commentary: The inspection function is possibly the most critical activity in the entire code enforcement process. At the various stages of construction, an inspector often provides the final check of the building for safety-related compliance. If necessary, the building official may accept inspection reports from approved agencies verifying compliance with the applicable provisions of the code. It is important that such agencies be highly qualified and reliable.

Required inspections (where applicable):

- Foundation
- Plumbing, mechanical, gas and electrical
- Floodplain
- Frame and masonry
- Fire-resistance-rated construction
- Others as required by the building official
- Final

The permit holder or authorized agent must notify the building official that the work is ready for inspection. Furthermore, the responsibility to make the work accessible and available for inspection rests with the person requesting the inspection.

Topic: Approval Required
Reference: IRC R109.4

Category: Administration
Subject: Inspections

Code Text: *Work shall not be done beyond the point indicated in each successive inspection without first obtaining the approval of the building official. The building official, upon notification, shall make the requested inspections and shall either indicate the portion of the construction that is satisfactory as completed, or shall notify the permit holder or an agent of the permit holder wherein the same fails to comply with the IRC. Any portions that do not comply shall be corrected and such portion shall not be covered or concealed until authorized by the building official.*

Discussion and Commentary: Each successive inspection must be approved prior to further work. This practice helps to control the concealment of any work that must be inspected, which would result in the removal of materials that block access.

The building official must establish a consistent and convenient procedure for notification of the appropriate persons when an inspection approval is not granted. The notification method should encourage an efficient and effective resolution to the issues under consideration.

Topic: Use and Occupancy

Reference: IRC R110.1

Category: Administration

Subject: Certificate of Occupancy

Code Text: *No building or structure shall be used or occupied, and no change in the existing occupancy classification of a building or structure or portion thereof shall be made until the building official has issued a certificate of occupancy therefor as provided herein.* See exception for work exempt from permits. *Issuance of a certificate of occupancy shall not be construed as an approval of a violation of the provisions of* the IRC *or of other ordinances of the jurisdiction.*

Discussion and Commentary: The certificate of occupancy is the tool by which the building official regulates and controls the uses and occupancies of the various buildings and structures within the jurisdiction. The code makes it unlawful to use or occupy a building unless a certificate of occupancy has been issued for that specific use. A temporary certificate of occupancy may be issued prior to completion of all of the work, but only in those portions of the building that can be safely occupied.

Certificate of Occupancy

(*Address of Structure*)

This (applicable portion of structure) has been inspected for compliance
with the laws and ordinances of (jurisdiction) and is hereby issued a
Certificate of Occupancy

Building permit number _____

Applicable edition of code _____

Sprinkler system provided _____

Special conditions _____

_____

_____

_____

Building Official _____

Name and address of owner _____

_____

The building official is permitted to suspend or revoke a certificate of occupancy based on any of the following reasons: (1) when the certificate is issued in error, (2) when incorrect information is supplied or (3) when the building is in violation of the code.

Topic: General Provisions **Category:** Administration
Reference: IRC R112 **Subject:** Board of Appeals

Code Text: *In order to hear and decide appeals of orders, decisions or determinations made by the building official relative to the application and interpretation of the IRC, there shall be and is hereby created a board of appeals. An application for appeal shall be based on a claim that the true intent of the code or the rules legally adopted thereunder have been incorrectly interpreted, the provisions of the code do not fully apply, or an equally good or better form of construction is proposed. The board shall have no authority to waive requirements of the IRC.*

Discussion and Commentary: Any aggrieved party with a material interest in the decision of the building official may appeal such a decision before a board of appeals. This provides a forum other than the court system in which the building official's action can be reviewed.

Members of the board of appeals shall be qualified by experience and training to rule on issues pertaining to building construction. Appendix B of the *International Building Code*® provides greater detail on administrative procedures and the necessary qualifications for board service.

Topic: Application　　　　　　　　　　**Category:** Referenced Standards
Reference: IRC Chapter 43　　　　　　**Subject:** Standards

Code Text: *This chapter lists the standards that are referenced in various sections of the IRC. The standards are listed herein by the promulgating agency of the standard, the standard identification, the effective date and title, and the section or sections of the IRC that reference the standard. The application of the referenced standard shall be as specified in Section R102.4.*

Discussion and Commentary: Limited in their scope, standards define more precisely the general provisions in the code. The *International Residential Code* references several hundred standards, each addressing a specific aspect of building design or construction. In general terms, the standards referenced by the IRC are primarily materials, testing and installation standards.

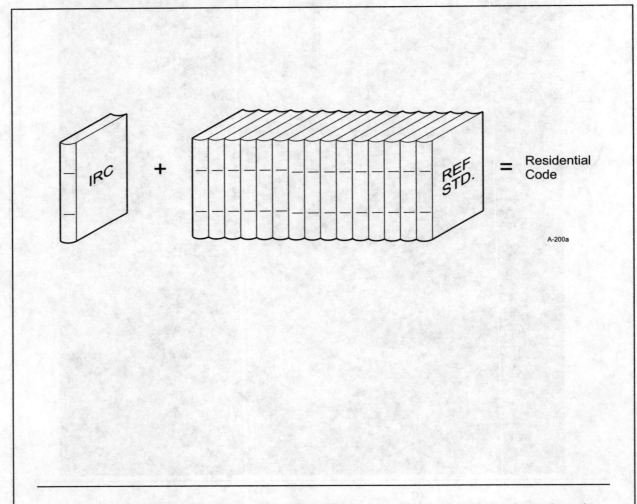

The codes and standards referenced in the IRC are an extension of the code, but only to the degree prescribed by the IRC. Where there is a conflict between the provisions in the IRC and any referenced code or standard, the provisions of the IRC shall apply.

QUIZ

Study Session 1

1. The *International Residential Code* is applicable to townhouses a maximum of _____ stories in height and provided with separate means of egress.
   a. one
   b. two
   c. three
   d. four

2. If there is a conflict in the code between a general requirement and a specific requirement, the _____ requirement shall apply.
   a. general
   b. specific
   c. least restrictive
   d. most restrictive

3. Provisions of the appendices do not apply unless _____.
   a. specified in the code
   b. applicable to unique conditions
   c. specifically adopted
   d. relevant to fire or life safety

4. The _____ is used in the code as the term to describe the individual in charge of the department of building safety.
   a. building official
   b. code official
   c. code administrator
   d. chief building inspector

5. The building official has the authority to _____ the provisions of the code.
   a. ignore
   b. waive
   c. modify
   d. interpret

6. Used materials may be utilized under which of the following conditions?
   a. they meet the requirements for new materials
   b. when approved by the building official
   c. used materials may never be used in new construction
   d. a representative sampling is tested for compliance

7. The building official has the authority to grant modifications to the code _____.
   a. for only those issues not affecting life safety or fire safety
   b. for individual cases where the strict letter of the code is impractical
   c. where the intent and purpose of the code cannot be met
   d. related only to administrative functions

8. The building official shall not grant modifications to any provision related to areas prone to flooding without _____.
   a. granting of a variance by the board of appeals
   b. approval of the jurisdiction's governing body
   c. the presentation of an approved design certificate
   d. submission by a registered design professional

9. Tests performed by _____ may be required by the building official where there is insufficient evidence of code compliance.
   a. the owner
   b. the contractor
   c. an approved agency
   d. a design professional

10. A permit is not required for the construction of a one-story detached accessory structure when it has a maximum floor area of _____ square feet.
    a. 100
    b. 120
    c. 150
    d. 200

11. Prefabricated swimming pools are subject to a building permit where they have a minimum depth of _____.
    a. 12 inches
    b. 24 inches
    c. 30 inches
    d. 36 inches

12. Unless an extension is authorized, a permit becomes invalid when work does not commence within _____ after permit issuance.
    a. 90 days
    b. 180 days
    c. one year
    d. two years

13. The building permit, or a copy of the permit, shall be kept _____ until completion of the project.
   a. at the job site
   b. by the permit applicant
   c. by the contractor
   d. by the owner

14. When a building permit is issued, the construction documents shall be approved _____.
   a. in writing or by stamp
   b. and two sets returned to the applicant
   c. and stamped as "Accepted as Reviewed"
   d. pending payment of the plan review fee

15. Unless otherwise mandated by state or local laws, the approved construction documents shall be retained by the building official for a minimum of _____ from the date of completion of the permitted work.
   a. 90 days
   b. 180 days
   c. one year
   d. two years

16. Where applicable, which one of the following inspections is not specifically identified by the *International Residential Code* as a required inspection?
   a. foundation inspection
   b. frame inspection
   c. fire-resistive-construction inspection
   d. energy efficiency inspection

17. Whose duty is it to provide access to work in need of inspection?
   a. the permit holder
   b. the owner or owner's agent
   c. the contractor
   d. the person requesting the inspection

18. The certificate of occupancy shall contain all of the following information except:
   a. the name and address of the owner
   b. the name of the building official
   c. the edition of the code under which the permit was issued
   d. the maximum occupant load

19. A temporary certificate of occupancy is valid for what period of time?
   a. 30 days
   b. 60 days
   c. 180 days
   d. a period set by the building official

20. The board of appeals is not authorized to rule on an appeal based on a claim that _____.
   a. the provisions of the code do not fully apply
   b. a code requirement should be waived
   c. the rules have been incorrectly interpreted
   d. a better form of construction is provided

21. The membership of the board of appeals _____
   a. shall include a jurisdictional member
   b. must consist of at least 5 members
   c. shall be knowledgeable of building construction
   d. must include an engineer or an architect

22. The notice of a code violation shall be served to _____.
   a. the permit holder
   b. the person responsible for the work
   c. the owner or owner's agent
   d. the general contractor

23. When a stop work order is issued, it shall be given to any of the following individuals except for the _____.
   a. owner
   b. owner's agent
   c. permit applicant
   d. person doing the work

24. Which of the following standards is applicable for factory-built fireplaces?
   a. ASTM A 951–98
   b. CPSC 16 CFR 1404
   c. NFPA 8501–97
   d. UL 127–99

25. ACI 318-02 is a reference standard addressing _____.
   a. structural concrete
   b. wood construction
   c. structural steel buildings
   d. gypsum board

26. All of the following issues are specifically identified for achieving the purpose of the *International Residential Code*, except for _____.
   a. affordability
   b. structural strength
   c. energy conservation
   d. usability and accessibility

18

27.   A permit is not required for the installation of a window awning provided the awning projects a maximum of _____ inches from the exterior wall.
    a. 30
    b. 36
    c. 48
    d. 54

28.   Where a self-contained refrigeration system contains a maximum of _____ pound(s) of refrigerant, a permit is not required.
    a. 1
    b. 2
    c. 5
    d. 10

29.   It is the duty of the _____ or their agent to notify the building official that work is ready for inspection.
    a. permit holder
    b. owner
    c. contractor
    d. design professional

30.   What is the role of the building official in relationship to the board of appeals?
    a. advisor only
    b. ex officio member
    c. full voting member
    d. procedural reviewer

INTERNATIONAL RESIDENTIAL CODE
Study Session 2
Sections R301 and R302—Building Planning I

OBJECTIVE: To gain an understanding of the design criteria to be used in the application of the prescriptive provisions of the *International Residential Code*, including wind and seismic limitations, snow loads and live loads, as well as the methods for addressing buildings located in very close proximity to property lines.

CODE REFERENCE: Sections R301 and R302, 2003 *International Residential Code*

KEY POINTS:

- What is the definition of light-framed construction? Under what conditions must such construction be designed in accordance with accepted engineering practice?
- What design standards are identified as acceptable alternatives to the prescriptive structural provisions of the IRC?
- At what wind speed must the structural system be designed based on a source other than the IRC? What design standards are acceptable for residential wind design?
- What is the most common design wind speed in the United States? Where are the hurricane-prone regions?
- How is the appropriate wind exposure category determined? Which category is assumed unless the site meets the definition of another category?
- What are the two methods of establishing wind speed? How do they compare to each other? Which method is used in the IRC?
- Buildings constructed in which Seismic Design Categories are subject to the seismic provisions of the IRC? Which category requires design to the requirements of the IBC?
- In which Seismic Design Categories are irregular buildings required to be designed? How is an irregular building determined?
- What alternate methods are permitted in the determination of a site's Seismic Design Category?
- What is the maximum snow load permitted when using the conventional construction provisions of the IRC?
- What are the limitations on story height for the various types of construction methods?
- What is a dead load? A live load?
- What is the design live load to be used for sleeping rooms? For rooms other than sleeping rooms?
- How is an attic defined? How does the minimum live load for attic areas differ based on roof slope?
- What is the maximum deflection permitted for floor structural members? For roof rafters with a ceiling attached directly to the bottom of the rafters?
- What is fire separation distance? How is it measured?
- At what fire separation distance from an interior property line do exterior walls not need a fire-resistive rating? If a rated wall is required, what is the minimum required rating of the wall?
- How far must a projection be located from an interior property line?
- At what fire separation distance are openings permitted in exterior walls?
- Under what condition may the exterior wall of an accessory structure located on the property line have no fire-resistance rating?
- How are penetrations of exterior walls required to have a fire-resistance rating to be addressed?

Topic: Design
Reference: IRC R301.1

Category: Building Planning
Subject: Design Criteria

Code Text: *The requirements of the IRC are based on platform and balloon-frame construction for light-frame buildings. The requirements for concrete and masonry buildings are based on a balloon framing system. When a building of otherwise conventional construction contains structural elements exceeding the limits of Section R301 or otherwise, not conforming to the IRC, these elements shall be designed in accordance with accepted engineering practice.*

Discussion and Commentary: Light-framed construction is a type of construction whose vertical and horizontal structural elements are primarily formed by a system of repetitive wood or light-gage steel framing members. Platform construction is defined as a method of construction by which floor framing bears on load-bearing walls that are not continuous through the story levels or floor framing. In balloon framing, the wall studs extend beyond the floor line.

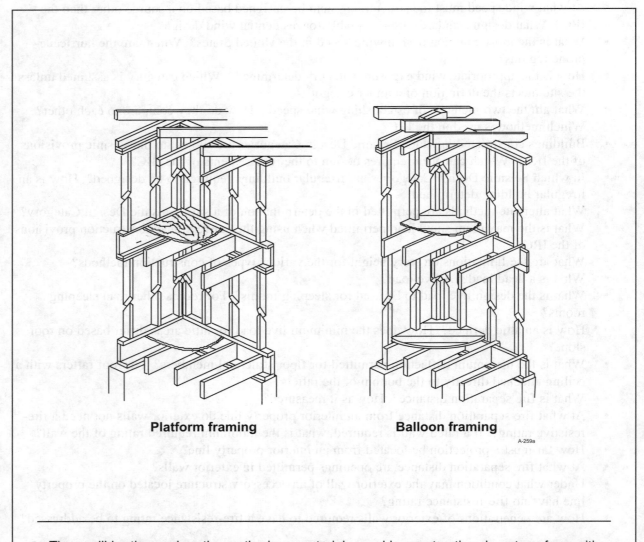

Platform framing   **Balloon framing**

A-259a

There will be times when the methods or materials used in construction do not conform with the prescriptive structural provisions of the IRC. In such situations, it is acceptable to use an engineered solution to satisfy the requirements of the code.

Topic: Alternative Provisions
Reference: IRC R301.1.1

Category: Building Planning
Subject: Design Criteria

Code Text: *As an alternative to the requirements in Section R301.1 the following standards are permitted subject to the limitations of* the IRC *and the limitations therein. Where engineered design is used in conjunction with these standards the design shall comply with the International Building Code. 1) American Forest and Paper Association (AF&PA) Wood Frame Construction Manual (WFCM) and 2) American Iron and Steel Institute (AISI), Standard for Cold-Formed Steel Framing—Prescriptive Method for One-and Two-family Dwellings (COFS/PM).*

Discussion and Commentary: Although the *International Residential Code* is intended to address most structural conditions encountered in residential construction, it may be advantageous to utilize more comprehensive prescriptive design methods that are set forth in the various industry standards.

Where utilizing the design criteria of either the *Wood Frame Construction Manual* or the *Cold-Formed Steel Design Manual*, they must be in accordance with the IBC. The IBC becomes the design basis in those cases where the prescriptive provisions are not used.

Topic: Climatic and Geographic Criteria   **Category:** Building Planning
Reference: IRC R301.2   **Subject:** Design Criteria

Code Text: *Buildings shall be constructed in accordance with the provisions of the IRC as limited by the provisions of Section R301. Additional criteria shall be established by the local jurisdiction and set forth in Table R301.2(1).*

Discussion and Commentary: Table R301.2(1) is designed so that jurisdictions recognize and use certain climatic and geographic design criteria that vary from location to location. Communities are directed to complete the table with a variety of factors. The code references the table in many of its provisions, mandating completion of the table in order to understand the code requirements. Footnotes to the table provide guidance on how and where the determination of the appropriate information is obtained.

TABLE R301.2(1)
CLIMATIC AND GEOGRAPHIC DESIGN CRITERIA

GROUND SNOW LOAD	WIND SPEED[e] (mph)	SEISMIC DESIGN CATEGORY[g]	SUBJECT TO DAMAGE FROM				WINTER DESIGN TEMP[f]	ICE SHIELD UNDERLAYMENT REQUIRED[i]	FLOOD HAZARDS[h]	AIR FREEZING INDEX[j]	MEAN ANNUAL TEMP[k]
			Weathering[a]	Frost line depth[b]	Termite[c]	Decay[d]					

For SI: 1 pound per square foot = 0.0479 kN/m^2, 1 mile per hour = 1.609 km/h.

a. Weathering may require a higher strength concrete or grade of masonry than necessary to satisfy the structural requirements of this code. The weathering column shall be filled in with the weathering index (i.e., "negligible," "moderate" or "severe") for concrete as determined from the Weathering Probability Map [Figure R301.2(3)]. The grade of masonry units shall be determined from ASTM C 34, C 55, C 62, C 73, C 90, C 129, C 145, C 216 or C 652.

b. The frost line depth may require deeper footings than indicated in Figure R403.1(1). The jurisdiction shall fill in the frost line depth column with the minimum depth of footing below finish grade.

c. The jurisdiction shall fill in this part of the table with "very heavy," "moderate to heavy," "slight to moderate," or "none to slight" in accordance with Figure R301.2(6) depending on whether there has been a history of local damage.

d. The jurisdiction shall fill in this part of the table with "moderate to severe," "slight to moderate," or "none to slight" in accordance with Figure R301.2(7) depending on whether there has been a history of local damage.

e. The jurisdiction shall fill in this part of the table with the wind speed from the basic wind speed map [Figure R301.2(4)]. Wind exposure category shall be determined on a site-specific basis in accordance with Section R301.2.1.4.

f. The outdoor design dry-bulb temperature shall be selected from the columns of 97^1/$_2$-percent values for winter from Appendix D of the *International Plumbing Code*. Deviations from the Appendix D temperatures shall be permitted to reflect local climates or local weather experience as determined by the building official.

g. The jurisdiction shall fill in this part of the table with the Seismic Design Category determined from Section R301.2.2.1.

h. The jurisdiction shall fill in this part of the table with (a) the date of the jurisdiction's entry into the National Flood Insurance Program (date of adoption of the first code or ordinance for management of flood hazard areas), (b) the date(s) of the currently effective FIRM and FBFM, or other flood hazard map adopted by the community, as may be amended.

i. In accordance with Sections R905.2.7.1, R905.4.3, R905.5.3, R905.6.3, R905.7.3 and R905.8.3, for areas where the average daily temperature in January is 25°F (-4°C) or less, or where there has been a history of local damage from the effects of ice damming, the jurisdiction shall fill in this part of the table with "YES." Otherwise, the jurisdiction shall fill in this part of the table with "NO."

j. The jurisdiction shall fill in this part of the table with the 100-year return period air freezing index (BF-days) from Figure R403.3(2) or from the 100-year (99%) value on the National Climatic Data Center data table "Air Freezing Index- USA Method (Base 32° Fahrenheit)" at www.ncdc.noaa.gov/fpsf.html.

k. The jurisdiction shall fill in this part of the table with the mean annual temperature from the National Climatic Data Center data table "Air Freezing Index- USA Method (Base 32° Fahrenheit)" at www.ncdc.noaa.gov/fpsf.html.

The frost line depth should be established by the local jurisdiction based on past practice or available information. Since frost penetration is dependent on soil conditions and elevations as well as expected winter temperatures, it may vary greatly within a limited geographical area.

Topic: Wind Limitations
Reference: IRC R301.2.1

Category: Building Planning
Subject: Design Criteria

Code Text: *Buildings and portions thereof shall be limited by wind speed, as defined in Table R301.2(1), and construction methods in accordance with the IRC. Basic wind speeds shall be determined from Figure R301.2(4). Construction in regions where the basic wind speeds from Figure R301.2(4) equal or exceed 110 miles per hour (177.1 km/h) shall be designed in accordance with one of five methods. See five possible methods.*

Discussion and Commentary: With the exception of steel framing, the prescriptive provisions of the IRC apply only where the basic wind speed is less than 110 mph. Steel framing provisions are applicable in regions with wind speeds up to 130 mph, provided the site is not Exposure D. Where the basic wind speed exceeds the limitations of the code, structures must be designed.

wind

Steel 130 mph
Wood 110 mph

WIND LOAD PUSHES AGAINST END WALL

END WALLS PUSH/PULL ROOF THIS WAY

WIND PULLS END WALL THIS WAY

WALL HOLDS ROOF BACK

B-486

Wind loads are a major consideration in designing a structure's lateral force-resisting system. In addition, other building elements are affected as evidenced by provisions for roof tie-downs, for example. Windows, skylights and exterior doors must withstand the component and cladding pressures.

Topic: Wind Internal Pressure
Reference: IRC R301.2.1.2

Category: Building Planning
Subject: Design Criteria

Code Text: *Windows in buildings located in wind borne debris regions shall have glazed openings protected from windborne debris or the building shall be designed as a partially enclosed building in accordance with the* International Building Code. *See exception for use of precut wood structural panels with complying attachment hardware.*

Discussion and Commentary: Opening protection in windborne debris regions is usually in the form of permanent shutters or laminated glass that meet the requirements of the Large Missile Test of ASTM E 1996 and E 1886. The exception provides a prescriptive, and more economical, approach, which is permitted in one- and two-story buildings. This method of opening protection uses wood structural panels with a maximum span of 8 feet along with complying fasteners. The builder must precut these panels to fit each glazed opening and provide the necessary attachment hardware.

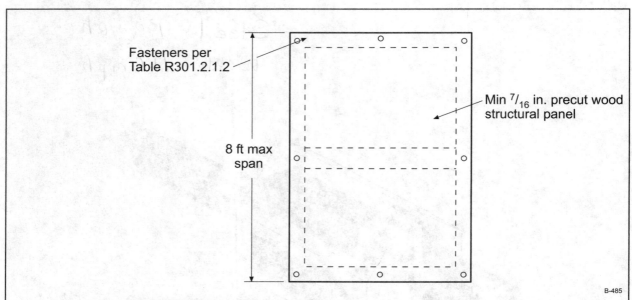

Fasteners per Table R301.2.1.2

Min $^7/_{16}$ in. precut wood structural panel

8 ft max span

B-485

TABLE R301.2.1.2
WINDBORNE DEBRIS PROTECTION FASTENING SCHEDULE
FOR WOOD STRUCTURAL PANELS[a,b,c,d]

FASTENER TYPE	FASTENER SPACING		
	Panel span ≤ 4 foot	4 foot < panel span ≤ 6 foot	6 foot < panel span ≤ 8 foot
$2^1/_2$" #6 Wood screws	16"	12"	9"
$2^1/_2$" #8 Wood screws	16"	16"	12"

For SI: 1 inch = 25.4 mm, 1 foot = 304.8 mm, 1 pound = 0.454 kg,
  1 mile per hour = 1.609 km/h.

a. This table is based on 110 mph wind speeds and a 33-foot mean roof height.
b. Fasteners shall be installed at opposing ends of the wood structural panel.
c. Nails shall be 10d common or 12d box nails.
d. Where screws are attached to masonry or masonry/stucco, they shall be attached utilizing vibration-resistant anchors having a minimum ultimate withdrawal capacity of 490 pounds.

If protection of glazed openings is not provided in one of the methods specified, the building must be designed per ASCE-7 as a "partially enclosed" structure. This approach accounts for the increased internal pressure that occurs as a result of a breach in the building envelope.

Topic: Wind Exposure Category
Reference: IRC R301.2.1.4

Category: Building Planning
Subject: Design Criteria

Code Text: *For each wind direction considered, an exposure category that adequately reflects the characteristics of ground surface irregularities shall be determined for the site at which the building or structure is to be constructed. For any given wind direction, the exposure in which a specific building or other structure is sited shall be assessed as being* in Exposure Category A, B, C or D.

Discussion and Commentary: In addition to wind speed, wind loading on structures is a function of the site's exposure category. The exposure category reflects the characteristics of ground surface irregularities and accounts for variations in ground surface roughness that arise from natural topography and vegetation as well as from constructed features.

Exposure B should be used unless the site meets the definition for another category of exposure. Exposure B includes urban and suburban areas, wooded areas and other terrain with numerous closely-spaced obstructions having the size of single-family dwellings or larger.

Topic: Seismic Provisions   **Category:** Building Planning
Reference: IRC R301.2.2   **Subject:** Design Criteria

Code Text: *The seismic provisions of* the IRC *shall apply to buildings constructed in Seismic Design Categories C, D_1, and D_2, as determined in accordance with* Section R301.2.2. See exception for detached one- and two-family dwellings in Category C. *Buildings shall be assigned a Seismic Design Category in accordance with Figure R301.2(2).*

Discussion and Commentary: Earthquakes generate internal forces in a structure due to inertia. These forces can cause a building to be distorted and severely damaged. The objective of earthquake resistant construction is to resist these forces and the resulting distortions. Buildings are deemed to be located in one of several seismic design categories, ranging from Category A to Category E. Buildings in Categories A and B as well as detached one- and two-family dwellings in Category C perform satisfactorily when constructed in accordance with the basic prescriptive provisions of the IRC. On the other hand, Category E buildings must be designed in accordance with the seismic provisions of the *International Building Code.*

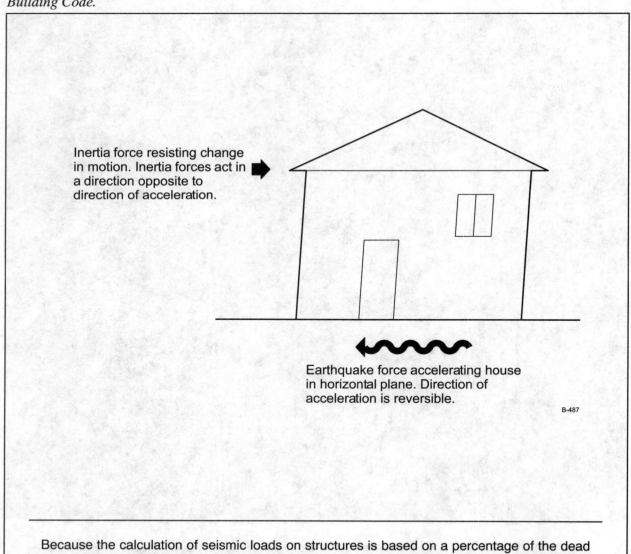

Inertia force resisting change in motion. Inertia forces act in a direction opposite to direction of acceleration.

Earthquake force accelerating house in horizontal plane. Direction of acceleration is reversible.

B-487

Because the calculation of seismic loads on structures is based on a percentage of the dead load, the IRC prescriptive seismic provisions are based on an assumed dead load. The building dead loads of Section R301.2.2.4 are reflected in the prescriptive design tables throughout the code.

Topic: Irregular Buildings
Reference: IRC R301.2.2.7

Category: Building Planning
Subject: Design Criteria

Code Text: *Concrete construction complying with Section R611 or R612 and conventional light-frame construction shall not be used in irregular portions of structures in Seismic Design Categories C, D_1 and D_2. Only such irregular portions of structures shall be designed in accordance with accepted engineering practice to the extent such irregular features affect the performance of the conventional framing system. A portion of a building shall be considered to be irregular when one or more of seven specific conditions exist. See seven possible conditions.*

Discussion and Commentary: Conventional light-frame construction typically allows cantilevers and offsets within certain limits in order to accommodate common design features and options. These features are not well suited to resisting earthquake loads. This becomes more of a concern in areas of higher seismic hazard.

- Portions of buildings are considered irregular where one or more of seven listed conditions exist:

  - Exterior shear wall lines or braced wall panels not in one plane vertically from foundation to uppermost story where required (exception for limited cantilevers and setbacks)

  - Section of floor or roof not laterally supported by shear walls or braced wall lines on all edges (exception for limited extension beyond shear wall or braced wall line)

  - End of braced wall panel occurs over an opening below and ends more than 1 foot from edge of opening (exception for opening with complying header)

  - Opening in a floor or roof exceeds 12 feet or exceeds 50 percent of least floor or roof dimension

  - Portions of floor level vertically offset (exceptions for framing directly supported by continuous foundations or where floor framing lapped or tied together)

  - Shear walls and braced wall lines do not occur in two perpendicular directions

  - Shear walls or braced wall lines constructed of dissimilar bracing systems on any story above grade

The IRC, consistent with the IBC, identifies building features that are considered irregular and are not permitted under conventional light-frame construction provisions for the higher seismic design categories (C, D_1 and D_2).

Topic: Floodplain Construction
Reference: IRC R301.2.4

Category: Building Planning
Subject: Design Criteria

Code Text: *Buildings and structures constructed in flood hazard areas (including A or V Zones) as established in Table R301.2(1) shall be designed and constructed in accordance with Section R323.* See exception for buildings in floodways designated by FIRM, FBFM or other flood hazard map adopted by the jurisdiction.

Discussion and Commentary: Application of the flood provisions of the code cannot prevent or eliminate all future flood damage. Rather, the provisions represent a reasonable balance of the knowledge and awareness of flood hazards, methods to guide development to less hazard-prone locations, methods of design and construction intended to resist flood damage, and each community's and landowner's reasonable expectations of land use.

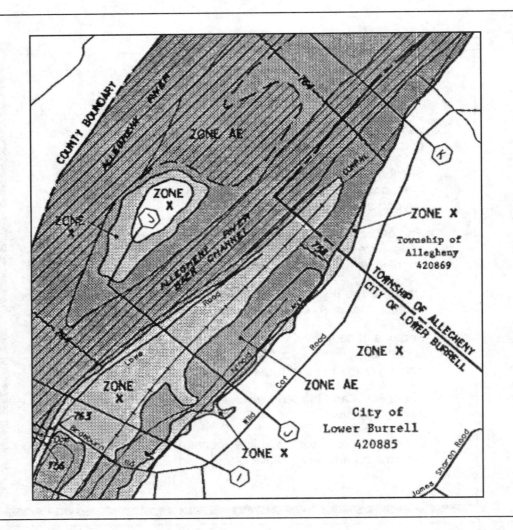

The flood hazard areas shown on FEMA maps are determined using the base flood, which is defined as have a 1-percent chance of occurring in any given year. The maps do not show the worst case flood nor the "flood of record," which usually is the most severe in history.

Topic: Snow Loads and Roof Loads   **Category:** Building Planning
Reference: IRC R301.2.3, R301.6   **Subject:** Design Criteria

Code Text: *Wood framed construction, cold-formed steel framed construction and masonry and concrete construction in regions with ground snow loads 70 psf (3.35 kN/m²) or less, shall be in accordance with Chapters 5, 6 and 8. Buildings in regions with ground snow loads greater than 70 psf (3.35 kN/m²) shall be designed in accordance with accepted engineering practice. Roofs shall be designed for the live load indicated in Table R301.6 or the snow load indicated in Table R301.2(1), whichever is greater.*

Discussion and Commentary: The prescriptive tables for floor, wall and roof construction are limited to buildings constructed in areas having a maximum 70 psf snow load. Structures in regions exceeding this limitation require an engineered design for all elements that carry snow loads. The basic 20 psf live load may be reduced based on the tributary area supported by any structural member of the roof, but no reduction is permitted for snow loads.

TABLE R301.6
MINIMUM ROOF LIVE LOADS IN POUNDS-FORCE PER SQUARE FOOT OF HORIZONTAL PROJECTION

ROOF SLOPE	TRIBUTARY LOADED AREA IN SQUARE FEET FOR ANY STRUCTURAL MEMBER		
	0 to 200	201 to 600	Over 600
Flat or rise less than 4 inches per foot (1:3)	20	16	12
Rise 4 inches per foot (1:3) to less than 12 inches per foot (1:1)	16	14	12
Rise 12 inches per foot (1:1) and greater	12	12	12

For SI: 1 square foot = 0.0929 m², 1 pound per square foot = 0.0479 kN/m², 1 inch per foot = 0.0833 mm/m.

For roofs, both the live load and snow load must be considered, with the more restrictive loading condition used. The live load takes into account that individuals and materials may be present on the roof during roofing operations.

Topic: Story Height
Reference: IRC R301.3

Category: Building Planning
Subject: Design Criteria

Code Text: *Buildings constructed in accordance with these provisions shall be limited to story heights of not more than the following: 1) for wood wall framing, the laterally unsupported bearing wall stud height permitted by Table R602.3(5) plus a height of floor framing not to exceed sixteen inches (see exception for braced walls); 2) for steel wall framing, a stud height of 10 feet, plus a height of floor framing not to exceed 16 inches; 3) for masonry walls, a maximum bearing wall clear height of 12 feet plus a height of floor framing not to exceed 16 inches (see exception for gable end walls); and 4) for insulating concrete form walls, the maximum bearing wall height per story as permitted by Section 611 tables plus a height of floor framing not to exceed 16 inches.*

Discussion and Commentary: This section identifies the maximum heights for use of the IRC's prescriptive provisions, with an allowance for a floor system up to sixteen inches in depth.

TABLE R602.3(5)
SIZE, HEIGHT AND SPACING OF WOOD STUDS[a]

STUD SIZE (inches)	BEARING WALLS					NONBEARING WALLS	
	Laterally unsupported stud height[a] (feet)	Maximum spacing when supporting roof and ceiling only (inches)	Maximum spacing when supporting one floor, roof and ceiling (inches)	Maximum spacing when supporting two floors, roof and ceiling (inches)	Maximum spacing when supporting one floor only (inches)	Laterally unsupported stud height[a] (feet)	Maximum spacing (inches)
2 × 3[b]	—	—	—	—	—	10	16
2 × 4	10	24	16	—	24	14	24
3 × 4	10	24	24	16	24	14	24
2 × 5	10	24	24	—	24	16	24
2 × 6	10	24	24	16	24	20	24

For SI: 1 inch = 25.4 mm.

a. Listed heights are distances between points of lateral support placed perpendicular to the plane of the wall. Increases in unsupported height are permitted where justified by analysis.

b. Shall not be used in exterior walls.

These requirements are height-limiting solely due to the prescriptive structural limitations of the code. Such story heights may be increased through the engineered design provisions in the *International Building Code*.

2003 IRC Study Companion

Topic: Live Loads
Reference: IRC R301.5

Category: Building Planning
Subject: Design Criteria

Code Text: *The minimum uniformly distributed live load shall be as provided in Table R301.5.*

Discussion and Commentary: Table R301.4 provides the minimum loads based on the use of a particular area or portion of a structure. These loads must be considered for the design of corresponding structural elements of any residence constructed under the code. For instance, bedrooms (sleeping rooms) require the use of a 30 psf live load, whereas all other rooms must be designed for a 40 psf uniform live load. Exterior balconies require a higher design load (60 psf) than decks (40 psf), simply because of the lack of redundancy in structural supports. An exterior balcony is defined as "an exterior floor projecting from and supported by a structure without additional independent supports." A deck is "supported on at least two opposing sides by an adjoining structure and/or posts, piers, or other independent supports."

TABLE R301.5
MINIMUM UNIFORMLY DISTRIBUTED LIVE LOADS
(In pounds per square foot)

USE	LIVE LOAD
Attics with storage[b]	20
Attics without storage[b]	10
Decks[e]	40
Exterior balconies	60
Fire escapes	40
Guardrails and handrails[d]	200
Guardrails in-fill components[f]	200
Passenger vehicle garages[a]	50[a]
Rooms other than sleeping rooms	40
Sleeping rooms	30
Stairs	40[c]

For SI: 1 pound per square foot = 0.0479 kN/m², 1 square inch = 645 mm², 1 pound = 4.45 N.

a. Elevated garage floors shall be capable of supporting a 2,000-pound load applied over a 20-square-inch area.

b. No storage with roof slope not over 3 units in 12 units.

c. Individual stair treads shall be designed for the uniformly distributed live load or a 300-pound concentrated load acting over an area of 4 square inches, whichever produces the greater stresses.

d. A single concentrated load applied in any direction at any point along the top.

e. See Section R502.2.1 for decks attached to exterior walls.

f. Guard in-fill components (all those except the handrail), balusters and panel fillers shall be designed to withstand a horizontally applied normal load of 50 pounds on an area equal to 1 square foot. This load need not be assumed to act concurrently with any other live load requirement.

Attic design loads are based on the slope of the roof. Where it is assumed that there is insufficient clearance or headroom to accumulate significant storage (slope of 3:12 or less), only 10 psf is mandated. For higher slopes where storage is possible, a 20 psf dead load is to be used.

Topic: Exterior Walls
Reference: IRC R302.1

Category: Building Planning
Subject: Location on Lot

Code Text: *Exterior walls with a fire separation distance less than 3 feet (914 mm) shall have not less than a one-hour fire-resistive rating with exposure from both sides. Projections shall not extend to a point closer than 2 feet (610 mm) from the line used to determine the fire separation distance. See exception allowing 4-inch projection for detached garages. Projections extending into the fire separation distance shall have not less than one-hour fire-resistive construction on the underside.*

Discussion and Commentary: To reduce the potential for fire spread from one residential structure to another due to their close proximity, the code provides a choice as to the method for addressing radiant heat transfer between structures. If the building is located at least 3 feet from a property line, the atmospheric separation mitigates the concern. Where the building is located closer to the property line, a fire-resistive barrier of one hour is mandated to restrict the spread of heat and flame.

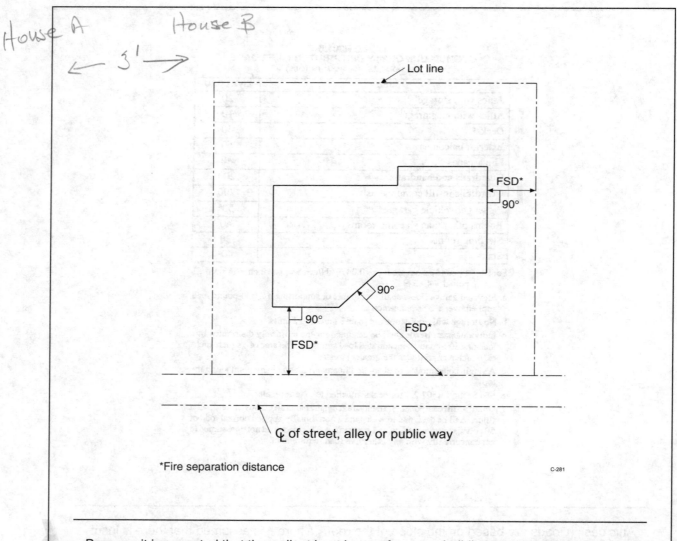

*Fire separation distance

C-281

Because it is expected that the radiant heat impact from one building across a property line to a neighboring building will be in a relatively direct manner, any exterior wall located at an angle of 90 degrees or more from the property line is not regulated.

Topic: Openings in Exterior Walls
Reference: IRC R302.2

Category: Building Planning
Subject: Location on Lot

Code Text: *Openings shall not be permitted in the exterior wall of a dwelling or accessory building with a fire separation distance less than 3 feet (914 mm).* See exceptions for foundation vents and those walls perpendicular to property lines.

Discussion and Commentary: Unprotected openings, typically doors and windows, in exterior walls of adjacent buildings provide an even greater opportunity for the spread of fire and heat from one structure to another than do the exterior walls. Therefore, there cannot be openings, including those in accessory structures, in exterior walls located in close proximity to an interior property line. There is no option, such as the use of fire windows or fire shutters, addressed by the code to allow for the protection of the openings.

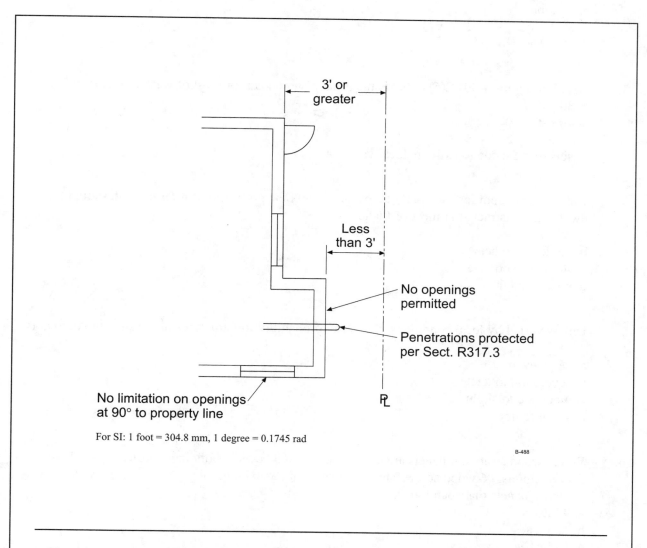

3' or greater

Less than 3'

No openings permitted

Penetrations protected per Sect. R317.3

No limitation on openings at 90° to property line

PL

For SI: 1 foot = 304.8 mm, 1 degree = 0.1745 rad

B-488

Should any items such as piping or conduit penetrate an exterior wall located within 3 feet (914 mm) of the property line, the penetrations must be protected in accordance with Section R317.3 as either through penetrations or membrane penetrations.

QUIZ

Study Session 2

1. What is the weathering potential for all dwelling sites located in the state of Missouri?
    a. severe
    b. moderate
    c. negligible
    d. none

2. The basic wind speed to be used for most of the United States is _____.
    a. 70 mph
    b. 75 mph
    c. 85 mph
    d. 90 mph

3. Based on Figure R301.2(5), what is the ground snow load for most of western Colorado?
    a. 30 psf
    b. 40 psf
    c. 50 psf
    d. site-specific due to extreme local variations

4. Unless local conditions warrant otherwise, what is the probability for termite infestation for dwellings constructed in most of Idaho?
    a. very heavy
    b. moderate to heavy
    c. slight to moderate
    d. none to slight

5. Unless varied by local conditions, the decay probability for structures in the state of Arizona is _____.
    a. very severe
    b. severe to moderate
    c. moderate to slight
    d. slight to none

6. Where wood structural panels are used to protect windows in buildings located in windborne debris regions, #6 wood screws shall be located at a maximum of _____ on center to fasten those panels that span 8 feet.
    a. 9 inches
    b. 12 inches
    c. 16 inches
    d. 24 inches

7. A "3-second gust" velocity of 90 miles per hour in the IRC is to be converted to a "fastest mile" wind speed of _____ for use in reference documents using the "fastest mile" criteria.
   a. 75 miles per hour
   b. 90 miles per hour
   c. 100 miles per hour
   d. 110 miles per hour

8. What wind exposure category is appropriate for a dwelling located in a residential development in a suburban area?
   a. Exposure A
   b. Exposure B
   c. Exposure C
   d. Exposure D

9. The Seismic Design Category for a site having a calculated SDS of 0.63g is _____.
   a. B
   b. D_1
   c. D_2
   d. E

10. In Seismic Design Category C, anchored masonry veneer is limited to _____ in thickness.
    a. 4 inches
    b. 5 inches
    c. 6 inches
    d. 8 inches

11. For a dwelling constructed in Seismic Design Category D_2, what is the maximum dead load for 8-inch-thick masonry walls?
    a. 40 psf
    b. 65 psf
    c. 80 psf
    d. 85 psf

12. A portion of a building is considered irregular for seismic purposes when a floor opening, such as for a stairway, exceeds _____ of the least floor dimension.
    a. 15 percent
    b. 25 percent
    c. $33^1/_3$ percent
    d. 50 percent

13. Buildings constructed in regions where the ground snow load exceeds _____ must be designed in accordance with accepted engineering practice.
    a. 30 psf
    b. 50 psf
    c. 70 psf
    d. 90 psf

14. For rooms other than sleeping rooms, what is the minimum uniformly distributed live load that is to be used for the floor system?

   a. 20 psf
   b. 30 psf
   c. 40 psf
   d. 50 psf

15. A minimum uniformly distributed live load of 10 psf is required for attic areas having no storage, provide the roof slope does not exceed _____.

   a. 3:12
   b. 4:12
   c. 5:12
   d. 6:12

16. A minimum uniformly distributed live load of _____ shall be used for sleeping rooms.

   a. 10 psf
   b. 20 psf
   c. 30 psf
   d. 40 psf

17. Where no snow load is present, what is the minimum roof live load for a 240 square foot tributary loaded roof area having a slope of 8:12?

   a. 20 psf
   b. 16 psf
   c. 14 psf
   d. 12 psf

18. What is the maximum allowable deflection for floor joists?

   a. L/120
   b. L/180
   c. L/240
   d. L/360

19. The maximum allowable deflection permitted for 7:12-sloped rafters with no ceiling attached is _____.

   a. L/120
   b. L/180
   c. L/240
   d. L/360

20. Except for walls perpendicular to the property line, a minimum 1-hour fire-resistance rating is required for all exterior walls having a fire separation distance of less than _____.

   a. 3 feet
   b. 4 feet
   c. 5 feet
   d. 10 feet

21.   What is the minimum distance a roof projection on a dwelling shall be located from an interior property line?
        a. 0 feet; it may extend to the property line
        b. 12 inches
        c. 2 feet
        d. 3 feet

22.   Tool and storage sheds, playhouses and similar accessory structures having a maximum floor area of _____ are not required to have exterior wall protection based on location on the lot.
        a. 100 square feet
        b. 120 square feet
        c. 150 square feet
        d. 200 square feet

23.   Openings located less than 3 feet from an interior property line shall be protected with what minimum fire-protection rating?
        a. 0 hours; no fire-protection rating is required
        b. 20 minutes
        c. 45 minutes
        d. openings are not permitted

24.   Where located in an exterior wall having a fire separation distance of less than 3 feet, penetrations protected with an approved penetration firestop system shall have a minimum fire-resistance rating of _____.
        a. 20 minutes
        b. 30 minutes
        c. 45 minutes
        d. 1 hour

25.   Where a detached garage is located within 2 feet of a lot line, the maximum eave projection is limited to _____ inches.
        a. 4
        b. 6
        c. 8
        d. 12

26.   For wind design purposes, a building located along the shoreline in a hurricane prone region is assessed as Exposure _____.
        a. A
        b. B
        c. C
        d. D

27. Which of the following buildings is exempt from the seismic requirements of the code?
    a. a townhouse in SDC C
    b. a one-family dwelling in SDC C
    c. a townhouse in SDC D$_1$
    d. a two-family dwelling in SDC D$_1$

28. For a building located in Seismic Design Category (SDC) D$_1$, the maximum average dead load for a floor assembly is _____ psf.
    a. 5
    b. 10
    c. 15
    d. 20

29. Under the IRC, the maximum story height for steel wall framing is limited to _____, plus up to 16 inches for floor framing.
    a. 10 feet
    b. 12 feet
    c. 16 feet
    d. 22 feet

30. An in-fill panel for a guard shall be designed to withstand a minimum load of _____ applied on an area of 1 square foot.
    a. 50 pounds
    b. 200 pounds
    c. 40 psf
    d. 50 psf

INTERNATIONAL RESIDENTIAL CODE
Study Session 3
Sections R303 through R310—Building Planning II

OBJECTIVE: To develop an understanding of the health and safety criteria of the code, including light and ventilation; minimum room areas and ceiling height; sanitation; toilet, bath, and shower spaces; glazing, including safety glazing; carports and garages; emergency escape and rescue openings; and exits.

REFERENCE: Sections R303 through R310, 2003 *International Residential Code*

KEY POINTS:
- Where natural light is used to satisfy the minimum illumination requirements, how is the minimum required amount of glazing determined? Where artificial light is used, what illumination level is mandated?
- Where natural ventilation is used to satisfy the minimum ventilation requirements, what is the minimum required area of openable exterior openings?
- How must the ventilation and illumination of bathrooms occur?
- How must mechanical and gravity outside air intake openings be located in relationship to vents, chimneys, parking lots and other potential areas of a hazardous or noxious contaminant?
- Where must illumination be located in relationship to interior stairways? Exterior stairways?
- In what climatic areas must a heating system be provided? What performance level is mandated for the system?
- What is the minimum size of the largest habitable room in a dwelling unit?
- What is the minimum dimension permitted for a habitable room other than a kitchen?
- What is the minimum ceiling height permitted for a living room or bedroom? A hallway? Bathroom? Basement? Where can a reduction in such heights be acceptable?
- How much clear floor space is required in front of a water closet? In front of a shower opening? What is the minimum distance needed between the centerline of a water closet and the nearest adjoining obstruction such as a wall or shower compartment?
- In what manner must safety glazing be identified? Multipane assemblies?
- Where is polished wired glass permitted to be used as safety glazing?
- What specific locations in and adjacent to doors are subject to human impact and require safety glazing? In tub and shower areas? Adjacent to swimming pools, hot tubs, and spas? At stairways and stairway landings?
- When is sloped glazing considered a skylight? What glazing materials are permitted in skylights? When must a screen be installed below a skylight?
- How must a private garage be separated from a dwelling unit? What type of door is required? How should a duct penetration be addressed?
- How does a carport differ from a garage? What limitations are placed on carports?
- Where are escape and rescue openings required? What is the minimum size of such openings? Maximum sill height? What limitations are placed on the operation of the opening?
- When a window well serves an escape and rescue opening, what is its minimum size?
- How may a bulkhead enclosure be utilized as an escape and rescue opening?

Topic: Habitable Rooms
Reference: IRC R303.1

Code Text: *All habitable rooms shall be provided with aggregate glazing area of not less than 8 percent of the floor area of such rooms. Natural ventilation shall be through windows, doors, louvers or other approved openings to the outdoor air. The minimum openable area to the outdoors shall be 4 percent of the floor area being ventilated.* Exceptions allow the use of artificial light and mechanical ventilation.

Discussion and Commentary: A usable and sanitary interior environment depends on the inclusion of adequate light and ventilation for the habitable spaces within the dwelling unit. Traditionally, the use of natural light and, to some degree, natural ventilation has been mandated as the means for achieving such an environment. It has become increasingly more common to use artificial lighting and a mechanical ventilation system. These methods create additional design flexibility and functionality while maintaining a pleasant and sanitary living environment.

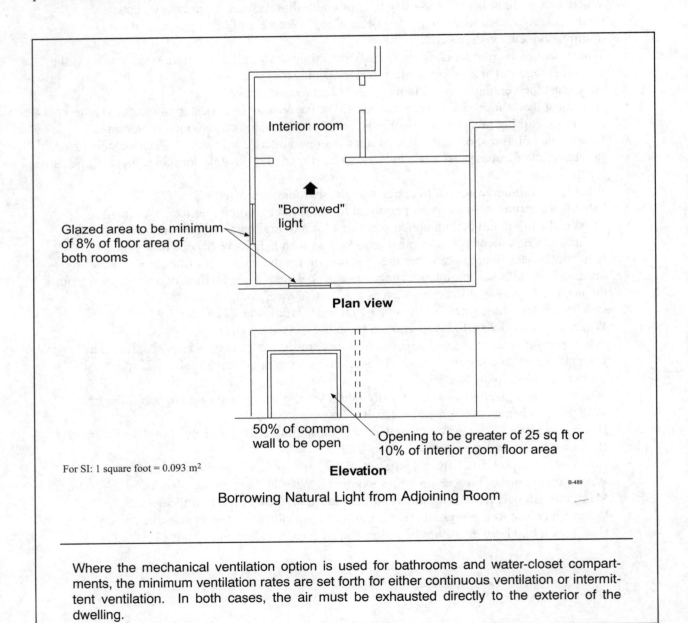

Interior room

"Borrowed" light

Glazed area to be minimum of 8% of floor area of both rooms

Plan view

50% of common wall to be open

Opening to be greater of 25 sq ft or 10% of interior room floor area

For SI: 1 square foot = 0.093 m²

Elevation

B-489

Borrowing Natural Light from Adjoining Room

Where the mechanical ventilation option is used for bathrooms and water-closet compartments, the minimum ventilation rates are set forth for either continuous ventilation or intermittent ventilation. In both cases, the air must be exhausted directly to the exterior of the dwelling.

Topic: Opening Location
Reference: IRC R303.4

Category: Building Planning
Subject: Light, Ventilation and Heating

Code Text: *Mechanical and gravity outside air intake openings shall be located a minimum of 10 feet (3048 mm) from any hazardous or noxious contaminant, such as vents, chimneys, plumbing vents, streets, alleys, parking lots and loading docks, except as otherwise specified in the IRC. Where a source of contaminant is located within 10 feet (3048 mm) of an intake opening, such opening shall be located a minimum of 2 feet (610 mm) below the contaminant source. Outside exhaust openings shall be located so as not to create a nuisance. Exhaust air shall not be directed onto walkways.*

Discussion and Commentary: In the context of this section, intake openings include windows, doors, combustion air intakes and similar openings that naturally or mechanically draw in air from the building exterior. The alternative to the 10-foot separation requirement, a 2-foot vertical separation distance, will allow noxious gases and contaminants to disperse into the atmosphere before they can be drawn into an air intake opening.

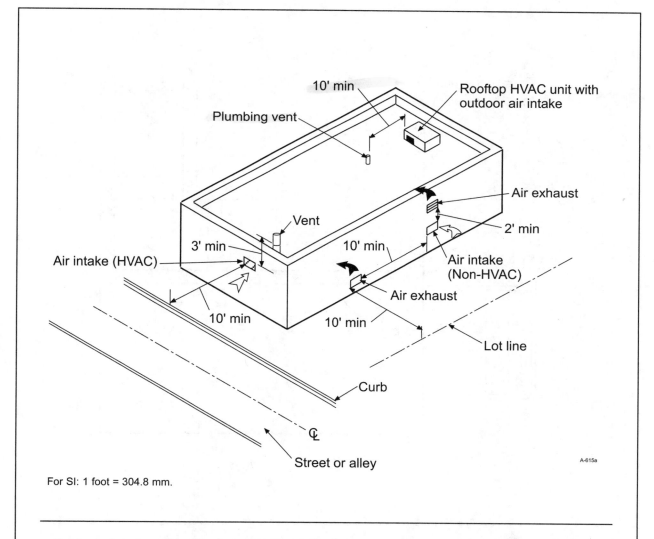

For SI: 1 foot = 304.8 mm.

Outdoor air exhaust and intake openings are regulated in the same manner as other openings in exterior walls. As such, they are not permitted in walls having a fire separation distance of less than 3 feet, except in exterior walls that are perpendicular to the lot line.

Topic: Stairway Illumination

Category: Building Planning

Reference: IRC R303.6

Subject: Light, Ventilation and Heating

Code Text: *Interior stairways shall be provided with an artificial light source located in the immediate vicinity of each landing of the stairway.* See exception where light source is located over each stairway section. *Exterior stairways shall be provided with an artificial light source located in the immediate vicinity of the top landing of the stairway.*

Discussion and Commentary: A stairway is one of the most hazardous areas of a dwelling unit. As such, the code highly regulates the design and construction of all stairways. In addition, adequate lighting must be provided to enable the stairway user to see the treads, their nosings and any obstructions that may be present. Stairway landings must be adequately lighted also.

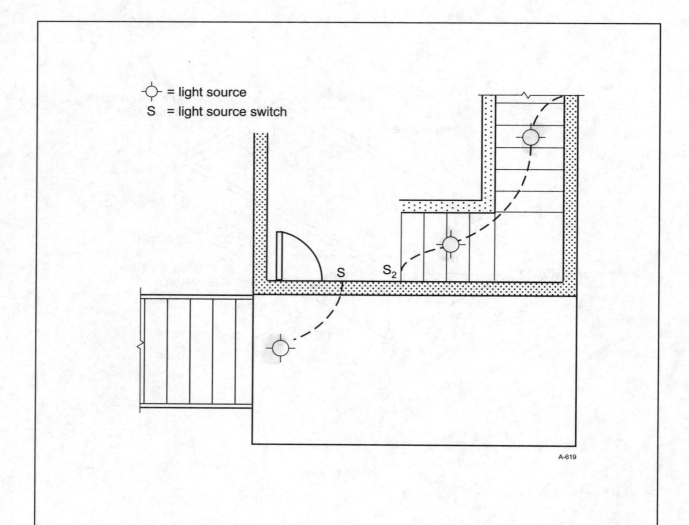

A-619

Unless the light sources are on continuously or automatically activated, interior stairway lights must be controlled from both the top and bottom of each stairway. For exterior stairway lighting, the control switch is to be located within the dwelling unit.

Topic: Minimum Areas and Dimensions
Reference: IRC R304

Category: Building Planning
Subject: Minimum Room Areas

Code Text: *Every dwelling unit shall have at least one habitable room that shall have not less than 120 square feet (11.2 m²) of gross floor area. Other habitable rooms shall have a floor area of not less than 70 square feet (6.5 m²). Habitable rooms shall not be less than 7 feet (2134 mm) in any horizontal dimension.* See exceptions for kitchens regarding minimum size and horizontal dimensions.

Discussion and Commentary: Acceptable sizes for habitable rooms have been established. Because habitable rooms are expected to be those spaces within a dwelling unit where most activities take place, they are the only areas regulated. Most habitable rooms need be only 7 feet by 10 feet to comply with the provisions; however, at least one larger room must be provided. It is seldom that any habitable room in today's typical dwelling unit would be designed with such a small floor area.

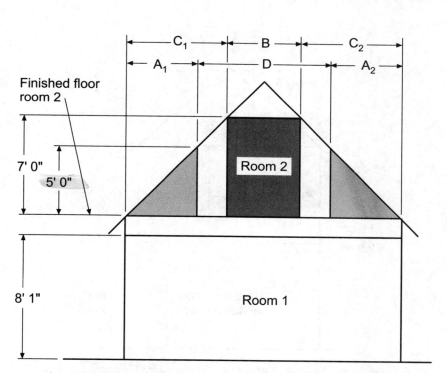

Area A - Floor area based on A_1 + A_2 times room depth does not contribute to the minimum required habitable area for Room 2

Area B ≤ 50% D; floor area based on B times room depth is not to exceed 50% of total floor area of D times room depth

For SI: 1 foot = 304.8 mm.

A-681

The minimum required floor area for any habitable room having a sloping ceiling, as would typically be encountered where an attic area is finished for use as a living or sleeping area, can be based on only those portions of the room with a ceiling height of at least 5 feet (1524 mm).

Topic: Minimum Height   **Category:** Building Planning
Reference: IRC R305.1   **Subject:** Ceiling Height

Code Text: *Habitable rooms, hallways, corridors, bathrooms, toilet rooms, laundry rooms and basements shall have a ceiling height of not less than 7 feet (2134 mm).* See four exceptions addressing ceilings with exposed beams, basements without habitable spaces, rooms with sloped ceilings and bathrooms. *The required height shall be measured from the finish floor to the lowest projection from the ceiling.*

Discussion and Commentary: For both safety reasons and usability by the occupants, the minimum ceiling height throughout occupiable areas of a dwelling unit is regulated. Most rooms that are commonly used by the occupants are included, other than closets and storage areas. Where basements are used for habitable purposes, they too must comply with the minimum height requirement of 7 feet (2134 mm).

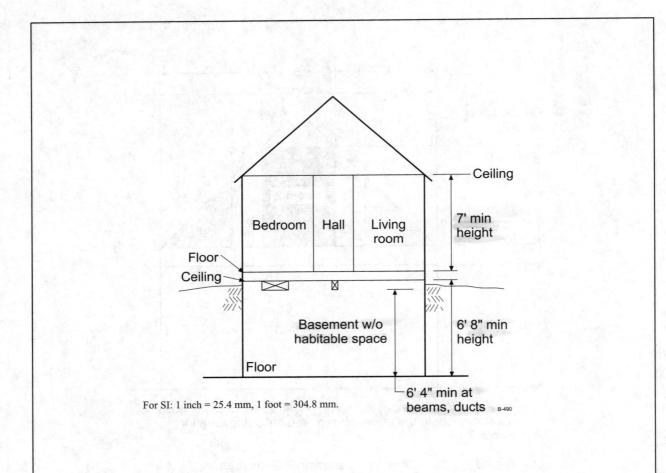

For SI: 1 inch = 25.4 mm, 1 foot = 304.8 mm.

Beams and girders spaced not less than 4 feet (1219 mm) on center may project not more than 6 inches (152 mm) below the required ceiling height.

Topic: Minimum Space Required
Reference: IRC R307

Category: Building Planning
Subject: Toilet, Bath and Shower Spaces

Code Text: *Fixtures shall be spaced as per Figure R307.2. Bathtub and shower floors and walls above bathtubs with installed shower heads and in shower compartments shall be finished with a nonabsorbent surface. Such wall surfaces shall extend to a height of not less than 6 feet (1829 mm) above the floor.* ✕

Discussion and Commentary: It is necessary to provide adequate clearances at and around bathroom fixtures to allow for ease of use by the occupants of the dwelling unit. The code addresses clear floor space and clearances for lavatories, water closets, bathtubs, and showers. In addition, the minimum permitted size for a shower is 30 inches by 30 inches (762 mm by 762 mm).

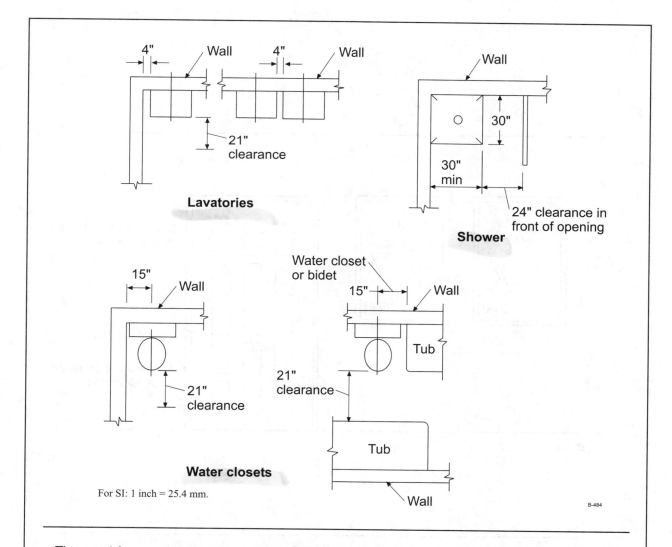

The need for nonabsorbent surfaces in shower areas subject to water splash is based on two concerns. For sanitation purposes, finish materials must be of a type that can be cleaned easily. Also, continued absorption of moisture will lead to deterioration of the structural components.

Topic: Safety Glazing Identification

Category: Building Planning

Reference: IRC R308.1

Subject: Glazing

Code Text: *Except as indicated in Section R308.1.1, each pane of glazing installed in hazardous locations as defined in Section R308.4 shall be provided with a manufacturer's or installer's label, designating the type and thickness of glass and the safety glazing standard with which it complies, which is visible in the final installation. The label shall be acid etched, sandblasted, ceramic-fired, embossed mark, or shall be of a type which once applied cannot be removed without being destroyed.* See exceptions for tempered spandrel glass and where certifications of compliance are provided.

Discussion and Commentary: Improper glazing installed in areas subject to human impact can create a serious hazard. Accordingly, it is critical that glazing in such locations be appropriately identified to help ensure that the proper glazing is in place.

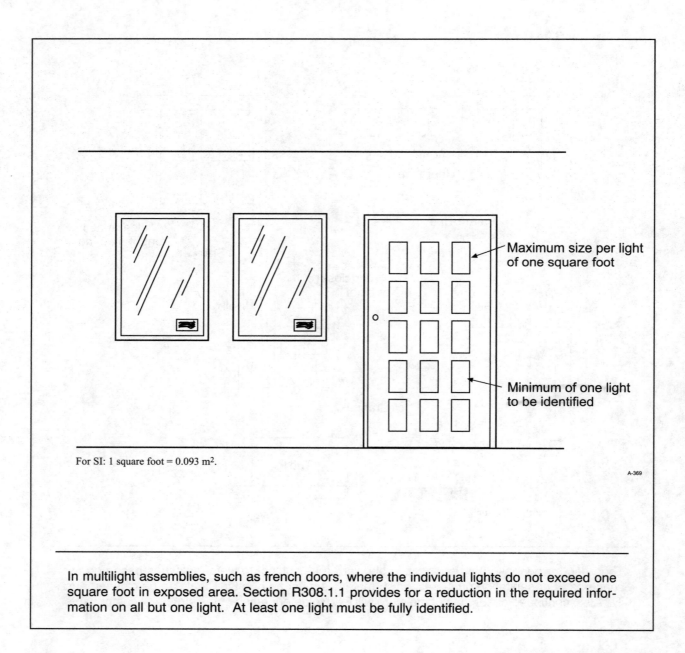

Maximum size per light of one square foot

Minimum of one light to be identified

For SI: 1 square foot = 0.093 m².

A-369

In multilight assemblies, such as french doors, where the individual lights do not exceed one square foot in exposed area. Section R308.1.1 provides for a reduction in the required information on all but one light. At least one light must be fully identified.

Topic: Hazardous Locations
Reference: IRC R308.4

Category: Building Planning
Subject: Glazing

Code Text: *The following shall be considered specific hazardous locations for the purposes of glazing:* here the code identifies eleven specific locations including glazing in and adjacent to doors, glazing in shower and bathtub enclosure walls, glazing in walls adjacent to pools and spas, and glazing enclosing stairway landings.

Discussion and Commentary: A number of glazing locations have been identified over the years as being particular hazards due to their frequent impact by building occupants. Glazing in doors is of particular concern due to the increased likelihood of accidental impact by individuals operating or opening the doors. In addition, a person may push against a glazed portion of the door to gain leverage. Large pieces of glass may also create a hazard where located close to a travel path, because it is possible to impact glazing where no barrier is provided.

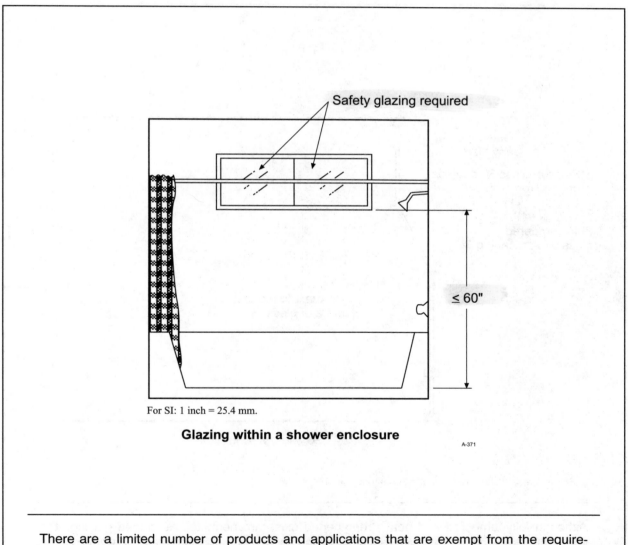

For SI: 1 inch = 25.4 mm.

Glazing within a shower enclosure

A-371

There are a limited number of products and applications that are exempt from the requirements for hazardous locations, including small openings in doors through which a 3-inch-diameter sphere will not pass, and specific decorative assemblies, such as leaded, faceted or carved glass.

Topic: Skylights and Sloped Glazing

Reference: IRC R308.6

Category: Building Planning

Subject: Glazing

Code Text: *The following types of glazing may be used (for skylights and sloped glazing): 1) laminated glass with a complying polyvinyl butyral interlayer; 2) fully tempered glass; 3) heat-strengthened glass; 4) wired glass; and 5) approved rigid plastics.*

Discussion and Commentary: A skylight or sloped glazing is considered *glass or other transparent or translucent glazing material installed at a slope of more than 15 degrees from vertical. Glazing materials in skylights, including unit skylights, solariums, sunrooms, roofs and sloped walls are included.* The requirements enhance the protection of a building's occupants from the possibility of falling glazing materials. Only certain glazing materials have the necessary characteristics to be permitted in an overhead installation. In addition, the provisions of the IRC address design loads normally attributed to roofs.

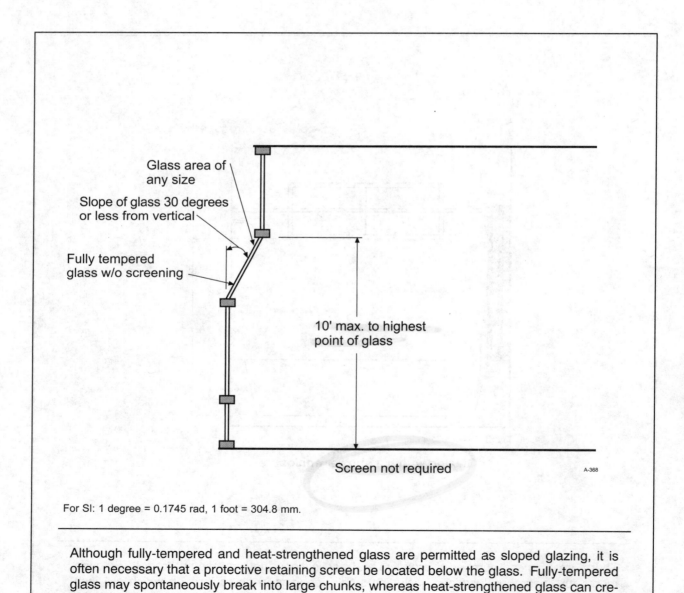

Glass area of any size

Slope of glass 30 degrees or less from vertical

Fully tempered glass w/o screening

10' max. to highest point of glass

Screen not required

A-368

For SI: 1 degree = 0.1745 rad, 1 foot = 304.8 mm.

Although fully-tempered and heat-strengthened glass are permitted as sloped glazing, it is often necessary that a protective retaining screen be located below the glass. Fully-tempered glass may spontaneously break into large chunks, whereas heat-strengthened glass can create falling shards.

Topic: Opening Protection
Reference: IRC R309.1

Category: Building Planning
Subject: Garages and Carports

Code Text: *Openings from a private garage directly into a room used for sleeping purposes shall not be permitted. Other openings between the garage and residence shall be equipped with solid wood doors not less than 1³/₈ inches (35 mm) in thickness, solid or honeycomb core steel doors not less than 1³/₈ inches (35 mm) thick, or 20-minute fire-rated doors.*

Discussion and Commentary: Although gypsum board or other approved material provides an adequate fire separation at the walls and/or ceiling between the garage and the dwelling unit, it is important that any other openings penetrating the separation be appropriately protected. Therefore, the type of door construction or fire rating of the door is mandated. Because the code covers only the door itself, it is not necessary to have a complete fire-rated door assembly, nor is it a requirement to provide a self-closing device. The level of separation provided by the door is consistent with that provided by the gypsum board installed on the garage side.

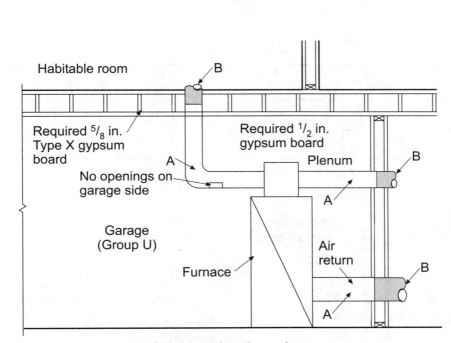

Cross section through garage

Ducts A - 0.019 in. (No. 26 Gage) galvanized steel with no openings into garage

Ducts B - any approved duct

For SI: 1 inch = 25.4 mm.

B-491c

Where any duct passes through the gypsum board membrane located on the garage side of the separation, the duct material must be of minimum No. 26 gage (0.48 mm) sheet steel. There must be no openings in the duct within the garage area.

Topic: Separation Required
Reference: IRC R309.2

Category: Building Planning
Subject: Garages and Carports

Code Text: *The garage shall be separated from the residence and its attic area by not less than 1/2-inch (12.7 mm) gypsum board applied to the garage side. Garages beneath habitable rooms shall be separated from all habitable rooms above by not less than 5/8-inch (15.9 mm) Type X gypsum board or equivalent. Where the separation is a floor-ceiling assembly, the structure supporting the separation shall also be protected by not less than 1/2-inch (12.7 mm) gypsum board or equivalent.*

Discussion and Commentary: It is not uncommon for a fire to start in a private garage attached directly to a dwelling unit. In many cases, the fire may grow and go unnoticed for a period of time, becoming a distinct hazard to the residence and its occupants. The code mandates a minimum level of fire separation from the garage to the dwelling unit to allow the residents time for escape. In addition, the separation may be adequate to restrict the spread of fire to the dwelling unit until the fire can be controlled and extinguished.

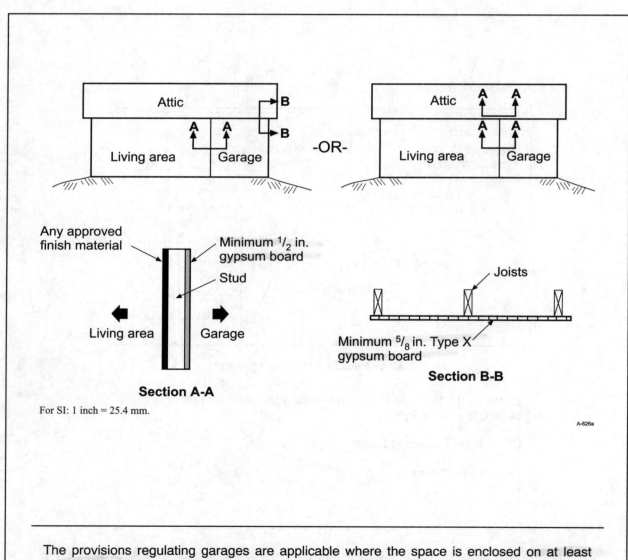

The provisions regulating garages are applicable where the space is enclosed on at least three sides. Where enclosed on two sides or fewer, the parking area is considered a carport, and the separation requirements do not apply.

Topic: Window Wells
Reference: IRC R310.1, R310.2

Category: Building Planning
Subject: Emergency Escape and Rescue Openings

Code Text: *Emergency escape and rescue openings with a finished sill height below the adjacent ground elevation shall be provided with a window well . . . . The minimum horizontal area of the window well shall be 9 square feet (0.84 m2), with a minimum horizontal projection and width of 36 inches (914 mm).* See exception for ladder encroachment into required dimensions. *The area of the window well shall allow the emergency and rescue opening to be fully opened.*

Discussion and Commentary: The window well provisions address those emergency escape and rescue openings that occur below grade. Simply applying the standard opening criteria to these window wells does not provide an adequate clear opening size to transition from the below-grade space to grade level. The increased cross-sectional dimension will be sufficient to provide for the escape of occupants or the entry of firefighters or other rescue personnel.

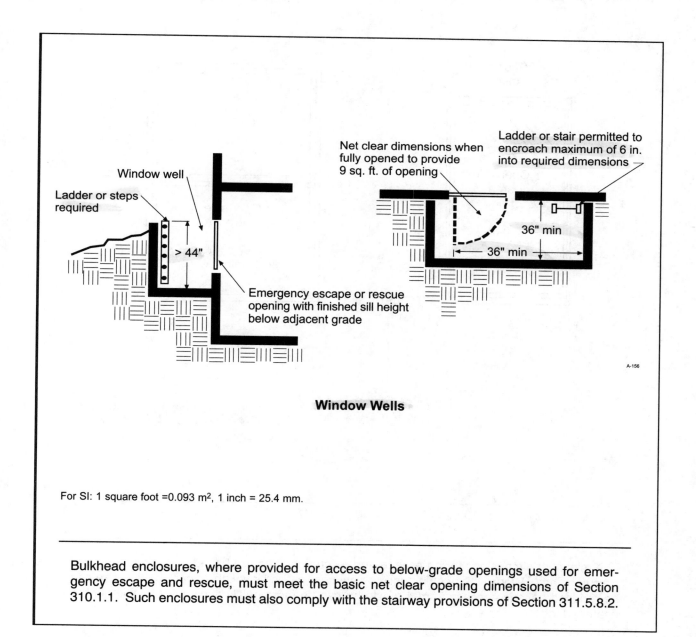

Window Wells

For SI: 1 square foot =0.093 m2, 1 inch = 25.4 mm.

Bulkhead enclosures, where provided for access to below-grade openings used for emergency escape and rescue, must meet the basic net clear opening dimensions of Section 310.1.1. Such enclosures must also comply with the stairway provisions of Section 311.5.8.2.

Topic: Openings Required
Reference: IRC R310.1

Category: Building Planning
Subject: Emergency Escape and Rescue Openings

Code Text: *Basements with habitable space and every sleeping room shall have at least one openable emergency escape and rescue opening. Where basements contain one or more sleeping rooms, emergency egress and rescue openings shall be required in each sleeping room, but shall not be required in adjoining areas of the basement. Where emergency escape and rescue openings are provided they shall have a sill height of not more than 44 inches (1118 mm) above the floor.*

Discussion and Commentary: Because so many deaths and injuries from fire occur as the result of occupants of residential buildings being asleep at the time of a fire, the code requires that basements containing any habitable space and all sleeping rooms have doors or windows that may be used for emergency escape or rescue. The concern is based on the fact that often a fire will have spread before the occupants are aware of the danger; thus, the normal means of escape will most likely be blocked.

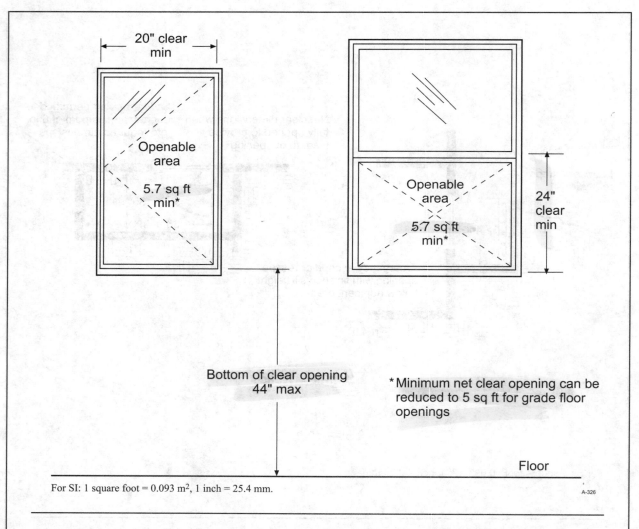

20" clear min

Openable area

5.7 sq ft min*

Openable area

5.7 sq ft min*

24" clear min

Bottom of clear opening 44" max

*Minimum net clear opening can be reduced to 5 sq ft for grade floor openings

Floor

For SI: 1 square foot = 0.093 m², 1 inch = 25.4 mm.

A-326

The minimum dimensions for an escape and rescue opening are based on both escape and rescue criteria. Where the opening occurs very near ground level, the clear opening size can be reduced because access from the exterior can be more easily accomplished without the use of a ladder.

QUIZ
Study Session 3

1. When natural light is used to satisfy the minimum light requirements for habitable rooms, the aggregate glazing area shall be a minimum of _____ of the floor area.
   a. 4 percent
   b. 5 percent
   c. 8 percent
   d. 10 percent

2. When natural ventilation is used to satisfy the minimum ventilation requirements for habitable rooms, the aggregate open area to the outdoors shall be a minimum of _____ of the floor area.
   a. 4 percent
   b. 5 percent
   c. 8 percent
   d. 10 percent

3. Where a whole-house mechanical ventilation system is installed, the natural ventilation requirements are not applicable, provided the system is capable of supplying a minimum of _____ of outside ventilation air per computed occupant.
   a. 15 cfm
   b. 20 cfm
   c. 30 cfm
   d. 50 cfm

4. Where an intermittent mechanical ventilation system is provided in a bathroom in lieu of exterior openings, the minimum required ventilation rate shall be _____.
   a. 15 cfm
   b. 20 cfm
   c. 30 cfm
   d. 50 cfm

5. At least one habitable room with a minimum floor area of _____ square feet shall be provided in every dwelling unit.
   a. 100
   b. 120
   c. 150
   d. 200

6. The floor area beneath a furred ceiling can be considered to be contributing to the minimum required habitable area for the room where it has a minimum height of _____ above the floor.
    a. 5 feet
    b. 6 feet, 4 inches
    c. 6 feet, 8 inches
    d. 7 feet

7. Other than in a kitchen, what is the minimum permitted horizontal dimension of any habitable room?
    a. 6 feet, 8 inches
    b. 7 feet, 0 inches
    c. 7 feet, 6 inches
    d. 8 feet, 0 inches

8. In general, the minimum ceiling height of all habitable rooms is _____.
    a. 6 feet, 8 inches
    b. 7 feet, 0 inches
    c. 7 feet, 6 inches
    d. 8 feet, 0 inches

9. Ceilings in basements without habitable spaces shall have a minimum ceiling height of _____ from the finished floor to beams, ducts, or other obstructions.
    a. 6 feet, 4 inches
    b. 6 feet, 6 inches
    c. 6 feet, 8 inches
    d. 7 feet, 0 inches

10. A minimum clearance of _____ shall be provided in front of a water closet.
    a. 21 inches
    b. 24 inches
    c. 30 inches
    d. 32 inches

11. The minimum distance between the centerline of a water closet and a side wall shall be _____.
    a. 15 inches
    b. 16 inches
    c. 18 inches
    d. 21 inches

12. Within a shower compartment, the walls shall be finished with a nonabsorbent surface to a minimum height of _____.
    a. 70 inches
    b. 72 inches
    c. 78 inches
    d. 84 inches

13.   Unless located a minimum of 60 inches above the floor or walking surface, glazing within a
        _____ arc of a door shall be considered to be installed in a hazardous location.
        a. 12-inch
        b. 24-inch
        c. 36-inch
        d. 60-inch

14.   Where located less than 5 feet above the walking surface, glazing in a wall enclosing an indoor
        hot tub shall be considered to be installed in a hazardous location unless the glazing is a mini-
        mum of _____ horizontally from the water's edge.
        a. 3 feet
        b. 4 feet
        c. 5 feet
        d. 10 feet

15.   Sloped glazing is glass or other glazing material installed at a minimum slope of _____.
        a. 15 degrees from the vertical
        b. 30 degrees from the vertical
        c. 15 degrees from the horizontal
        d. 30 degrees from the horizontal

16.   Curbs are not required for skylights installed on roofs having a minimum slope of _____.
        a. 2:12
        b. 3:12
        c. 4:12
        d. 5:12

17.   Where a solid wood door is installed as a permitted opening between the garage and the resi-
        dence, the minimum thickness of the door shall be _____.
        a. $1^1/_4$ inches
        b. $1^3/_8$ inches
        c. $1^5/_8$ inches
        d. $1^3/_4$ inches

18.   Any duct penetrating the wall or ceiling separating the dwelling from the garage shall be con-
        structed of a minimum _____ sheet steel or other approved material.
        a. No. 22 gage
        b. No. 24 gage
        c. No. 26 gage
        d. No. 28 gage

19. A garage shall be separated from a residence and its attic area by minimum _____ gypsum board.
   a. $^1/_2$-inch
   b. $^1/_2$-inch Type "X"
   c. $^5/_8$-inch
   d. $^5/_8$-inch Type "X"

20. Escape and rescue openings shall have a maximum sill height of _____ above the floor.
   a. 40 inches
   b. 42 inches
   c. 44 inches
   d. 48 inches

21. Escape and rescue openings when considered as grade floor openings shall have a minimum net clear opening of _____.
   a. 4.0 square feet
   b. 4.4 square feet
   c. 5.0 square feet
   d. 5.7 square feet

22. For an emergency escape and rescue opening, the minimum clear height shall be _____, and the minimum clear width shall be _____.
   a. 24 inches, 20 inches
   b. 24 inches, 28 inches
   c. 28 inches, 22 inches
   d. 28 inches, 24 inches

23. Where a window well is required in conjunction with an escape and rescue opening, the window well shall have a minimum net clear area of _____ square feet with a minimum horizontal dimension of _____.
   a. 5.7, 20 inches
   b. 5.7, 24 inches
   c. 9.0, 30 inches
   d. 9.0, 36 inches

24. A garage located beneath a habitable room shall be separated from the room by a minimum _____ gypsum board.
   a. $^1/_2$-inch
   b. $^1/_2$-inch Type X
   c. $^5/_8$-inch
   d. $^5/_8$-inch Type X

25. A 6-square-foot skylight of laminated glass shall have a minimum _____ polyvinyl butyral interlayer where located 14 feet above the walking surface.
    a. 0.015-inch
    b. 0.024-inch
    c. 0.030-inch
    d. 0.044-inch

26. A mechanical outside air intake opening shall be located a minimum of _____ feet from a plumbing vent.
    a. 3
    b. 5
    c. 10
    d. 12

27. The illumination source for interior stairs shall be capable of illuminating the treads and landings to a minimum level of _____ foot-candle(s).
    a. 1
    b. 5
    c. 8
    d. 10

28. A minimum ceiling height of _____ is required at the front clearance area of bathroom fixtures.
    a. 6 feet, 4 inches
    b. 6 feet, 6 inches
    c. 6 feet, 8 inches
    d. 7 feet, 0 inches

29. Glazing adjacent to a door to a storage closet is not required to be safety glazing provided the closet is a maximum of _____ in depth.
    a. 24 inches
    b. 30 inches
    c. 36 inches
    d. 42 inches

30. Glazing in a door shall be safety glazing where the opening allows the passage of a minimum _____ sphere.
    a. 3-inch
    b. $3^1/_2$-inch
    c. 4-inch
    d. 6-inch

INTERNATIONAL RESIDENTIAL CODE
Study Session 4
Sections R311 through R323—Building Planning III

OBJECTIVE: To obtain an understanding of the fire and life safety criteria of the code, including stairways, ramps and landings; handrails and guards; smoke alarms; foam plastic and insulation; flame spread and smoke density of wall and ceiling finishes; separation of dwelling units, including townhouse separations; protection against decay and termite infestation; and flood-resistant construction.

REFERENCE: Sections R311 through R323, 2003 *International Residential Code*

KEY POINTS:
- How should unenclosed space under stairs be addressed? What if it is enclosed but not accessible? What if it is accessible for use?
- What is the minimum required width of a hallway?
- How many exit doors from a dwelling are required? Under what conditions is an additional exit door required? What is the minimum required sized of an exit door?
- What is the minimum size required for a landing? Under what conditions is a landing not required on each side of an exterior door?
- How wide must a stairway be? Where is this width measured? What type of encroachments into the minimum required width are permitted? How far may they encroach? Where may they be located?
- What is the maximum allowable riser height in a stairway? Minimum allowable tread run? Within a flight of stairs, what is the maximum tolerance permitted between the smallest and greatest riser height? Between the smallest and greatest tread run?
- What are the various criteria for tread nosings? When are nosings not required?
- What is the minimum allowable headroom clearance at stairways and landings? How is this height measured?
- How is handrail height measured? What is the minimum height permitted? The maximum height? What sizes and shapes of handrails are acceptable?
- What types of special stairways are addressed? How do they differ in layout and size from traditional stairway design?
- How is a ramp defined? What is the maximum slope permitted for a ramp?
- When is a handrail required for a ramp? What size landings are required?
- When is a guard required? What is the minimum allowable height of the guard above the walking surface? How do the provisions differ for guards on open sides of stairs?
- How must intermediate rails be located?
- Where are smoke alarms required to be located? Do they need to be interconnected? How must they be powered? What must be done when alterations or repairs take place?
- How are foam plastics regulated? What are the restrictions on wall and ceiling finishes? Insulation materials?
- In a two-family dwelling, how must the units be separated? What minimum fire-resistive separation is mandated between townhouses? When is a common two-hour fire-resistance-rated wall acceptable? When are parapets required? How are penetrations of dwelling unit separation assemblies to be protected?
- What areas are considered subject to decay damage? Termite damage? How should such conditions be abated?
- What type of building constructed under the IRC requires some degree of accessibility?
- What are the general requirements for buildings constructed in flood hazard areas?

Topic: Exit Doors

Category: Building Planning

Reference: IRC R311

Subject: Means of Egress

Code Text: *Not less than one exit door conforming to the provisions of Section R311 shall be provided from each dwelling unit. The required exit door shall provide for direct access from the habitable portions of the dwelling to the exterior without requiring travel through a garage. The required exit door shall be a side-hinged door not less than 3 feet (914 mm) in width and 6 feet 8 inches (2032 mm) in height. Other doors shall not be required to comply with these minimum dimensions. All egress doors shall be readily openable from the side from which egress is to be made without the use of a key or special knowledge or effort.*

Discussion and Commentary: Openings to sleeping rooms from garages are not allowed, because a person might not awake in time if there were a hazard from carbon monoxide fumes or smoke in the garage. Where openings are permitted, they must maintain the integrity of the required separation between the garage and dwelling unit. If 20-minute fire-rated doors are utilized in lieu of the other options identified in the code, they are not required to be provided with a rated door frame or self-closing device.

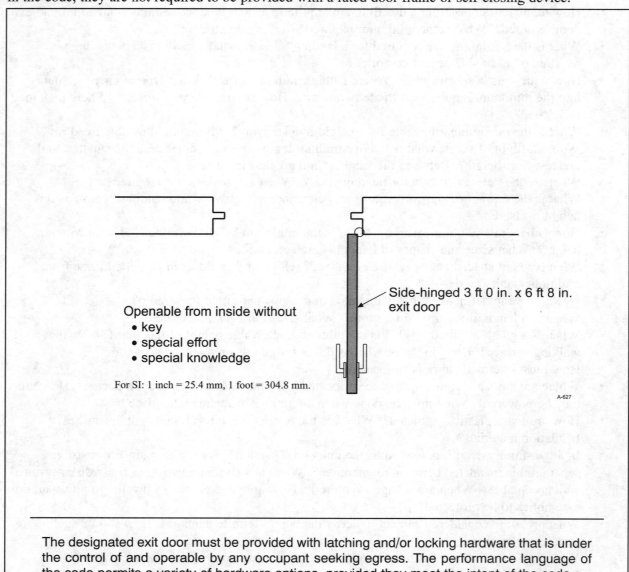

Openable from inside without
- key
- special effort
- special knowledge

For SI: 1 inch = 25.4 mm, 1 foot = 304.8 mm.

Side-hinged 3 ft 0 in. x 6 ft 8 in. exit door

A-627

The designated exit door must be provided with latching and/or locking hardware that is under the control of and operable by any occupant seeking egress. The performance language of the code permits a variety of hardware options, provided they meet the intent of the code.

2003 IRC Study Companion

Topic: Stairway
Reference: IRC R311.5.1

Category: Building Planning
Subject: Means of Egress

Code Text: *Stairways shall not be less than 36 inches (914 mm) in clear width at all points above the permitted handrail height and below the required headroom height. Handrails shall not project more than 4.5 inches (114 mm) on either side of the stairway and the minimum clear width of the stairway at and below the handrail height, including treads and landings, shall not be less than 31.5 inches (787 mm) where a handrail is installed on one side and 27 inches (686 mm) where handrails are provided on both sides. See exception for spiral stairways.*

Discussion and Commentary: Although a minimum width of 36 inches is required for stairways, the code is not as concerned about elements such as trim, stringers or other items that may be found below the level of the handrail, as long as they do not exceed the projection of the handrail(s). This width is based on the body's movements as a person walks on a stairway.

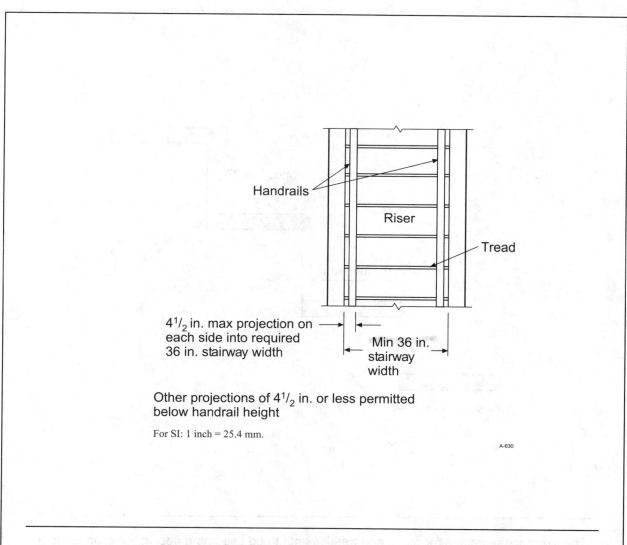

Handrails

Riser

Tread

$4^1/_2$ in. max projection on each side into required 36 in. stairway width

Min 36 in. stairway width

Other projections of $4^1/_2$ in. or less permitted below handrail height

For SI: 1 inch = 25.4 mm.

A-630

Typically, when a building code mandates a minimum width, it is expected that the width be the clear, net, usable, unobstructed width. In this case, however, limited projections into the minimum width are acceptable where located at or below the height of the handrail.

Topic: Stair Treads and Risers
Reference: IRC R311.5.3

Category: Building Planning
Subject: Means of Egress

Code Text: *The maximum riser height shall be 7³/₄ inches (196 mm). The greatest riser height within any flight of stairs shall not exceed the smallest by more than ³/₈ inch (9.5 mm). The minimum tread depth shall be 10 inches (254 mm). The greatest tread depth within any flight of stairs shall not exceed the smallest by more than ³/₈ inch (9.5 mm).*

Discussion and Commentary: Although the rise and run configuration is important in regulating stairway safety, another significant safety factor relative to stairways is the uniformity of risers and treads in any flight, which is the section of a stairway leading from one landing to the next. It is important that any variation that would interfere with the rhythm of the stair user be avoided. Special allowances are provided for winder treads.

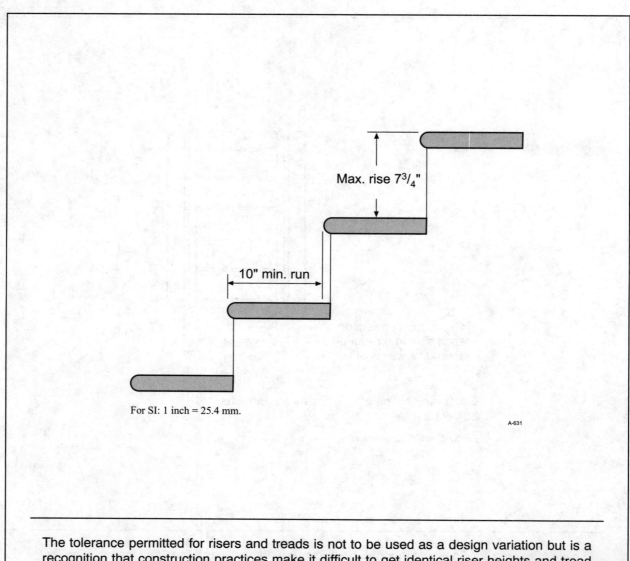

Max. rise 7³/₄"

10" min. run

For SI: 1 inch = 25.4 mm.

A-631

The tolerance permitted for risers and treads is not to be used as a design variation but is a recognition that construction practices make it difficult to get identical riser heights and tread dimensions in field construction or where changes occur in stair and floor coverings.

2003 IRC Study Companion

Topic: Special Stairways
Reference: IRC R311.5.3, R311.5.8

Category: Building Planning
Subject: Means of Egress

Code Text: *Winder treads shall have a minimum tread depth of 10 inches (254 mm) measured at a point 12 inches (305 mm) from the side where the treads are narrower. Winder treads shall have a minimum tread depth of 6 inches (152 mm) at any point. Spiral stairways are permitted, provided the minimum width shall be 26 inches (660 mm) with each tread having a 7 1/2-inches (190 mm) minimum tread depth at 12 inches (305 mm) from the narrower edge. All treads shall be identical, and the rise shall be not more than 9 1/2 inches (241 mm). Stairways serving bulkhead enclosures, not part of the required building egress, providing access from the outside grade level to the basement shall be exempt from the require-ments of Sections R311.4.3 (landings at doors) and R311.5 (stairways) where the maximum height from the basement finished floor level to grade adjacent to the stairway does not exceed 8 feet.*

Discussion and Commentary: Four special types of stairways are allowed for use in dwelling units, pro-vided they are constructed in compliance with the specific limitations mandated by the code.

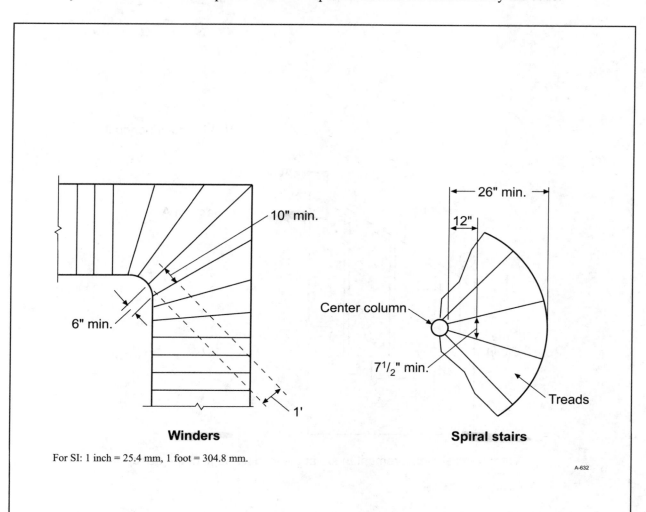

Winders

For SI: 1 inch = 25.4 mm, 1 foot = 304.8 mm.

Spiral stairs

A-632

It is obvious that winders, spiral stairs and circular stairways are not as easy to use as the tra-ditional stairway design. It is important that, while the rise and run may differ from the funda-mental criteria, the travel path be consistent in size and shape to reduce the hazard level.

Topic: Stairway Handrail
Reference: IRC R311.5.6

Category: Building Planning
Subject: Means of Egress

Code Text: *Handrails shall be provided on at least one side of each continuous run of treads or flight with four or more risers. Handrail height, measured vertically from the sloped plane adjoining the tread nosing, or finish surface of ramp slope, shall be not less than 34 inches (864 mm) and not more than 38 inches (965 mm). Handrails for stairways shall be continuous for the full length of the flight, from a point directly above the top riser of the flight to a point directly above the lowest riser of the flight. Handrail ends shall be returned or shall terminate in newel posts or safety terminals.* See exceptions.

Discussion and Commentary: Any stairway consisting of four or more risers must be provided with a handrail on at least one side. The rail is a proven safety feature for users of stairways, particularly those who fail to pay proper attention to stair use. Two types of handrail shapes are described, differing because of their perimeter measurement.

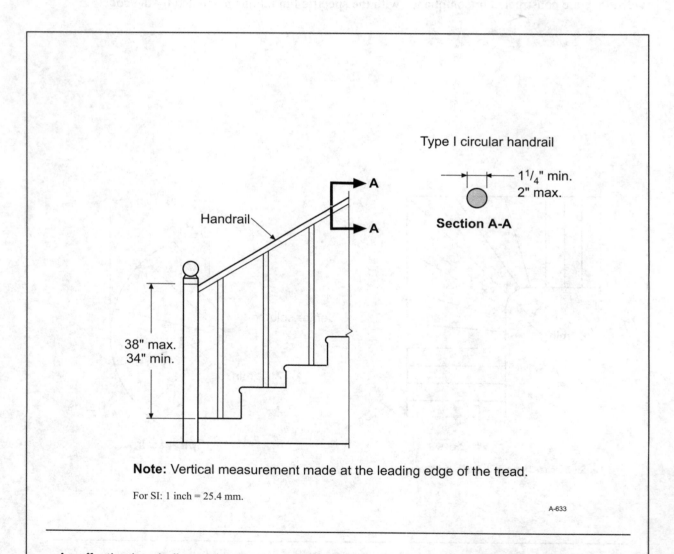

Type I circular handrail

$1\frac{1}{4}$" min.
2" max.

Section A-A

Handrail

38" max.
34" min.

Note: Vertical measurement made at the leading edge of the tread.

For SI: 1 inch = 25.4 mm.

A-633

An effective handrail must be of a size and shape that can be easily grasped by the majority of stair users. Whereas the minimum and maximum cross sections of $1\frac{1}{4}$ inches and 2 inches are specific for circular rails, other shapes with equivalent grasping surfaces may be acceptable.

Topic: Ramps
Reference: IRC R311.6

Category: Building Planning
Subject: Means of Egress

1:8

Code Text: *Ramps shall have a maximum slope of one unit vertical in eight units horizontal (12.5-percent slope). A minimum 3-foot-by-3-foot (914 mm by 914 mm) landing shall be provided: at the top and bottom of ramps, where doors open onto ramps, and where ramps change direction. Handrails shall be provided on at least one side of all ramps exceeding a slope of one unit vertical in 12 units horizontal (8.33-percent slope).*

Discussion and Commentary: A ramp is defined as a walking surface that has a running slope steeper than 1 unit vertical in 20 units horizontal (5-percent slope). For general use purposes as well as exiting, the maximum permitted slope is 1:8. Ramps with slopes of 1:12 to 1:20 do not require a handrail, as the rise or descent is gradual enough to provide a safe travel path.

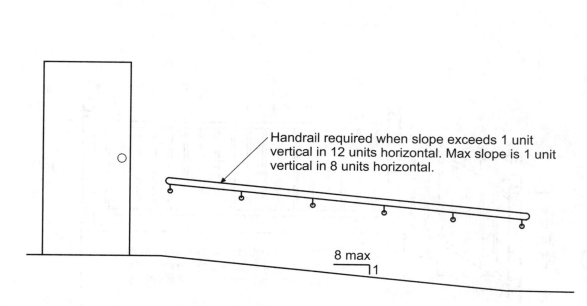

Handrail required when slope exceeds 1 unit vertical in 12 units horizontal. Max slope is 1 unit vertical in 8 units horizontal.

8 max
1

3 ft x 3 ft min landing required:
- At the top and bottom of ramp
- Where doors open onto ramp
- Where ramp changes direction

For SI: 1 foot = 304.8 mm.

A-629

The ramp provisions in the IRC differ significantly from those in the IBC because of the building code's emphasis on accessibility. The IRC requirements are simply to allow safe movement through the dwelling unit and do not address use by persons with physical disabilities.

Topic: Location and Height
Reference: IRC R312.1

Category: Building Planning
Subject: Guards

Code Text: *Porches, balconies or raised floor surfaces located more than 30 inches (762 mm) above the floor or grade below shall have guards not less than 36 inches (914 mm) in height. Open sides of stairs with a total rise of more than 30 inches (762 mm) above the floor or grade below shall have guards not less than 34 inches (864 mm) in height measured vertically from the nosing of the treads.*

Discussion and Commentary: Some form of protection is necessary at elevated floor areas with a vertical drop of more than 30 inches, as a fall from that height can potentially result in injury. Where required, the guard must be of an adequate height to prevent someone from falling over the edge of the protected area.

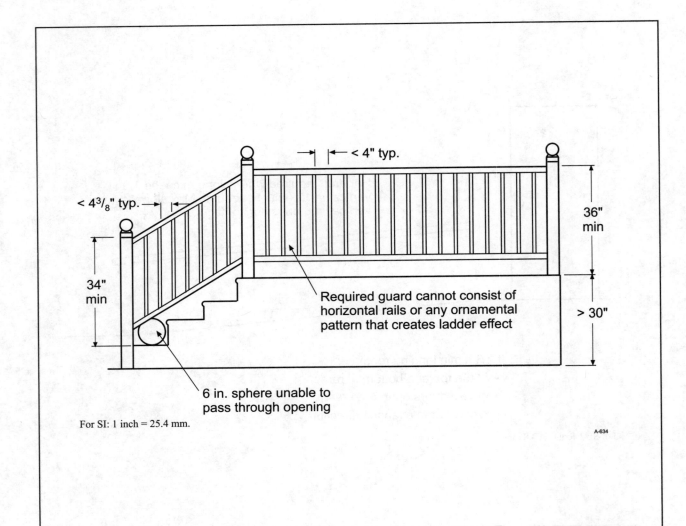

For SI: 1 inch = 25.4 mm.

A-634

Required at balconies, porches, mezzanines and landings, guards are also needed on the open sides of stairs where the vertical distance to the surface below exceeds 30 inches. In such situations, it is acceptable to use a complying handrail as the guard.

Topic: Opening Limitations
Reference: IRC R312.2

Category: Building Planning
Subject: Guards

Code Text: *Required guards on open sides of stairways, raised floor areas, balconies and porches shall have intermediate rails or ornamental closures which do not allow passage of a sphere 4 inches (102 mm) or more in diameter.* See exception for triangular openings formed by riser, tread, and bottom rail.

Discussion and Commentary: Guards must be constructed so that they not only prevent people from falling over them but also prevent children from crawling through them. The criteria spacing was chosen after many years of research and discussion. The chance of even a very small child being able to get through such a narrow opening is very low.

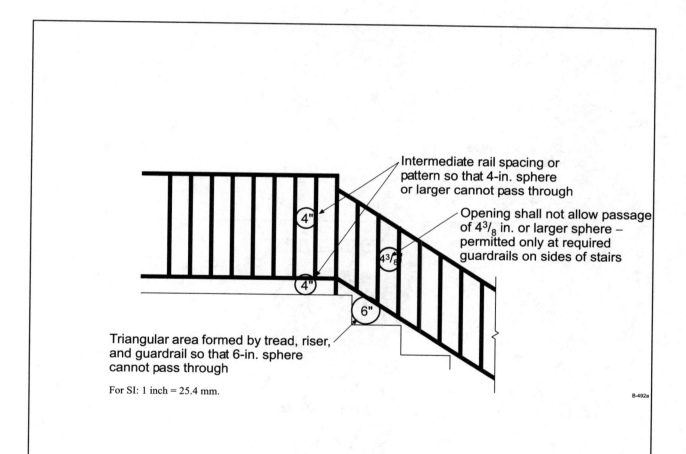

Intermediate rail spacing or pattern so that 4-in. sphere or larger cannot pass through

Opening shall not allow passage of 4³/₈ in. or larger sphere – permitted only at required guardrails on sides of stairs

Triangular area formed by tread, riser, and guardrail so that 6-in. sphere cannot pass through

For SI: 1 inch = 25.4 mm.

B-492a

In addition to limiting the opening size along the open sides of stairways, the code addresses the hazard created by open risers. Section R311.5.3.3 allows the use of open risers but only where the opening between the treads does not permit the passage of a 4-inch diameter sphere.

Topic: Location and Interconnection
Reference: IRC R313.1

Category: Building Planning
Subject: Smoke Alarms

Code Text: *Smoke alarms shall be installed in the following locations: 1) in each sleeping room, 2) outside of each separate sleeping area in the immediate vicinity of the bedrooms, and 3) on each additional story of the dwelling, including basements but not including crawl spaces and unhabitable attics. When more than one smoke alarm is required to be installed within an individual dwelling unit the alarm devices shall be interconnected in such a manner that the actuation of one alarm will activate all of the alarms in the individual unit.*

Discussion and Commentary: A majority of deaths from fire in residential structures occur because of the delay in detecting a fire. Therefore, smoke alarms are mandated throughout the typically occupied areas of a dwelling unit to not only detect the products of combustion but also to alert the occupants of the fire.

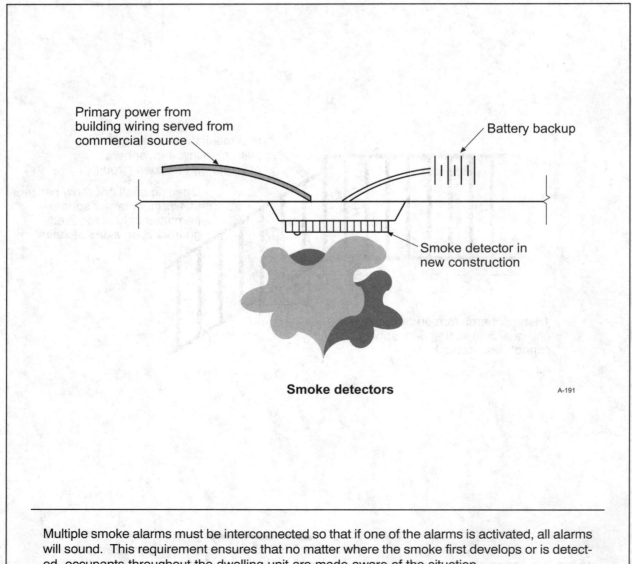

Smoke detectors   A-191

Multiple smoke alarms must be interconnected so that if one of the alarms is activated, all alarms will sound. This requirement ensures that no matter where the smoke first develops or is detected, occupants throughout the dwelling unit are made aware of the situation.

Topic: Thermal Barrier
Reference: IRC R314.1.2

Category: Building Planning
Subject: Foam Plastic

Code Text: *Foam plastic, except where otherwise noted, shall be separated from the interior of a building by minimum $^1/_2$-inch (12.7 mm) gypsum board or an approved finish material equivalent to a thermal barrier to limit the average temperature rise of the unexposed surface to no more than 250°F (121°C) after 15 minutes of fire exposure to the ASTM E 119 standard time temperature curve.*

Discussion and Commentary: Foam plastic must not be exposed to the building's interior during the early stages of a fire; rather, it must be isolated from the fire for a short period of time. The adequacy of the barrier is based on its ability to remain in place for 15 minutes.

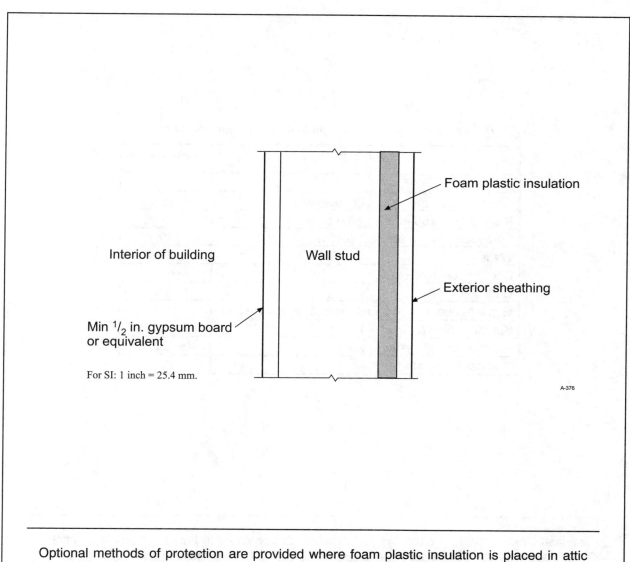

Interior of building

Wall stud

Foam plastic insulation

Exterior sheathing

Min $^1/_2$ in. gypsum board or equivalent

For SI: 1 inch = 25.4 mm.

A-376

Optional methods of protection are provided where foam plastic insulation is placed in attic areas or crawl spaces. The level of separation is reduced, but not eliminated, where the attic or crawl space provides entry only for the service of utilities.

Topic: Wall and Ceiling Finishes
Reference: IRC R315

Category: Building Planning
Subject: Flame Spread and Smoke Density

Code Text: *Wall and ceiling finishes shall have a flame-spread classification of not greater than 200.* See exceptions for trim, doors and windows, and very thin materials cemented to a wall or ceiling surface. *Wall and ceiling finishes shall have a smoke-developed index of not greater than 450. Tests shall be made in accordance with ASTM E 84.*

Discussion and Commentary: The control of interior finish materials is an important aspect of fire protection. The dangers of unregulated interior finish include both the rapid spread of fire and the contribution of additional fuel to the fire. Rapid fire spread presents a threat to the building occupants by limiting or denying their exit travel. This can be caused by either the rapid spread of the fire itself or by the production of large quantities of dense, black smoke that obscures the exit path or makes movement difficult.

Flame-spread ratings of typical construction materials

Material	Flame spread
Ceilings:	
Glass-fiber sound-absorbing blankets	15 to 30
Mineral-fiber sound-absorbing panels	10 to 25
Sprayed cellulose fibers (treated)	20
Walls:	
Brick or concrete block	0
Cork	175
Gypsum board (with paper surface on both sides)	10 to 25
Southern pine (untreated)	130 to 190
Plywood paneling (untreated)	75 to 275
Red oak (untreated)	100

A-635

The installation of materials that will not significantly contribute to fire conditions is permitted. Trim, door and window frames, and baseboards are not regulated, because of their limited quantity. Thin materials such as wallpaper are allowed, as they tend to perform similar to their backing.

Topic: Two-family Dwellings
Reference: IRC R317.1

Category: Building Planning
Subject: Dwelling Unit Separation

Code Text: *Dwelling units in two-family dwellings shall be separated from each other by wall and/or floor assemblies of not less than 1-hour fire-resistance rating when tested in accordance with ASTM E 119.* See reduction in rating to ¹/₂-hour for fully sprinklered buildings. *When floor assemblies are required to be fire-resistance-rated by Section R317.1, the supporting construction of such assemblies shall have an equal or greater fire-resistive rating.*

Discussion and Commentary: The code mandates a limited degree of fire separation to protect the occupants of one dwelling unit from the actions of their neighbor. Where the units are side by side, the separation wall must extend vertically to the underside of the roof deck. Where one unit is located over the other, the fire-rated floor/ceiling assembly must extend to and be tight against the exterior walls. The required fire-resistance-rating of one hour shall be verified by compliance with ASTM E 119, mandating an assembly constructed in accordance with its listing.

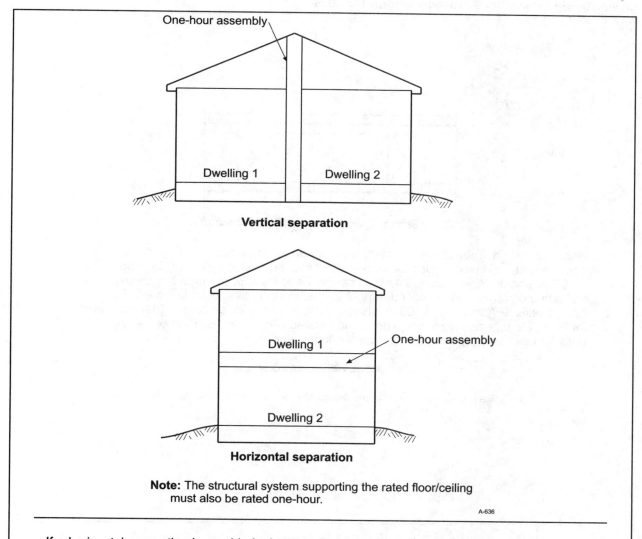

Note: The structural system supporting the rated floor/ceiling must also be rated one-hour.

A-636

If a horizontal separation is provided, elements such as bearing walls and columns that support the separation must also have a one-hour fire-resistive rating. It is of little value to provide a complying floor/ceiling assembly that quickly fails structurally because of unprotected supports.

Topic: Townhouse Separations

Reference: IRC R317.2

Category: Building Planning

Subject: Dwelling Unit Separation

Code Text: *Each townhouse shall be considered a separate building and shall be separated by fire-resistance-rated wall assemblies meeting the requirements of Section R302 for exterior walls.* See exception for permitted use of common 2-hour fire-resistance-rated wall. *The common wall for townhouses shall be continuous from the foundation to the underside of the roof sheathing, deck or slab and shall extend the full length of the common wall including walls extending through and separating attached accessory structures.*

Discussion and Commentary: The application of townhouse provisions has its base in the exterior wall requirements of Section R302, which deal with the building's location on the lot. In general, because the "exterior wall" of the townhouse is essentially being constructed with no fire separation distance where one townhouse adjoins another, that wall must have a minimum one-hour fire-resistive rating. The adjacent townhouse would have the same requirement; thus, two separate one-hour walls would be constructed side by side where one townhouse adjoins the other.

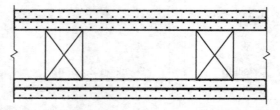

Construction type: gypsum wallboard, studs

Base layer $^5/_8$ in. Type X gypsum wallboard or veneer base applied at right angles to each side of 2 x 4 wood studs 24 in. on center with 6d coated nails, $1^7/_8$ in. long, 0.085 in. shank, $^1/_4$ in. heads, 24 in. on center. Face layer $^5/_8$ in. Type X gypsum wallboard or veneer base applied at right angles to studs over base layer with 8d coated nails, $2^3/_8$ in. long, 0.100 in. shank, $^1/_4$ in. heads, 8 in. on center. Stagger joints 24 in. on center each layer and side. Sound tested with studs 16 in. on center with nails for base layer spaced 6 in. on center (load-bearing).

Two-hour wall assembly

(Fire Resistance/Sound Control Design Manual, Fifteenth Edition, Gypsum Association)

For SI: 1 inch = 25.4 mm.

A-637

As an option, the construction of a single two-hour fire-resistive wall is permitted in lieu of two separate one-hour walls. This provides a similar level of separation between units. Where a two-hour wall is constructed, no plumbing or mechanical elements are permitted within the wall.

Topic: Location Required
Reference: IRC R319.1

Category: Building Planning
Subject: Protection Against Decay

Code Text: *In areas subject to decay damage as established by Table R301.2(1), the following locations shall require the use of an approved species and grade of lumber, pressure treated . . . or decay-resistant heartwood of redwood, black locust, or cedars.* See seven specific locations where such lumber is required.

Discussion and Commentary: For those portions of a wood-framed structure that are subject to damage by decay, lumber must be pressure preservatively treated or of a species of wood having a natural resistance to decay. Crawl spaces and unexcavated areas under a building usually contain moisture-laden air. Foundation walls and floor slabs-on-grade absorb moisture from the ground and, by capillary action, move it to framing members to which they are in contact. Other locations also have similar conditions where damage to the wood is quite possible unless adequate clearance is maintained or the appropriate type of material is used.

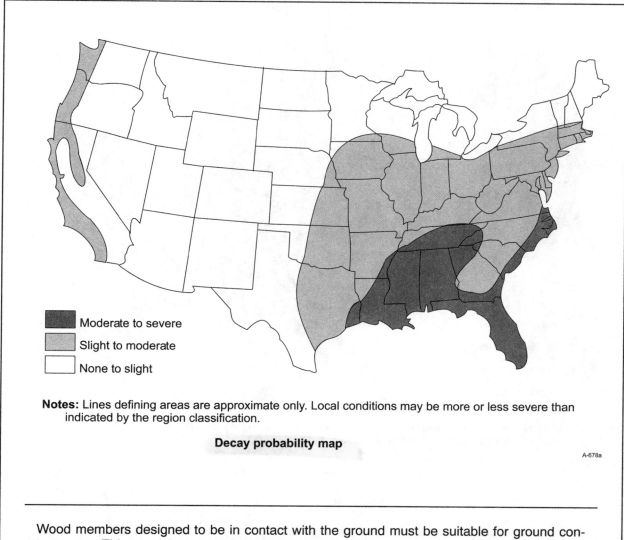

■ Moderate to severe
▨ Slight to moderate
□ None to slight

Notes: Lines defining areas are approximate only. Local conditions may be more or less severe than indicated by the region classification.

Decay probability map

A-678a

Wood members designed to be in contact with the ground must be suitable for ground contact use. This provision applies to all wood members that support permanent structures designed for human occupancy. Untreated wood is permitted where it is used entirely below groundwater level.

Topic: Subterranean Termite Control
Reference: IRC R320.1

Category: Building Planning
Subject: Protection Against Termites

Code Text: *In areas favorable to termite damage as established by Table R301.2(1), methods of protection shall be by chemical soil treatment, pressure preservatively treated wood in accordance with the AWPA standards listed in Section R319.1, naturally termite-resistant wood or physical barriers (such as metal or plastic termite shields), or any combination of these methods.*

Discussion and Commentary: Figure R301.2(6) depicts the geographical areas where termite damage to structures is probable. In those areas, the structure must be protected from termite damage in an appropriate manner. Although there are a number of approved methods, the most common is soil poisoning. Alternatives include the use of pressure preservatively treated wood and termite shields over perimeter walls. Often, a combination of these methods is necessary to establish the desired level of protection.

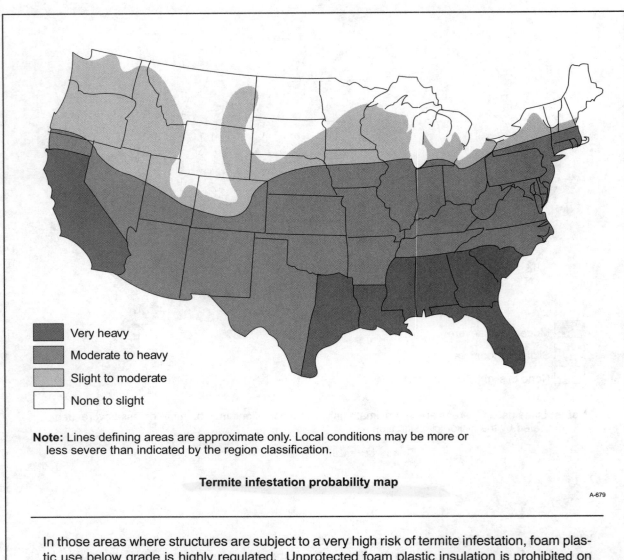

Very heavy

Moderate to heavy

Slight to moderate

None to slight

Note: Lines defining areas are approximate only. Local conditions may be more or less severe than indicated by the region classification.

Termite infestation probability map

A-679

In those areas where structures are subject to a very high risk of termite infestation, foam plastic use below grade is highly regulated. Unprotected foam plastic insulation is prohibited on the exterior face of foundation walls and similar locations unless the structure is noncombustible or of treated wood.

Topic: Scope

Reference: IRC R322, IBC 1107.6.3

Category: Building Planning

Subject: Accessibility

Code Text: *Where there are four or more dwelling units or sleeping units in a single structure, the provisions of Chapter 11 of the* International Building Code *for Group R-3 shall apply. In occupancies in Group R-3 where there are four or more dwelling units or sleeping units intended to be occupied as a residence in a single structure, every dwelling unit and sleeping unit intended to be occupied as a residence shall be a Type B unit.* See exception for reduction in number of Type B units.

Discussion and Commentary: Although most structures constructed under the provisions of the *International Residential Code* are exempt from the accessibility provisions, structures containing four or more townhouses will be regulated. The required Type B units are described in ICC A117.1 and are consistent with the design and construction requirements of the federal Fair Housing Act.

A-327a

The technical requirements for Type B dwelling units are found in Section 1003 of ICC A117.1. The issues addressed by the accessibility standard include the primary entrance, accessible route, walking surfaces, doors, ramps, elevators and lifts, operable parts, bathrooms and kitchens.

QUIZ

Study Session 4

1. The minimum width of a hallway shall be _____.
   a. 32 inches
   b. 36 inches
   c. 42 inches
   d. 44 inches

2. Every landing shall have a minimum length of _____ measured in the direction of travel.
   a. 30 inches
   b. 32 inches
   c. 36 inches
   d. 42 inches

3. The maximum slope of a ramp shall be one unit vertical in _____ units horizontal.
   a. five
   b. eight
   c. ten
   d. twelve

4. Handrails are not required on ramps having a maximum slope of _____.
   a. 1:5
   b. 1:8
   c. 1:12
   d. 1:15

5. Where a door opens onto a ramp, a minimum _____ landing shall be provided.
   a. 32-inch-by-32-inch
   b. 36-inch-by-36-inch
   c. 36-inch-by-60-inch
   d. 60-inch-by-60-inch

6. Stairways shall be a minimum _____ wide at all points above the permitted handrail height and below the required headroom height.
   a. 30 inches
   b. 34 inches
   c. 36 inches
   d. 38 inches

7.   Where handrails are provided on both sides of a stairway, what is the minimum required clear width at and below the handrail height?
       a. 27 inches
       b. 29 inches
       c. $31^1/_2$ inches
       d. 36 inches

8.   Stairways shall have a maximum riser height of _____ and a minimum tread run of _____.
       a. $8^1/_4$ inches, 9 inches
       b. 8 inches, 9 inches
       c. $7^3/_4$ inches, 10 inches
       d. $7^1/_2$ inches, 10 inches

9.   The maximum variation between the greatest riser height and the smallest riser height within any flight of stairs shall be _____.
       a. $^1/_4$ inch
       b. $^3/_8$ inch
       c. $^1/_2$ inch
       d. $^5/_8$ inch

10.   Unless a minimum tread depth of _____ is provided, a minimum $^3/_4$-inch nosing is required at the leading edge of all treads.
       a. $9^1/_2$ inches
       b. 10 inches
       c. 11 inches
       d. 12 inches

11.   Unless the stair has a total rise of no more than 30 inches, the opening between treads at open risers shall be such that a minimum _____ sphere shall not pass through.
       a. 3-inch
       b. 4-inch
       c. 6-inch
       d. no limitation is mandated

12.   Along all portions of a stairway, including at landings, the minimum headroom shall be _____.
       a. 6 feet, 6 inches
       b. 6 feet, 8 inches
       c. 6 feet, 9 inches
       d. 7 feet, 0 inches

13.   What is the maximum riser height permitted for spiral stairways?
       a. $7^3/_4$ inches
       b. 8 inches
       c. $8^1/_4$ inches
       d. $9^1/_2$ inches

14. Stairway handrails shall be located a minimum of _____ and a maximum of _____ measured vertically from the nosing of the treads.
   a. 30 inches, 34 inches
   b. 30 inches, 38 inches
   c. 34 inches, 38 inches
   d. 34 inches, 42 inches

15. Where a stairway handrail is located adjacent to a wall, a minimum clearance of _____ shall be provided between the wall and the handrail.
   a. $1^1/_4$ inches
   b. $1^1/_2$ inches
   c. 2 inches
   d. $3^1/_2$ inches

16. A Type 1 handrail having a circular cross section shall have a minimum diameter of _____ and a maximum diameter of _____.
   a. $1^1/_4$ inches, 2 inches
   b. $1^1/_4$ inches, $2^5/_8$ inches
   c. $1^1/_2$ inches, 2 inches
   d. $1^1/_2$ inches, $2^5/_8$ inches

17. A porch, balcony or similar raised floor surface more than 30 inches above the floor or grade below shall be provided with a guard having a minimum height of _____.
   a. 32 inches
   b. 34 inches
   c. 36 inches
   d. 42 inches

18. Required guards shall be provided with intermediate rails or ornamental closures such that a minimum _____ sphere cannot pass through.
   a. 3-inch
   b. 4-inch
   c. 6-inch
   d. 9-inch

19. Which one of the following areas in a dwelling unit does not specifically require the installation of a smoke alarm?
   a. basement
   b. kitchen
   c. habitable attic
   d. sleeping room

20. In general, foam plastic shall be separated from the interior of a dwelling by minimum
_____.
   a. $^3/_8$-inch gypsum board
   b. $^1/_2$-inch gypsum board
   c. $^5/_8$-inch gypsum board
   d. $^5/_8$-inch Type "X" gypsum board

21. Regulated wall and ceiling finishes shall have a maximum flame-spread classification of
_____.
   a. 25
   b. 75
   c. 200
   d. 450

22. What is the minimum required separation for a wall between dwelling units in a nonsprinklered
two-family dwelling?
   a. $^1/_2$-inch gypsum board on each side
   b. 30-minute fire-resistance-rated
   c. 1-hour fire-resistance-rated
   d. 2-hour fire-resistance-rated

23. When a common wall is used to separate townhouses, it shall have a minimum _____ fire-
resistance rating.
   a. 1-hour
   b. 2-hour
   c. 3-hour
   d. 4-hour

24. Unless they are made of pressure preservatively treated or approved decay-resistant lumber,
wood sill plates that rest on masonry or concrete exterior walls shall be located a minimum of
_____ from exposed ground.
   a. 6 inches
   b. 8 inches
   c. 12 inches
   d. 18 inches

25. In the establishment of a design flood elevation, the depth of peak elevation of flooding for a
_____ flood is used.
   a. 10-year
   b. 20-year
   c. 50-year
   d. 100-year

26. Walls and the under stair surface that enclose accessible space under a stairway shall be protected on the enclosed side by minimum _____ gypsum board.
    a. $^1/2$-inch
    b. $^1/2$-inch Type X
    c. $^5/8$-inch
    d. $^5/8$-inch Type X

27. Where foam plastic is spray applied to a sill plate without a thermal barrier, it shall have a maximum density of _____ pcf.
    a. 1.0
    b. 1.5
    c. 2.0
    d. 3.0

28. The maximum width of a Type II handrail above the recess shall be _____ inches.
    a. $1^1/2$
    b. 2
    c. $2^5/8$
    d. $2^3/4$

29. Handrails shall be provided on at least one side of a stairway having a minimum of _____ risers.
    a. 2
    b. 3
    c. 4
    d. 5

30. What is the minimum required size for an exit door from a dwelling unit?
    a. 32 inches by 78 inches
    b. 32 inches by 80 inches
    c. 36 inches by 78 inches
    d. 36 inches by 80 inches

INTERNATIONAL RESIDENTIAL CODE
Study Session 5
Chapter 4—Foundations

OBJECTIVE: To gain an understanding of the requirements relating to foundation systems, including soil evaluation; footing size, depth and slope; foundation design and reinforcing; frost-protected shallow foundations; wood foundations; insulating form foundation walls; foundation drainage; waterproofing and dampproofing; and under-floor spaces.

REFERENCE: Chapter 4, 2003 *International Residential Code*

KEY POINTS:
- How must lots be graded? What is the minimum required slope of the grade away from the foundation?
- When are soils tests required? When are presumptive load-bearing values to be used?
- What criteria determines the minimum size of footings? What is the minimum required footing thickness? Minimum projection distance? Minimum depth?
- What additional conditions must be met for footings located in Seismic Design Categories D_1 and D_2? Where must any required reinforcing be located?
- What is the maximum slope permitted for the top surface of footings? The bottom surface?
- In what manner must the sill plate be connected to the foundation wall? What is the minimum size and spacing for anchor bolts? How must the connection be made in Seismic Design Categories D_1 and D_2?
- How shall buildings be located in respect to an adjacent ascending slope? Descending slope?
- How are foundations and floor slabs for buildings located on expansive soils to be regulated?
- What is the minimum required height of a foundation above a street gutter or other point of drainage discharge?
- What are the provisions for the installation of wood foundation systems?
- How does a frost-protected shallow foundation work? What are the primary conditions for compliance?
- What criteria are necessary to address plain concrete and masonry foundation walls? Reinforced concrete and masonry foundation walls?
- What special conditions are required for foundation walls of buildings constructed in Seismic Design Categories D_1 and D_2? Where must reinforcement be located?
- When are pier foundations acceptable? What are the minimum requirements?
- What is an insulating concrete form (ICF) foundation wall system? What are the different types of ICF systems? How are they regulated?
- When are drains required around foundation walls? What drainage methods are acceptable?
- When must a foundation wall be dampproofed? Waterproofed? What methods are used for dampproofing and waterproofing?
- What is the minimum required size for under-floor ventilation openings? Where shall such openings be located? How may the required area of openings be reduced?
- How shall access to an under-floor space be provided?
- When is a drainage system needed for an under-floor space?

Topic: Drainage
Reference: IRC R401.3

Category: Foundations
Subject: General Requirements

Code Text: *Surface drainage shall be diverted to a storm sewer conveyance or other approved point of collection so as to not create a hazard. Lots shall be graded so as to drain surface water away from foundation walls. The grade away from foundation walls shall fall a minimum of 6 inches (152 mm) within the first 10 feet (3048 mm).* See exception for use of drains or swales.

Discussion and Commentary: Proper site drainage is an important element in preventing wet basements, damp crawl spaces, eroded banks and the possible failure of a foundation system. Drainage patterns should result in adequate slopes to approved drainage devices that are capable of carrying concentrated runoff. It is often necessary to use gutters and downspouts to direct roof water to the appropriate drainage points.

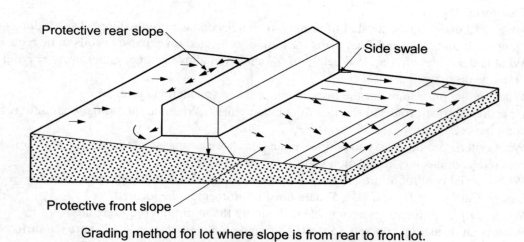

Grading method for lot where slope is from rear to front lot.
Drainage swales are located at rear and sides of dwelling.

B-493

Drains or swales are effective measures where the site does not allow for the necessary fall away from the structure. Minimum slope gradients for swales are often based on ground frost and moisture conditions. The permeability of soil and type of drainage devices are also important.

Topic: Soil Tests   **Category:** Foundations
Reference: IRC R401.4   **Subject:** General Requirements

Code Text: *In areas likely to have expansive, compressible, shifting or other unknown soil characteristics, the building official shall determine whether to require a soil test to determine the soil's characteristics at a particular location. This test shall be made by an approved agency using an approved method. In lieu of a complete geotechnical evaluation, the load-bearing values in Table R401.4.1 shall be assumed.*

Discussion and Commentary: Expansive soils, soil instability and increased lateral pressure due to a high water table or to surcharge loads from adjacent structures are special conditions that must be considered in the design of a foundation system. Where such conditions occur, they are beyond the scope of the prescriptive provisions of the code for foundation systems. In some cases, a soil test may be required to evaluate the soil's characteristics.

TABLE R401.4.1
PRESUMPTIVE LOAD-BEARING VALUES OF
FOUNDATION MATERIALS[a]

CLASS OF MATERIAL	LOAD-BEARING PRESSURE (pounds per square foot)
Crystalline bedrock	12,000
Sedimentary and foliated rock	4,000
Sandy gravel and/or gravel (GW and GP)	3,000
Sand, silty sand, clayey sand, silty gravel and clayey gravel (SW, SP, SM, SC, GM and GC)	2,000
Clay, sandy clay, silty clay, clayey silt, silt and sandy silt (CI, ML, MH and CH)	1,500[b]

For SI: 1 pound per square foot = 0.0479 kN/m^2.

a. When soil tests are required by Section R401.4, the allowable bearing capacities of the soil shall be part of the recommendations.

b. Where the building official determines that in-place soils with an allowable bearing capacity of less than 1,500 psf are likely to be present at the site, the allowable bearing capacity shall be determined by a soils investigation.

Where the bearing capacity of the soil has not been determined by geotechnical analysis such as borings, field load tests, laboratory tests and/or engineering analysis, it is a common practice to use presumptive bearing values for the design of the foundation system.

Topic: Wood and Concrete
Reference: IRC R402

Category: Foundations
Subject: Materials

Code Text: *All lumber and plywood* used in wood foundation systems *shall be treated in accordance with AWPA C22, and shall bear the label of an accredited agency showing 0.60 retention. Concrete shall have a minimum specified compressive strength as shown in Table R402.2. Concrete subject to weathering as indicated in Table R301.2(1) shall be air entrained as specified in Table R402.2.*

Discussion and Commentary: Freezing and thawing cycles can be the most destructive weathering factors for concrete. For nonair entrained concrete, deterioration is caused by the water freezing in the cement matrix, in the aggregate, or both. Studies have documented that concrete provided with proper air entrainment is highly resistant to this deterioration.

TABLE R402.2
MINIMUM SPECIFIED COMPRESSIVE STRENGTH OF CONCRETE

TYPE OR LOCATIONS OF CONCRETE CONSTRUCTION	MINIMUM SPECIFIED COMPRESSIVE STRENGTH[a] (f'_c)		
	Weathering potential[b]		
	Negligible	Moderate	Severe
Basement walls, foundations and other concrete not exposed to the weather	2,500	2,500	2,500[c]
Basement slabs and interior slabs on grade, except garage floor slabs	2,500	2,500	2,500[c]
Basement walls, foundation walls, exterior walls and other vertical concrete work exposed to the weather	2,500	3,000[d]	3,000[d]
Porches, carport slabs and steps exposed to the weather, and garage floor slabs	2,500	3,000[d,e]	3,500[d,e]

For SI: 1 pound per square inch = 6.895 kPa.

a. At 28 days psi.

b. See Table R301.2(1) for weathering potential.

c. Concrete in these locations that may be subject to freezing and thawing during construction shall be air-entrained concrete in accordance with Footnote d.

d. Concrete shall be air entrained. Total air content (percent by volume of concrete) shall not be less than 5 percent or more than 7 percent.

e. See Section R402.2 for minimum cement content.

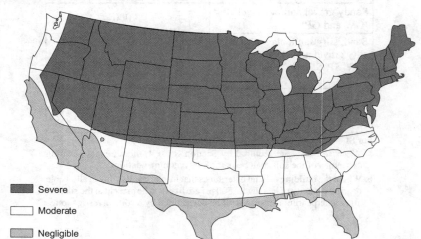

■ Severe

□ Moderate

▨ Negligible

a. Alaska and Hawaii are classified as severe and negligible, respectively.
b. Lines defining areas are approximate only. Local conditions may be more or less severe than indicated by region classification. A severe classification is where weather conditions result in significant snowfall combined with extended periods during which there is little or no natural thawing causing deicing salts to be used extensively.

Weathering probability map for concrete

A-680

Performance of wood foundation systems is dependent on the use of properly treated materials. Verification of the proper materials is provided by requiring identification showing the approval of an accredited inspection agency.

Topic: Minimum Size
Reference: IRC R403.1.1

Category: Foundations
Subject: Footings

Code Text: *Minimum sizes for concrete and masonry footings shall be as set forth in Table R403.1 and Figure R403.3.1(1). The footing width, W, shall be based on the load-bearing value of the soil in accordance with Table R401.4.1. Spread footings shall be at least 6 inches (152 mm) in thickness. Footing projections, P, shall be at least 2 inches (51 mm) and shall not exceed the thickness of the footing.*

Discussion and Commentary: To avoid overstressing the footing, minimum size requirements are established based on construction type, number of stories and soil load-bearing value. The projection limitation is critical for 6-inch-thick footings supported by poor soil conditions. Excessive projections could result in the footing being cracked in the same plane as the foundation wall, which could occur if the allowable stress in the concrete is exceeded.

TABLE R403.1
MINIMUM WIDTH OF CONCRETE OR MASONRY FOOTINGS (inches)[a]

	LOAD-BEARING VALUE OF SOIL (psf)			
	1,500	2,000	3,000	≥ 4,000
Conventional light-frame construction				
1-story	12	12	12	12
2-story	15	12	12	12
3-story	23	17	12	12
4-inch brick veneer over light frame or 8-inch hollow concrete masonry				
1-story	12	12	12	12
2-story	21	16	12	12
3-story	32	24	16	12
8-inch solid or fully grouted masonry				
1-story	16	12	12	12
2-story	29	21	14	12
3-story	42	32	21	16

For SI: 1 inch = 25.4 mm, 1 pound per square foot = 0.0479 kN/m².

a. Where minimum footing width is 12 inches, a single wythe of solid or fully grouted 12-inch nominal concrete masonry units is permitted to be used.

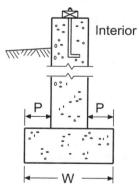

Interior

Basement or crawl space
with concrete wall and
spread footing

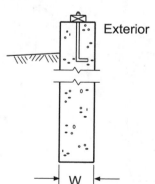

Exterior

Basement or crawl space
with foundation wall
bearing directly on soil

A-641a

In buildings constructed in Seismic Design Categories D₁ or D₂, footing reinforcement is mandated. Interconnection of the stem wall and its supporting footing is necessary to resist the tendency to slip during an earthquake. Various methods address different foundation systems.

Topic: Minimum Depth
Reference: IRC R403.1.4

Category: Foundations
Subject: Footings

Code Text: *All exterior footings shall be placed at least 12 inches (305 mm) below the undisturbed ground. Except where otherwise protected from frost, foundation walls, piers and other permanent supports of buildings and structures shall be protected from frost by one or more of the following methods: 1) extending below the frost line specified in Table R301.2(1); 2) constructing in accordance with Section R403.3 (frost protected shallow foundations); 3) constructing in accordance with ASCE 32-01 (Design and Construction of Frost Protected Shallow Foundations); and 4) erected on solid rock.* See exceptions for small freestanding accessory structures and decks not supported by a dwelling.

Discussion and Commentary: The volume changes (frost heave) that take place during freezing and thawing cycles produce excessive stresses in the foundations and create extensive damage to the supported walls. Thus, foundations must be extended below the depth of frost penetration.

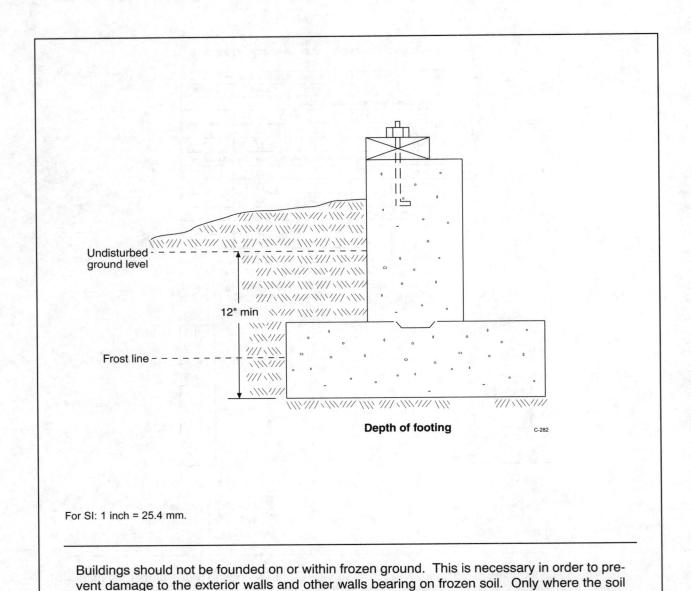

Undisturbed ground level

12" min

Frost line

Depth of footing

C-282

For SI: 1 inch = 25.4 mm.

Buildings should not be founded on or within frozen ground. This is necessary in order to prevent damage to the exterior walls and other walls bearing on frozen soil. Only where the soil is maintained in a permanently frozen condition can such construction be permitted.

Topic: Minimum Slope
Reference: IRC R403.1.5

Category: Foundations
Subject: Footings

Code Text: *The top surface of footings shall be level. The bottom surface of footings shall not have a slope exceeding one unit vertical in 10 units horizontal (10-percent slope). Footings shall be stepped where it is necessary to change the elevation of the top surface of the footings or where the slope of the bottom surface of the footings will exceed one unit vertical in ten units horizontal (10-percent slope).*

Discussion and Commentary: Although the code places no restriction on a stepped foundation, there is a recommended overlap of the top of the foundation wall beyond the step in the foundation. It should be larger than the vertical step in the foundation wall at that point. This recommendation is based on possible crack propagation at an angle of 45 degrees.

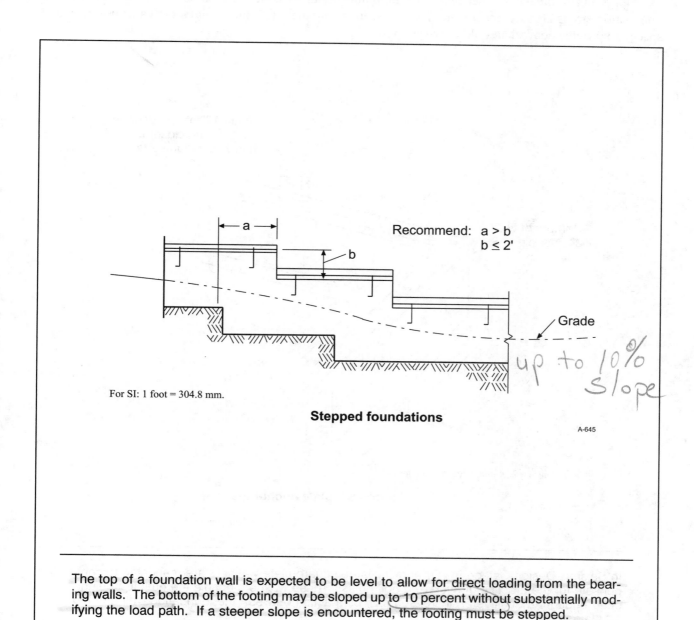

Recommend: a > b
b ≤ 2'

Grade

up to 10% slope

For SI: 1 foot = 304.8 mm.

Stepped foundations

A-645

The top of a foundation wall is expected to be level to allow for direct loading from the bearing walls. The bottom of the footing may be sloped up to 10 percent without substantially modifying the load path. If a steeper slope is encountered, the footing must be stepped.

Topic: Foundation Anchorage

Reference: IRC R403.1.6

Category: Foundations

Subject: Footings

Code Text: *When braced wall panels are supported directly on continuous foundations, the wall wood sill plate or cold-formed steel bottom track shall be anchored to the foundation in accordance with Section R403.1.6. The wood sole plate at exterior walls on monolithic slabs and wood sill plate shall be anchored to the foundation with anchor bolts spaced a maximum of 6 feet (1829 mm) on center. There shall be a minimum of two bolts per plate section with one bolt located not more than 12 inches (305 mm) or less than seven bolt diameters from each end of the plate section. Bolts shall be at least 1/2 inch (12.7 mm) in diameter and shall extend a minimum of 7 inches (178 mm) into masonry or concrete. See exception for foundation anchor straps.*

Discussion and Commentary: To prevent walls and floors from shifting under lateral loads, anchorage to the supporting foundation is needed. The minimum required connection is supplied by anchor bolts. Foundation straps are also acceptable if installed in accordance with the manufacturer's instructions and spaced to provide equivalent hold-down strength.

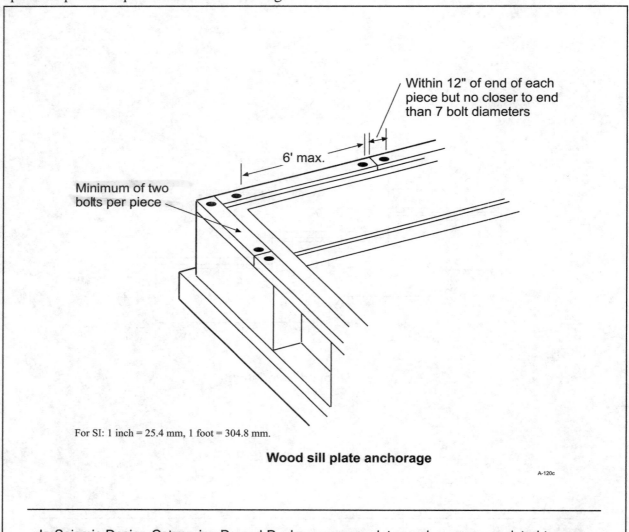

Within 12" of end of each piece but no closer to end than 7 bolt diameters

6' max.

Minimum of two bolts per piece

For SI: 1 inch = 25.4 mm, 1 foot = 304.8 mm.

Wood sill plate anchorage

A-120c

In Seismic Design Categories D_1 and D_2, large square plate washers are mandated to compensate for the practice of oversized bolt holes. These washers enable the nuts to be tightened enough to achieve an increased clamping action between the foundation wall and the sill plate.

Topic: Foundation Elevation
Reference: IRC R403.1.7.3

Category: Foundation
Subject: Footings

Code Text: *On graded sites, the top of any exterior foundation shall extend above the elevation of the street gutter at point of discharge or the inlet of an approved drainage device a minimum of 12 inches (305 mm) plus 2 percent. Alternate elevations are permitted subject to the approval of the building official, provided it can be demonstrated that required drainage to the point of discharge and away from the structure is provided at all locations on the site.*

Discussion and Commentary: Where natural drainage away from a building is not available, the site must be graded so that water will not drain toward, or accumulate at, the exterior foundation wall. A prescriptive elevation is set forth that will ensure positive drainage to a street gutter or other drainage point; however, any other method that moves water away from the building can be accepted by the building official.

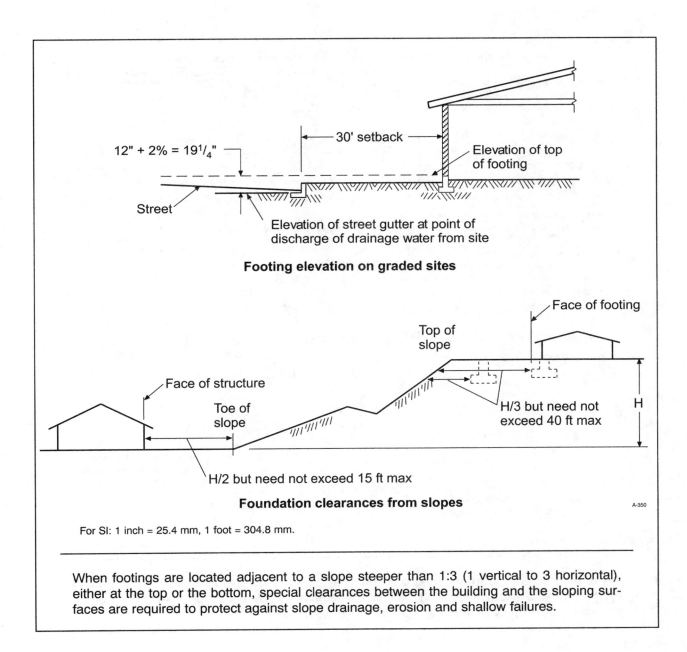

Footing elevation on graded sites

Foundation clearances from slopes

A-350

For SI: 1 inch = 25.4 mm, 1 foot = 304.8 mm.

When footings are located adjacent to a slope steeper than 1:3 (1 vertical to 3 horizontal), either at the top or the bottom, special clearances between the building and the sloping surfaces are required to protect against slope drainage, erosion and shallow failures.

Topic: Frost Protected Foundations
Reference: IRC R403.3

Category: Foundations
Subject: Footings

Code Text: *For buildings where the monthly mean temperature of the building is maintained at a minimum of 64°F (18°C), footings are not required to extend below the frost line when protected from frost by insulation in accordance with Figure R404.3(1) and Table R403.3.* Not permitted for unheated areas such as porches, garages and crawl spaces.

Discussion and Commentary: As a fundamental rule, footings must be placed below the frost line. However, the use of a frost protected foundation is an acceptable alternative, allowing placement of the footing/foundation above the frost line. This foundation system uses insulation to reduce the heat loss at the slab edge. By holding heat from the dwelling in the ground under the foundation, the insulation eliminates the potential for freezing, and thus the consequences of freeze/thaw conditions are avoided.

TABLE R403.3
MINIMUM INSULATION REQUIREMENTS FOR FROST-PROTECTED FOOTINGS IN HEATED BUILDINGS[a]

AIR FREEZING INDEX (°F-days)[b]	VERTICAL INSULATION R-VALUE[c,d]	HORIZONTAL INSULATION R-VALUE[c,e]		HORIZONTAL INSULATION DIMENSIONS PER FIGURE R403.3(1) (inches)		
		Along walls	At corners	A	B	C
1,500 or less	4.5	NR	NR	NR	NR	NR
2,000	5.6	NR	NR	NR	NR	NR
2,500	6.7	1.7	4.9	12	24	40
3,000	7.8	6.5	8.6	12	24	40
3,500	9.0	8.0	11.2	24	30	60
4,000	10.1	10.5	13.1	24	36	60

For SI: 1 inch = 25.4 mm, °C = [(°F)-32]/1.8.

a. Insulation requirements are for protection against frost damage in heated buildings. Greater values may be required to meet energy conservation standards. Interpolation between values is permissible.

b. See Figure R403.3(2) for Air Freezing Index values.

c. Insulation materials shall provide the stated minimum R-values under long-term exposure to moist, below-ground conditions in freezing climates. The following R-values shall be used to determine insulation thicknesses required for this application: Type II expanded polystyrene—2.4R per inch; Type IV extruded polystyrene—4.5R per inch; Type VI extruded polystyrene—4.5R per inch; Type IX expanded polystyrene—3.2R per inch; Type X extruded polystyrene—4.5R per inch. NR denotes "not required."

d. Vertical insulation shall be expanded polystyrene insulation or extruded polystyrene insulation.

e. Horizontal insulation shall be extruded polystyrene insulation.

Insulation detail

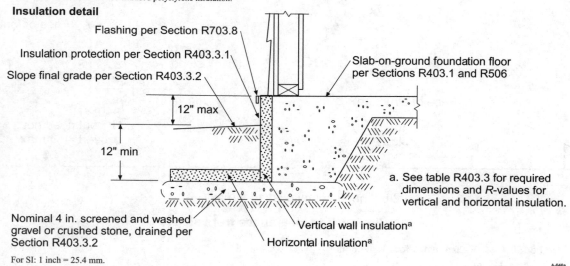

For SI: 1 inch = 25.4 mm.

A-646a

It is often necessary to extend the insulation away from the foundation horizontally. By doing so, it is probable that the insulating material will be damaged by landscape work or other activities unless an approved method of protection is provided.

Topic: Concrete Foundation Walls
Reference: IRC R404.1.2, R404.1.3

Category: Foundations
Subject: Foundation Walls

Code Text: *Concrete foundation walls shall be constructed as set forth in Tables R404.1.1(1), R404.1.1(2), R404.1.1(3) and R404.1.1(4), and shall also comply with the provisions of Section R404.1.2. A design in accordance with accepted engineering practice shall be provided for concrete or masonry foundation walls when any of the following conditions exist: 1) Walls are subject to hydrostatic pressure from groundwater. 2) Walls supporting more than 48 inches (1219 mm) of unbalanced backfill that do not have permanent lateral support at the top and bottom.*

Discussion and Commentary: Foundation walls are usually designed and constructed to carry the vertical loads from the structure above, resist wind and any lateral forces transmitted to the foundation, and sustain earth pressures exerted against the wall, including any forces that may be imposed by frost action.

TABLE R404.1.1(2)
REINFORCED CONCRETE AND MASONRY[a] FOUNDATION WALLS

MAXIMUM WALL HEIGHT (feet)	MAXIMUM UNBALANCED BACKFILL HEIGHT[e] (feet)	MINIMUM VERTICAL REINFORCEMENT SIZE AND SPACING[b, c] FOR 8-INCH NOMINAL WALL THICKNESS		
		Soil classes[d]		
		GW, GP, SW and SP soils	GM, GC, SM, SM-SC and ML soils	SC, MH, ML-CL and inorganic CL soils
6	5	#4 at 48″ o.c.	#4 at 48″ o.c.	#4 at 48″ o.c.
	6	#4 at 48″ o.c.	#4 at 40″ o.c.	#5 at 48″ o.c.
7	4	#4 at 48″ o.c.	#4 at 48″ o.c.	#4 at 48″ o.c.
	5	#4 at 48″ o.c.	#4 at 48″ o.c.	#4 at 40″ o.c.
	6	#4 at 48″ o.c.	#5 at 48″ o.c.	#5 at 40″ o.c.
	7	#4 at 40″ o.c.	#5 at 40″ o.c.	#6 at 48″ o.c.
8	5	#4 at 48″ o.c.	#4 at 48″ o.c.	#4 at 40″ o.c.
	6	#4 at 48″ o.c.	#5 at 48″ o.c.	#5 at 40″ o.c.
	7	#5 at 48″ o.c.	#6 at 48″ o.c.	#6 at 40″ o.c.
	8	#5 at 40″ o.c.	#6 at 40″ o.c.	#6 at 24″ o.c.
9	5	#4 at 48″ o.c.	#4 at 48″ o.c.	#5 at 48″ o.c.
	6	#4 at 48″ o.c.	#5 at 48″ o.c.	#6 at 48″ o.c.
	7	#5 at 48″ o.c.	#6 at 48″ o.c.	#6 at 32″ o.c.
	8	#5 at 40″ o.c.	#6 at 32″ o.c.	#6 at 24″ o.c.
	9	#6 at 40″ o.c.	#6 at 24″ o.c.	#6 at 16″ o.c.

For SI: 1 inch = 25.4 mm, 1 foot = 304.8 mm.

a. Mortar shall be Type M or S and masonry shall be laid in running bond.

b. Alternative reinforcing bar sizes and spacings having an equivalent cross-sectional area of reinforcement per lineal foot of wall shall be permitted provided the spacing of the reinforcement does not exceed 72 inches.

c. Vertical reinforcement shall be Grade 60 minimum. The distance from the face of the soil side of the wall to the center of vertical reinforcement shall be at least 5 inches.

d. Soil classes are in accordance with the Unified Soil Classification System. Refer to Table R405.1.

e. Unbalanced backfill height is the difference in height of the exterior and interior finish ground levels. Where an interior concrete slab is provided, the unbalanced backfill height shall be measured from the exterior finish ground level to the top of the interior concrete slab.

In lieu of an engineered foundation wall system, the IRC provides prescriptive tables for both reinforced and unreinforced masonry and concrete foundation walls. If reinforced walls are under consideration, three different wall widths (8 inch, 10 inch and 12 inch) are addressed.

Topic: Insulating Concrete Form Foundation Walls
Reference: IRC R404.4

Category: Foundations
Subject: Foundation Walls

Code Text: *Insulating concrete form (ICF) foundation walls shall be designed and constructed in accordance with the provisions of* Section R404.4 *or in accordance with the provisions of ACI 318.*

Discussion and Commentary: An insulating concrete form (ICF) system uses stay-in-place forms of rigid foam plastic insulation, a hybrid of cement and foam insulation, a hybrid of cement and wood chips, or other insulating material for constructing cast-in-place concrete walls. Three types of ICF systems are prescriptively addressed in the IRC; waffle-grid, flat-wall and screen grid. The use of ICF systems is limited to buildings of limited size and height located in the lower seismic design categories and wind speed areas.

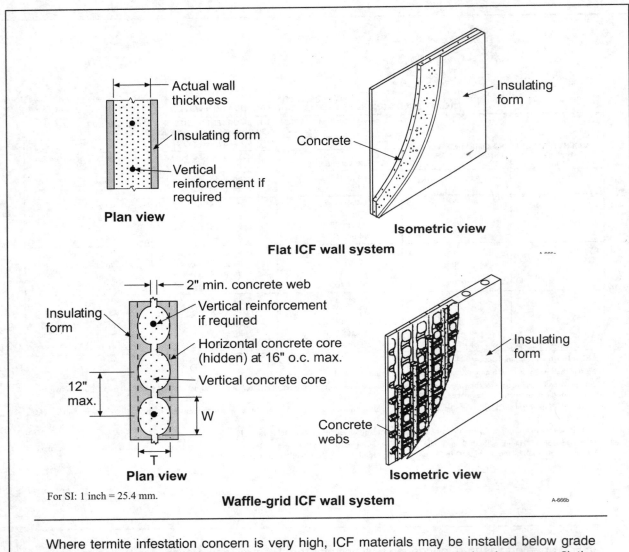

For SI: 1 inch = 25.4 mm.

Where termite infestation concern is very high, ICF materials may be installed below grade only if 1) a method is established to protect the foam plastic from termite damage or 2) the structure is of noncombustible or pressure preservatively treated wood construction.

Topic: Concrete or Masonry Foundations
Reference: IRC R405.1

Category: Foundations
Subject: Foundation Drainage

Code Text: *Drains shall be provided around all concrete or masonry foundations that retain earth and enclose habitable or usable spaces located below grade.* See exception for well-drained ground or sand-gravel mixture soils. *Drainage tiles, gravel or crushed stone drains, perforated pipe or other approved systems or materials shall be installed at or below the area to be protected and shall discharge by gravity or mechanical means into an approved drainage system.*

Discussion and Commentary: To allow free groundwater that may be present adjacent to the foundation wall to be removed, drains are usually placed around houses to remove water and prevent leakage into habitable or usable spaces below grade. Drainage tiles are extremely important in areas having moderate to heavy rainfall and soils with a low percolation rate.

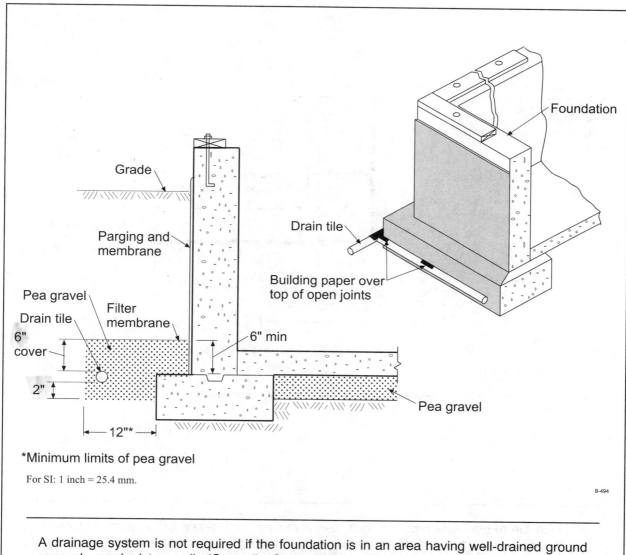

*Minimum limits of pea gravel

For SI: 1 inch = 25.4 mm.

B-494

A drainage system is not required if the foundation is in an area having well-drained ground or sand-gravel mixture soils (Group I). Such soils provide good drainage characteristics and include gravel sand mixtures, gravelly sands, silty gravels and silty sand.

Topic: Where Required

Reference: IRC R406.1, R406.2

Category: Foundations

Subject: Waterproofing and Dampproofing

Code Text: *Except where required to be waterproofed by Section R406.2, foundation walls that retain earth and enclose habitable or usable spaces located below grade shall be dampproofed from the top of the footing to the finished grade. In areas located where a high water table or other severe soil-water conditions are known to exist, exterior foundation walls that retain earth and enclose habitable or usable spaces located below grade shall be waterproofed with a membrane extending from the top of the footing to the finished grade.*

Discussion and Commentary: Dampproofing installations generally consist of the application of one or more coatings of impervious compounds that are intended to prevent the passage of water vapor through walls under slight pressure. Waterproofing installations consist of the application of a combination of sealing materials and impervious coatings used to prevent the passage of moisture in either a vapor or liquid form under conditions of significant hydrostatic pressure.

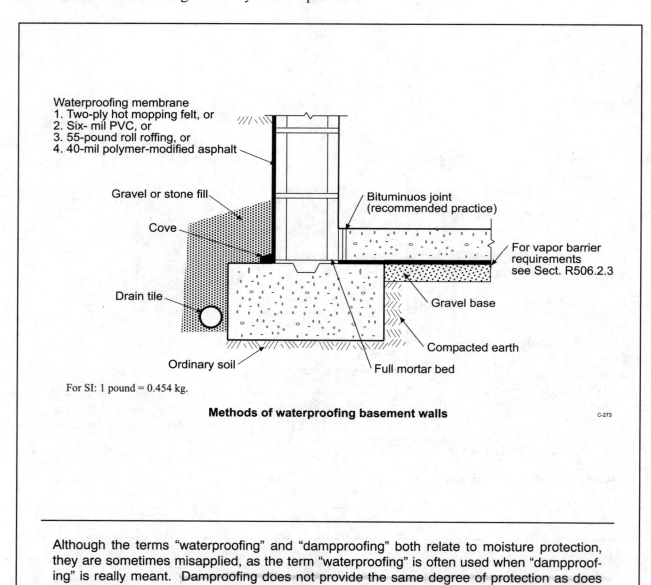

Waterproofing membrane
1. Two-ply hot mopping felt, or
2. Six- mil PVC, or
3. 55-pound roll roffing, or
4. 40-mil polymer-modified asphalt

Gravel or stone fill

Cove

Bituminuos joint
(recommended practice)

For vapor barrier
requirements
see Sect. R506.2.3

Gravel base

Drain tile

Compacted earth

Ordinary soil

Full mortar bed

For SI: 1 pound = 0.454 kg.

Methods of waterproofing basement walls

C-273

Although the terms "waterproofing" and "dampproofing" both relate to moisture protection, they are sometimes misapplied, as the term "waterproofing" is often used when "dampproofing" is really meant. Dampproofing does not provide the same degree of protection as does waterproofing.

Topic: Ventilation
Reference: IRC R408

Category: Foundations
Subject: Under-floor Space

Code Text: *The under-floor space between the bottom of the floor joists and the earth under any building (except space occupied by a basement or cellar) shall be provided with ventilation openings through foundation walls or exterior walls. The minimum net area of ventilation openings shall not be less than 1 square foot for each 150 square feet (0.67m² for each 100 m²) of under-floor space area.* See five exceptions for reduction or elimination of ventilation to the exterior

Discussion and Commentary: To control condensation in crawl space areas and thus reduce the chance of dry rot, natural ventilation of such spaces by reasonably distributed openings through exterior foundation walls is required. Susceptibility of a structure to condensation is a function of the geographical location and the climatic conditions.

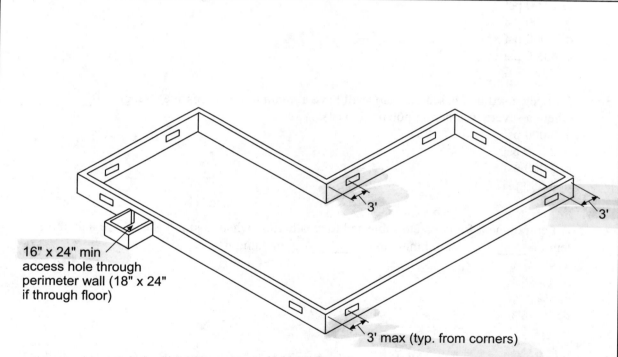

16" x 24" min access hole through perimeter wall (18" x 24" if through floor)

3'

3'

3' max (typ. from corners)

Screened openings (mesh opening ≥ 1/8 in.) through foundation wall to crawl space
Total net clear area of openings equals the crawl space area divided by 150

For SI: 1 inch = 25.4 mm, 1 foot = 304.8 mm.

A-650

The installation of a covering material over the ventilation opening keeps rodents and vermin from entering the crawl space. The code lists a variety of materials including corrosion-resistant wire mesh having a least dimension of 1/8 inch.

QUIZ

Study Session 5

1. Unless grading is prohibited by physical barriers, lots shall be graded away from the foundation with a minimum fall of _____ within the first _____
   a. 6 inches, 5 feet
   b. 6 inches, 10 feet
   c. 12 inches, 5 feet
   d. 12 inches, 10 feet

2. In the absence of a complete geotechnical evaluation to determine the soil's characteristics, clayey sand material shall be assumed to have a presumptive load-bearing value of _____.
   a. 1,500 psf
   b. 2,000 psf
   c. 3,000 psf
   d. 4,000 psf

3. Concrete used in a basement slab shall have a minimum compressive strength of _____ where a severe weathering potential exists.
   a. 2,000 psi
   b. 2,500 psi
   c. 3,000 psi
   d. 3,500 psi

4. Air entrainment for concrete subjected to weathering, when required, shall have a total air content of _____ minimum and _____ maximum.
   a. 4%, 7%
   b. 4%, 8%
   c. 5%, 7%
   d. 5%, 8%

5. A two-story conventional light-frame dwelling is to be constructed in an area where the load-bearing value of the soil is 2,000 psf. What is the minimum required width of the concrete footings?
   a. 8 inches
   b. 11 inches
   c. 12 inches
   d. 15 inches

6. The minimum required thickness for concrete spread footings is _____.
   a. 4 inches
   b. 6 inches
   c. 7 inches
   d. 8 inches

7. What are the minimum and maximum footing projections for a concrete footing having a thickness of 8 inches?
   a. 2 inches, 4 inches
   b. 2 inches, 8 inches
   c. 4 inches, 8 inches
   d. 8 inches, 12 inches

8. Where a permanent wood foundation basement wall system is used, the wall shall be supported by a minimum _____ footing plate resting on gravel or crushed stone fill a minimum of _____ in width.
   a. 2-inch-by-6-inch, 12 inches
   b. 2-inch-by-8-inch, 16 inches
   c. 2-inch-by-12-inch, 16 inches
   d. 2-inch-by-12-inch, 24 inches

9. For concrete foundation systems constructed in Seismic Design Category D_2, foundations with stem walls shall be provided with a minimum of _____ bar(s) at the top of the wall and _____ bar(s) at the bottom of the footing.
   a. one #4, one #4
   b. one #5, one #5
   c. one #5, two #4
   d. two #4, two #5

10. All exterior footings shall be placed a minimum of _____ below the undisturbed ground.
    a. 6 inches
    b. 8 inches
    c. 9 inches
    d. 12 inches

11. The maximum permitted slope for the bottom surface of footings shall be _____.
    a. 1:8
    b. 1:10
    c. 1:12
    d. 1:20

12. Anchor bolts used to attach wood sill plates to foundation walls shall be spaced a maximum of _____ on center.
    a. 4 feet
    b. 5 feet
    c. 6 feet
    d. 7 feet

13. In Seismic Design Categories D_1 and D_2, anchor bolts shall be a minimum of _____ in diameter and extend a minimum of _____ into masonry foundations.
    a. $1/2$ inch, 7 inches
    b. $1/2$ inch, 15 inches
    c. $5/8$ inch, 7 inches
    d. $5/8$ inch, 15 inches

14. On graded sites, the top of any exterior foundation wall shall extend a minimum of _____ above the elevation of the street gutter or other approved drainage discharge point.
    a. 6 inches plus 1 percent
    b. 6 inches plus 2 percent
    c. 12 inches plus 1 percent
    d. 12 inches plus 2 percent

15. In order to use frost protected shallow foundations, the monthly mean temperature of the building must be maintained at a minimum of _____.
    a. 60°F
    b. 64°F
    c. 68°F
    d. 70°F

16. A dwelling located in an area with an air freezing index of 3,500 is constructed with a frost-protected shallow foundation. What is the minimum required horizontal insulation R-value at the corners of the foundation?
    a. 8.0
    b. 8.6
    c. 11.2
    d. 13.1

17. A 9-foot-high plain masonry foundation wall is subjected to 7 feet of unbalanced backfill. If the soil class is GC, what is the minimum required nominal wall thickness?
    a. 8 inches
    b. 10 inches
    c. 12 inches
    d. 12 inches solid grouted

18. An 8-foot-high, 8-inch-thick reinforced concrete foundation wall is subjected to 6 feet of unbalanced backfill. If the soil class is GP, what minimum vertical reinforcement is required?
    a. #4 @ 48 inches on center
    b. #5 @ 48 inches on center
    c. #5 @ 40 inches on center
    d. #6 @ 48 inches on center

19. Where masonry veneer is used, concrete and masonry foundations shall extend a minimum of _____ above adjacent finished grade.
a. 4 inches
b. 6 inches
c. 8 inches
d. 12 inches

20. Unless the foundation wall supports a maximum of _____ of unbalanced backfill, backfill shall not be placed against the wall until the wall has sufficient strength and has been anchored to the floor above or has been sufficiently braced.
a. 2 feet
b. 4 feet
c. 6 feet
d. 8 feet

21. Where wood studs having an F_b value of 1320 are used in a wood foundation wall, the minimum stud size shall be _____ and the maximum stud spacing shall be _____ on center.
a. 2 inches by 4 inches, 24 inches
b. 2 inches by 4 inches, 16 inches
c. 2 inches by 6 inches, 24 inches
d. 2 inches by 6 inches, 16 inches

22. What is the maximum height of backfill permitted against a wood foundation wall not designed by AF&PA Report # 7?
a. 3 feet
b. 4 feet
c. 6 feet
d. 7 feet

23. Required foundation drains of gravel or crushed stone shall extend a minimum of _____ beyond the outside edge of the footing and _____ above the top of the footing.
a. 6 inches, 4 inches
b. 6 inches, 6 inches
c. 12 inches, 4 inches
d. 12 inches, 6 inches

24. As a general rule, the under-floor space between the bottom of floor joists and the earth shall be provided with ventilation openings sized for a minimum net area of 1 square foot for each _____ square feet of under-floor area.
a. 100
b. 120
c. 150
d. 200

25.   Access to an under-floor space through a perimeter wall shall be provided by a minimum _____ access opening.
       a. 16-inch-by-20-inch
       b. 16-inch-by-24-inch R 408.3
       c. 18-inch-by-24-inch
       d. 18-inch-by-30-inch

26.   Frost protection is not required for a foundation that supports a freestanding accessory structure that has a maximum floor area of _____ square feet and a maximum eave height of _____ feet.
       a. 200, 8
       b. 200, 10
       c. 400, 10
       d. 400, 12

27.   Gravel used as a footing for a wood foundation shall be washed, well graded and have a maximum stone size of _____ inch.
       a. $1/4$
       b. $1/2$
       c. $3/4$
       d. 1

28.   Horizontal insulation used in a frost protected shallow foundation system shall be protected against damage if it is located less than _____ inches below the ground surface.
       a. 12
       b. 15
       c. 18
       d. 24

29.   Where waterproofing of a masonry foundation wall is necessary due to the presence of a high water table, a minimum membrane thickness of _____ is required if the waterproofing material is polymer-modified asphalt.
       a. 6-mil
       b. 16-mil
       c. 30-mil
       d. 40-mil

30.   Ventilating openings providing under-floor ventilation shall be located so that at least one such opening is installed a maximum of _____ feet of each corner of the building.
       a. 2
       b. 3
       c. 5
       d. 6

INTERNATIONAL RESIDENTIAL CODE
Study Session 6
Chapter 5—Floors

OBJECTIVE: To develop an understanding of the provisions regulating floor construction, including the sizing and installation of floor joists and girders; drilling and notching of floor framing members; engineered wood products such as trusses, glue-laminated members and I-joists; draftstopping; floor sheathing spans and installation; steel floor framing; and concrete slab-on-ground floors.

REFERENCE: Chapter 5, 2003 *International Residential Code*

KEY POINTS:

- How are the maximum allowable spans of floor joists determined? What design live loads are to be used for joists supporting sleeping rooms? For areas other than sleeping rooms? For attics accessed by a permanent stairway?
- What is the general limitation for cantilever spans of floor joists? What spans are permitted for cantilevered floor joists supporting a bearing wall and a roof? Supporting an exterior balcony?
- How shall bearing partitions be supported by floor joists? What method is acceptable where piping or vents are installed at the point of support?
- How must bearing partitions perpendicular to floor joists be located?
- What conditions regulate the sizing of girders for exterior bearing walls? For interior bearing walls?
- How much bearing is required for the ends of floor joists supported by wood or metal? Supported by concrete or masonry? What other methods are acceptable?
- What methods of connection are permitted where joists frame from opposite sides of a bearing wall or girder?
- How are floor joists to be restrained to resist twisting and lateral movement?
- What are the limitations for notches in floor joists? Bored holes?
- When are single headers and single trimmers permitted at floor openings? When are framing anchors or joist hangers mandated?
- How are wood floor trusses regulated? What type of information must be provided on the truss design drawings?
- In what type of floor system is draftstopping required? At what intervals? What materials are permitted for draftstopping?
- How is wood structural panel sheathing identified? How does the panel span rating reflect the maximum allowable floor span? How is the allowable span for particleboard determined?
- What special concerns are addressed for pressure preservatively treated wood basement floors and floors on ground?
- What are the limitations for using the prescriptive provisions for steel floor framing?
- What materials are permitted as steel load-bearing members? How must they be identified? What are the fundamental fastening requirements for steel-to-steel floor connections? Floor to foundation or bearing wall connections? How are the maximum allowable spans determined?
- What is the minimum permitted thickness for concrete slab-on-ground floors? How must the site be prepared?
- How is any fill below a concrete slab-on-ground floor regulated? When is a base course needed? When is a vapor barrier required?

Topic: Allowable Joist Spans
Reference: IRC R502.3

Category: Floors
Subject: Wood Floor Framing

Code Text: *Table R502.3(1) shall be utilized to determine the maximum allowable span of floor joists that support sleeping areas and attics that are accessed by means of a fixed stairway provided that the design live load does not exceed 30 psf (1.44kN/m^2). Table R502.3.1(2) shall be utilized to determine the maximum allowable span of floor joists that support all areas of the building, other than sleeping and attics, provided that the design live load does not exceed 40 psf (1.92 kN/m^2).*

Discussion and Commentary: The joist span tables are based on common lumber species and grades, joist spacing and design loads. Although a separate span table is available for sleeping rooms based on 30 psf, it is common to use the 40 psf table for the entire dwelling unit. Both tables assume a maximum design dead load of 10 psf.

TABLE R502.3.1(2)
FLOOR JOIST SPANS FOR COMMON LUMBER SPECIES (Residential living areas, live load=40 psf, L/Δ=360)

JOIST SPACING (inches)	SPECIES AND GRADE		DEAD LOAD = 10 psf				DEAD LOAD = 20 psf			
			2x6	2x8	2x10	2x12	2x6	2x8	2x10	2x12
			Maximum floor joist spans							
			(ft.- in.)	(ft.- in.)	(ft.- in.)	(ft.- in.)	(ft.- in.)	(ft.- in.)	(ft.- in.)	(ft.- in.)
12	Douglas fir-larch	SS	11- 4	15- 0	19- 1	23- 3	11- 4	15- 0	19- 1	23- 3
	Douglas fir-larch	#1	10-11	14- 5	18- 5	22- 0	10-11	14- 2	17- 4	20- 1
	Douglas fir-larch	#2	10- 9	14- 2	17- 9	20- 7	10- 6	13- 3	16- 3	18-10
	Douglas fir-larch	#3	8- 8	11- 0	13- 5	15- 7	7-11	10- 0	12- 3	14- 3
	Hem-fir	SS	10- 9	14- 2	18- 0	21-11	10- 9	14- 2	18- 0	21-11
	Hem-fir	#1	10- 6	13-10	17- 8	21- 6	10- 6	13-10	16-11	19- 7
	Hem-fir	#2	10- 0	13- 2	16-10	20- 4	10- 0	13- 1	16- 0	18- 6
	Hem-fir	#3	8- 8	11- 0	13- 5	15- 7	7-11	10- 0	12- 3	14- 3
	Southern pine	SS	11- 2	14- 8	18- 9	22-10	11- 2	14- 8	18- 9	22-10
	Southern pine	#1	10-11	14- 5	18- 5	22- 5	10-11	14- 5	18- 5	22- 5
	Southern pine	#2	10- 9	14- 2	18- 0	21- 9	10- 9	14- 2	16-11	19-10
	Southern pine	#3	9- 4	11-11	14- 0	16- 8	8- 6	10-10	12-10	15- 3
	Spruce-pine-fir	SS	10- 6	13-10	17- 8	21- 6	10- 6	13-10	17- 8	21- 6
	Spruce-pine-fir	#1	10- 3	13- 6	17- 3	20- 7	10- 3	13- 3	16- 3	18-10
	Spruce-pine-fir	#2	10- 3	13- 6	17- 3	20- 7	10- 3	13- 3	16- 3	18-10
	Spruce-pine-fir	#3	8- 8	11- 0	13- 5	15- 7	7-11	10- 0	12- 3	14- 3
16	Douglas fir-larch	SS	10- 4	13- 7	17- 4	21- 1	10- 4	13- 7	17- 4	21- 0
	Douglas fir-larch	#1	9-11	13- 1	16- 5	19- 1	9- 8	12- 4	15- 0	17- 5
	Douglas fir-larch	#2	9- 9	12- 7	15- 5	17-10	9- 1	11- 6	14- 1	16- 3
	Douglas fir-larch	#3	7- 6	9- 6	11- 8	13- 6	6-10	8- 8	10- 7	12- 4
	Hem-fir	SS	9- 9	12-10	16- 5	19-11	9- 9	12-10	16- 5	19-11
	Hem-fir	#1	9- 6	12- 7	16- 0	18- 7	9- 6	12- 0	14- 8	17- 0
	Hem-fir	#2	9- 1	12- 0	15- 2	17- 7	8-11	11- 4	13-10	16- 1
	Hem-fir	#3	7- 6	9- 6	11- 8	13- 6	6-10	8- 8	10- 7	12- 4
	Southern pine	SS	10- 2	13- 4	17- 0	20- 9	10- 2	13- 4	17- 0	20- 9
	Southern pine	#1	9-11	13- 1	16- 9	20- 4	9-11	13- 1	16- 4	19- 6
	Southern pine	#2	9- 9	12-10	16- 1	18-10	9- 6	12- 4	14- 8	17- 2
	Southern pine	#3	8- 1	10- 3	12- 2	14- 6	7- 4	9- 5	11- 1	13- 2
	Spruce-pine-fir	SS	9- 6	12- 7	16- 0	19- 6	9- 6	12- 7	16- 0	19- 6
	Spruce-pine-fir	#1	9- 4	12- 3	15- 5	17-10	9- 1	11- 6	14- 1	16- 3
	Spruce-pine-fir	#2	9- 4	12- 3	15- 5	17-10	9- 1	11- 6	14- 1	16- 3
	Spruce-pine-fir	#3	7- 6	9- 6	11- 8	13- 6	6-10	8- 8	10- 7	12- 4

The span tables account for a uniform load condition. They also permit isolated concentrated loads such as nonbearing partitions. The use of additional joists and other adequate supports may be necessary where larger concentrated loads are encountered.

Topic: Floor Cantilevers
Reference: IRC R502.3.3

Category: Floors
Subject: Allowable Joist Spans

Code Text: *Floor cantilever spans shall not exceed the nominal depth of the wood floor joist. Floor cantilevers constructed in accordance with Table R502.3.3(1) shall be permitted when supporting a light-frame bearing wall and roof only. Floor cantilevers supporting an exterior balcony are permitted to be constructed in accordance with Table R502.3.3(2).*

Discussion and Commentary: Under the prescriptive limitations of the IRC, there are three conditions where the maximum allowable span of cantilevered floor joists can be determined. First, regardless of loading conditions, a maximum span equal to the depth of the floor joists is permitted. This requirement allows a very short cantilevered element. The second condition considers the support of both a single-story bearing wall and roof. The third situation is based on the cantilevered support of an exterior balcony with no roof load.

TABLE R502.3.3(2)
CANTILEVER SPANS FOR FLOOR JOISTS SUPPORTING EXTERIOR BALCONY[a, b, e, f]

| Member Size | Spacing | Maximum Cantilever Span (Uplift Force at Backspan Support in Lbs.)[c, d] | | |
| | | Ground Snow Load | | |
		≤30 psf	50 psf	70 psf
2 × 8	12"	42" (139)	39" (156)	34" (165)
2 × 8	16"	36" (151)	34" (171)	29" (180)
2 × 10	12"	61" (164)	57" (189)	49" (201)
2 × 10	16"	53" (180)	49" (208)	42" (220)
2 × 10	24"	43" (212)	40" (241)	34" (255)
2 × 12	16"	72" (228)	67" (260)	57" (268)
2 × 12	24"	58" (279)	54" (319)	47" (330)

For SI: 1 inch = 25.4 mm, 1 pound per square foot = 0.0479 kN/m².

a. Spans are based on No. 2 Grade lumber of Douglas fir-larch, hem-fir, southern pine, and spruce-pine-fir for repetitive (3 or more) members.

b. Ratio of backspan to cantilever span shall be at least 2:1.

c. Connections capable of resisting the indicated uplift force shall be provided at the backspan support.

d. Uplift force is for a backspan to cantilever span ratio of 2:1. Tabulated uplift values are permitted to be reduced by multiplying by a factor equal to 2 divided by the actual backspan ratio provided (2/backspan ratio).

e. A full-depth rim joist shall be provided at the cantilevered end of the joists. Solid blocking shall be provided at the cantilevered support.

f. Linear interpolation shall be permitted for ground snow loads other than shown.

The ratio of back span to cantilever span must be a minimum of 3:1 where the cantilevered floor joists support a bearing wall and a roof. Where only supporting an exterior balcony, the minimum ratio must be 2:1. Connections must be capable of resisting the indicated uplift force.

Topic: Joists Under Bearing Partitions
Reference: IRC R502.4

Category: Floors
Subject: Wood Floor Framing

Code Text: *Joists under parallel bearing partitions shall be of adequate size to support the load. Double joists, sized to adequately support the load, that are separated to permit the installation of piping or vents shall be full depth solid blocked with lumber not less than 2 inches (51 mm) in nominal thickness spaced not more than 4 feet (1219 mm) on center. Bearing partitions perpendicular to joists shall not be offset from supporting girders, walls or partitions more than the joist depth unless such joists are of sufficient size to carry the additional load.*

Discussion and Commentary: It is necessary to double floor joists under bearing partitions in order to provide support for the additional load from above. The multiple joists create a beam-type member that is capable of properly transmitting the loads to the bearing points at the joist ends. Therefore, a beam of equivalent size may be substituted for the double joist.

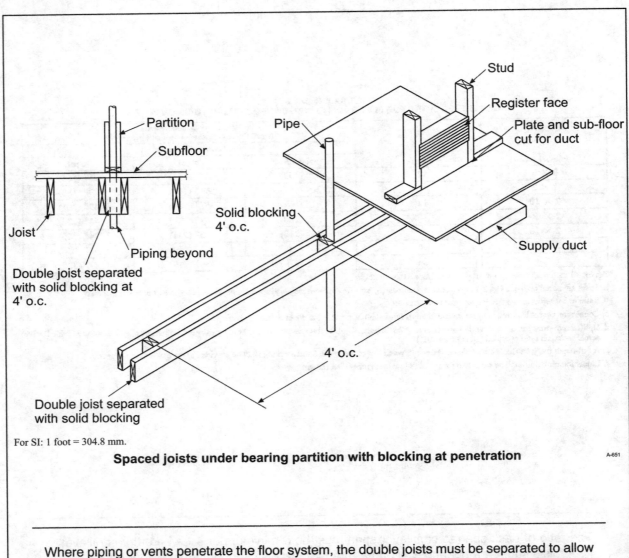

For SI: 1 foot = 304.8 mm.

Spaced joists under bearing partition with blocking at penetration

A-651

Where piping or vents penetrate the floor system, the double joists must be separated to allow the pipe or vent to pass through. Solid blocking between the joists at 4-foot intervals is required so that the two joists will function as a combined member without twisting.

Topic: Allowable Girder Spans
Reference: IRC R502.5

Category: Floors
Subject: Wood Floor Framing

Code Text: *The allowable spans of girders fabricated of dimension lumber shall not exceed the values set forth in Tables R502.5(1) and R502.5(2).*

Discussion and Commentary: Table R502.5(1) provides the maximum span of girders in exterior walls based on several criteria. The loads supported by the header, building width, built-up member size and ground snow load are needed to use the prescriptive table. The built-up members listed vary as to depth and number of members. The more commonly encountered condition of girders for interior bearing walls is addressed in Table R502.5(2). In using this table, only the loading condition, building span, and built-up member size are needed to apply the table.

# of jack studs

TABLE R502.5(2)
GIRDER SPANS[a] AND HEADER SPANS[a] FOR INTERIOR BEARING WALLS
(Maximum spans for Douglas fir-larch, hem-fir, southern pine and spruce-pine-fir[b] and required number of jack studs)

| HEADERS AND GIRDERS SUPPORTING | SIZE | BUILDING WIDTH[c] (feet) | | | | | |
| | | 20 | | 28 | | 36 | |
		Span	NJ[d]	Span	NJ[d]	Span	NJ[d]
One floor only	2-2×4	3-1	1	2-8	1	2-5	1
	2-2×6	4-6	1	3-11	1	3-6	1
	2-2×8	5-9	1	5-0	2	4-5	2
	2-2×10	7-0	2	6-1	2	5-5	2
	2-2×12	8-1	2	7-0	2	6-3	2
	3-2×8	7-2	1	6-3	1	5-7	2
	3-2×10	8-9	1	7-7	2	6-9	2
	3-2×12	10-2	2	8-10	2	7-10	2
	4-2×8	5-10	1	5-1	2	4-6	2
	4-2×10	10-1	1	8-9	1	7-10	2
	4-2×12	11-9	1	10-2	2	9-1	2
Two floors	2-2×4	2-2	1	1-10	1	1-7	1
	2-2×6	3-2	2	2-9	2	2-5	2
	2-2×8	4-1	2	3-6	2	3-2	2
	2-2×10	4-11	2	4-3	2	3-10	3
	2-2×12	5-9	2	5-0	3	4-5	3
	3-2×8	5-1	2	4-5	2	3-11	2
	3-2×10	6-2	2	5-4	2	4-10	2
	3-2×12	7-2	2	6-3	2	5-7	3
	4-2×8	4-2	2	3-7	2	3-2	2
	4-2×10	7-2	2	6-2	2	5-6	2
	4-2×12	8-4	2	7-2	2	6-5	2

FOOTNOTES TO TABLE R502.5(2)

For SI: 1 inch = 25.4 mm, 1 foot = 304.8 mm.

a. Spans are given in feet and inches.

b. Tabulated values assume #2 grade lumber.

c. Building width is measured perpendicular to the ridge. For widths between those shown, spans are permitted to be interpolated.

d. NJ - Number of jack studs required to support each end. Where the number of required jack studs equals one, the headers are permitted to be supported by an approved framing anchor attached to the full-height wall stud and to the header.

In addition to providing the maximum allowable girder spans, the table sets forth the number of jack studs. Typically, two jack studs are required to support the ends of the girder, providing at least 3 inches of bearing. Only under limited conditions is a single jack stud permitted.

Topic: Bearing

Reference: IRC R502.6

Category: Floors

Subject: Wood Floor Framing

Code Text: *The ends of each joist, beam or girder shall have not less than 1.5 inches (38 mm) of bearing on wood or metal and not less than 3 inches (76 mm) on masonry or concrete except where supported on a 1-inch-by-4-inch (25.4 mm by 102 mm) ribbon strip and nailed to the adjacent stud or by the use of approved joist hangers.*

Discussion and Commentary: Joists, beams and girders are typically provided with bearing surfaces. The required bearing length for masonry and concrete is twice that needed for wood or metal. As an alternative, bearing may be accomplished through the use of a ribbon strip, provided each member is directly connected to an adjoining stud. Approved joist hangers are also acceptable means of support in lieu of direct bearing.

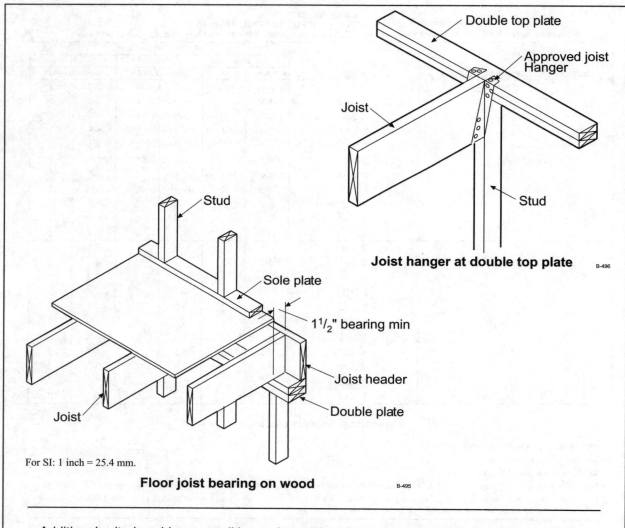

Joist hanger at double top plate   B-496

For SI: 1 inch = 25.4 mm.

Floor joist bearing on wood   B-495

Additional criteria address conditions where joists frame over the top of a beam, girder or bearing wall. A minimum 3-inch (76 mm) lap is required unless a splice having equivalent strength is installed. Framing anchors or ledgers are mandated when framing into girder sides.

Topic: Lateral Restraint at Supports
Reference: IRC R502.7

Category: Floors
Subject: Wood Floor Framing

Code Text: *Joists shall be supported laterally at the ends by full-depth solid blocking not less than 2 inches (51 mm) nominal in thickness; or by attachment to a header, band, or rim joist, or to an adjoining stud; or shall be otherwise provided with lateral support to prevent rotation. In Seismic Design Categories D_1 and D_2, lateral restraint shall also be provided at each intermediate support.*

Discussion and Commentary: The potential for rotation of floor joists is reduced by providing complying blocking at the end points. This is typically accomplished at exterior walls by nailing the rim joist directly to the joist ends. Full-depth blocking may be used where the joist ends bear at interior bearing walls. In addition, any other approved methods that can be shown to prevent the rotation of the joists are acceptable.

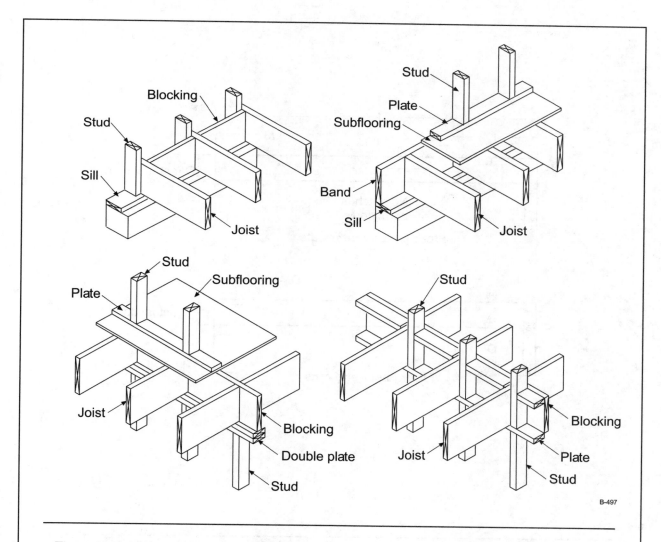

The potential for very deep members (2 in. x 14 in. or greater) to buckle laterally is mitigated by requiring solid blocking or diagonal bridging located at maximum intervals of 8 feet. As an alternative, a continuous 1 in. x 3 in. strip can be nailed to the bottoms of the joists at similar intervals.

Topic: Notching
Reference: IRC R502.8

Category: Floors
Subject: Wood Floor Framing

Code Text: *Notches in solid lumber joists, rafters and beams shall not exceed one-sixth of the depth of the member, shall not be longer than one-third of the depth of the member and shall not be located in the middle one-third of the span. Notches at the ends of the member shall not exceed one-fourth the depth of the member. The tension side of members 4 inches (102 mm) or greater in nominal thickness shall not be notched except at the ends of the members.*

Discussion and Commentary: Notches and bored holes potentially reduce the structural strength of framing members to unacceptable levels. Therefore, prescriptive limits on the size and location of both are set forth. The specified limitations are only applicable to solid sawn lumber members. Where engineered wood products such as trusses and I-joists are notched or have bored holes, the effects of the penetrations must be considered in the member design.

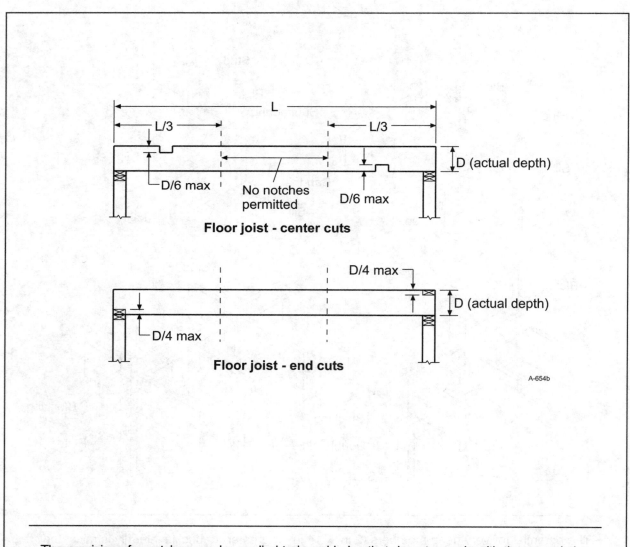

A-654b

The provisions for notches can be applied to bored holes that do not comply with the prescriptive provisions of the code. For example, a bored hole is not allowed within 2 inches of the bottom of a joist. If so located, it may be analyzed as a notch to determine compliance.

Topic: Bored Holes
Reference: IRC R502.8

Category: Floors
Subject: Wood Floor Framing

Code Text: *The diameter of holes bored or cut into members shall not exceed one-third the depth of the member. Holes shall not be closer than 2 inches (51 mm) to the top or bottom of the member, or to any other hole located in the member. Where the member is also notched, the hole shall not be closer than 2 inches (51 mm) to the notch.*

Discussion and Commentary: Where holes are drilled or bored in joists and other wood framing members, they tend to have the same effect as notches. The bored hole reduces the structural strength of the member. Holes cannot be oversized and must be located a minimum distance from the edges of the members. By placing any bored hole within the center portion of the structural member, the effect on the bending stresses is much less of a concern.

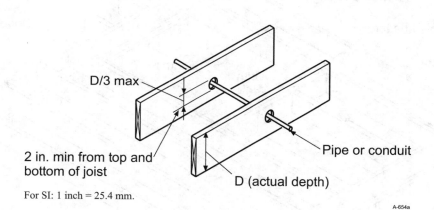

D/3 max

2 in. min from top and bottom of joist

For SI: 1 inch = 25.4 mm.

Pipe or conduit

D (actual depth)

A-654a

Although the code does not limit the number of bored holes within a member, it does restrict how far a bored hole can be located from a notch. Common sense should dictate that as few holes as necessary should be bored, with the holes well-distributed along the member span.

Topic: Framing of Openings   **Category:** Floors
Reference: IRC R502.10   **Subject:** Wood Floor Framing

Code Text: *Openings in floor framing shall be framed with a header and trimmer joists. When the header joist span does not exceed 4 feet (1219 mm), the header joist may be a single member the same size as the floor joist. When the header joist span exceeds 4 feet (1219 mm), the trimmer joists and the header joist shall be doubled and of sufficient cross section to support the floor joists framing into the header. Approved hangers shall be used for the header joist to trimmer joist connections when the header joist span exceeds 6 feet (1829 mm). Tail joists over 12 feet (3658 mm) long shall be supported at the header by framing anchors or on ledger strips not less than 2 inches by 2 inches (51 mm by 51 mm).*

Discussion and Commentary: Where larger floor openings are necessary to provide passage of such things as stairways or chases, the adequate transfer of floor loads to their supporting members must be provided.

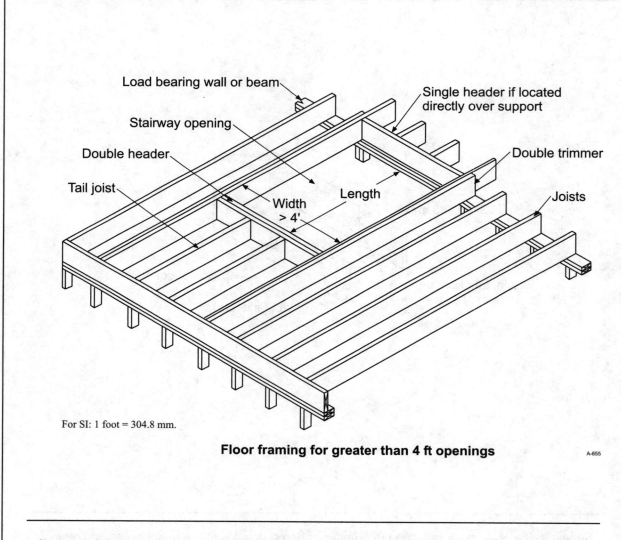

For SI: 1 foot = 304.8 mm.

Floor framing for greater than 4 ft openings

A-655

In many situations, nails are insufficient to transfer vertical loads. Therefore, positive connections must be used when the span of the header or tail joists exceeds a specified length.

Topic: Wood Trusses
Reference: IRC R502.11

Category: Floors
Subject: Wood Floor Framing

Code Text: *Wood trusses shall be designed in accordance with approved engineering practice. The design and manufacture of metal plate connected wood trusses shall comply with ANSI/TPI 1. Trusses shall be braced to prevent rotation and provide lateral stability in accordance with the requirements specified in the construction documents for the building and on the individual truss design drawings. Truss members and components shall not be cut, notched, spliced or otherwise altered in any way without the approval of a registered design professional.*

Discussion and Commentary: There are no prescriptive provisions for wood trusses in the IRC. All trusses must be designed, and any modification or alteration must be approved by an architect or engineer.

Minimum Truss Design Drawing Information

1. Slope or depth, span, and spacing.
2. Location of all joints.
3. Required bearing widths.
4. Design loads as applicable.
   - 4.1 Top chord live load (including snow loads).
   - 4.2 Top chord dead load.
   - 4.3 Bottom chord live load.
   - 4.4 Bottom chord dead load.
   - 4.5 Concentrated loads and their points of application.
   - 4.6 Controlling wind and earthquake loads.
5. Adjustments to lumber and joint connector design values for conditions of use.
6. Each reaction force and direction.
7. Joint connector type and description (e.g., size, thickness or gage); and the dimensioned location of each joint connector except where symmetrically located relative to the joint interface.
8. Lumber size, species and grade for each member.
9. Connection requirements for:
   - 9.1 Truss-to-truss girder.
   - 9.2 Truss ply-to ply.
   - 9.3 Field splices.
10. Calculated deflection ratio and/or maximum description for live and total load.
11. Maximum axial compression forces in the truss members to enable the building designer to design the size, connections and anchorage of the permanent continuous lateral bracing. Forces shall be shown on the truss drawing or on supplemental documents.
12. Required permanent truss member bracing location.

Truss design drawings must be prepared by a registered design professional and submitted to the building official and approved prior to installation of the trusses. The drawings shall be provided with the shipment of trusses to the site and must contain at least the information specified in Section R502.11.4.

Topic: Draftstopping

Category: Floors

Reference: IRC R502.12

Subject: Wood Floor Framing

Code Text: *When there is usable space both above and below the concealed space of a floor/ceiling assembly, draftstops shall be installed so that the area of the concealed space does not exceed 1,000 square feet (92.9 m².). Draftstopping shall divide the concealed space into approximately equal areas. Where the assembly is enclosed by a floor membrane above and a ceiling membrane below, draftstopping shall be provided in floor/ceiling assemblies under the following circumstances: 1) ceiling is suspended under the floor framing, or 2) floor framing is constructed of truss-type open-web or perforated members.*

Discussion and Commentary: Draftstopping, like fireblocking, is only mandated in combustible construction. Although the construction of draftstops is not as substantial as that for fireblocking, its integrity is just as critical and must be maintained.

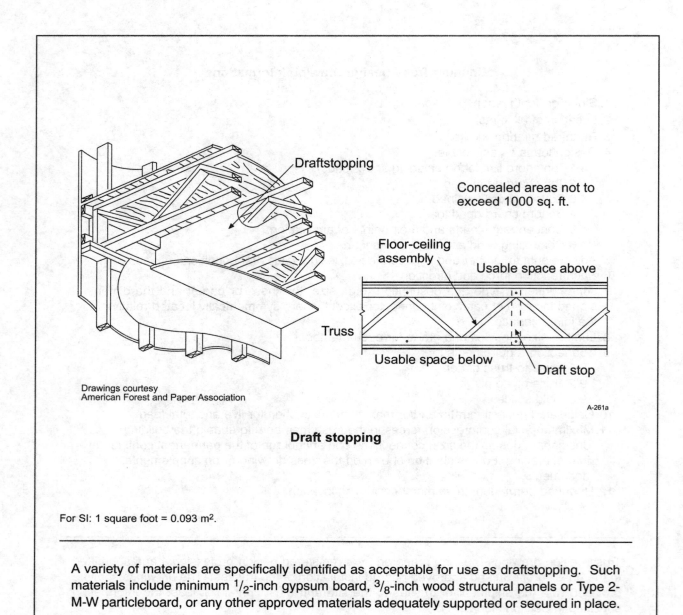

Drawings courtesy
American Forest and Paper Association

A-261a

Draft stopping

For SI: 1 square foot = 0.093 m².

A variety of materials are specifically identified as acceptable for use as draftstopping. Such materials include minimum $1/2$-inch gypsum board, $3/8$-inch wood structural panels or Type 2-M-W particleboard, or any other approved materials adequately supported or secured in place.

2003 IRC Study Companion

Topic: Wood Structural Panels
Reference: IRC R503.2
Category: Floors
Subject: Floor Sheathing

Code Text: *Where used as subflooring or combination subfloor underlayment, wood structural panels shall be of one of the grades specified in Table R503.2.1.1(1). The maximum allowable span for wood structural panels used as subfloor or combination subloor underlayment shall be as set forth in Table R503.2.1.1(1).*

Discussion and Commentary: The maximum spans for wood structural panel floor sheathing are limited by the stress and deflection imposed by the design live loads. The edges of the panels between floor supports are prevented from moving relative to each other by use of tongue and groove panel edges, the addition of wood blocking or by the use of an approved underlayment or structural finished floor system.

TABLE R503.2.1.1(1)
ALLOWABLE SPANS AND LOADS FOR WOOD STRUCTURAL PANELS FOR ROOF AND SUBFLOOR SHEATHING AND COMBINATION SUBFLOOR UNDERLAYMENT[a,b,c]

SPAN RATING	MINIMUM NOMINAL PANEL THICKNESS (inch)	MAXIMUM SPAN (inches)[d]		LOAD (pounds per square foot, at maximum span)		MAXIMUM SPAN (inches)
		With edge support	Without edge support	Total load	Live load	
Sheathing[e]			Roof[f]			Subfloor[j]
12/0	$5/16$	12	12	40	30	0
16/0	$5/16$	16	16	40	30	0
20/0	$5/16$	20	20	40	30	0
24/0	$3/8$	24	20[g]	40	30	0
24/16	$7/16$	24	24	50	40	16
32/16	$15/32, 1/2$	32	28	40	30	16[h]
40/20	$19/32, 5/8$	40	32	40	30	20[h,i]
48/24	$23/32, 3/48$	48	36	45	35	24
60/32	$7/8$	60	48	45	35	32
Underlayment, C-C plugged, single floor[e]			Roof[f]			Combination subfloor underlayment[k]
16 o.c.	$19/32, 5/8$	24	24	50	40	16[i]
20 o.c.	$19/32, 5/8$	32	32	40	30	20[i,j]
24 o.c.	$23/32, 3/4$	48	36	35	25	24
32 o.c.	$7/8$	48	40	50	40	32
48 o.c.	$1 3/32, 1 1/8$	60	48	50	40	48

For SI: 1 inch = 25.4 mm, 1 pound per square foot = 0.0479 kN/m².

a. The allowable total loads were determined using a dead load of 10 psf. If the dead load exceeds 10 psf, then the live load shall be reduced accordingly.

b. Panels continuous over two or more spans with long dimension perpendicular to supports. Spans shall be limited to values shown because of possible effect of concentrated loads.

c. Applies to panels 24 inches or wider.

d. Lumber blocking, panel edge clips (one midway between each support, except two equally spaced between supports when span is 48 inches), tongue-and-groove panel edges, or other approved type of edge support.

e. Includes Structural 1 panels in these grades.

f. Uniform load deflection limitation: $1/180$ of span under live load plus dead load, $1/240$ of span under live load only.

g. Maximum span 24 inches for $15/32$- and $1/2$-inch panels.

h. Maximum span 24 inches where $3/4$-inch wood finish flooring is installed at right angles to joists.

i. Maximum span 24 inches where 1.5 inches of lightweight concrete or approved cellular concrete is placed over the subfloor.

j. Unsupported edges shall have tongue-and-groove joints or shall be supported with blocking unless minimum nominal $1/4$-inch thick underlayment with end and edge joints offset at least 2 inches or 1.5 inches of lightweight concrete or approved cellular concrete is placed over the subfloor, or $3/4$-inch wood finish flooring is installed at right angles to the supports. Allowable uniform live load at maximum span, based on deflection of $1/360$ of span, is 100 psf.

k. Unsupported edges shall have tongue-and-groove joints or shall be supported by blocking unless nominal $1/4$-inch-thick underlayment with end and edge joints offset at least 2 inches or $3/4$-inch wood finish flooring is installed at right angles to the supports. Allowable uniform live load at maximum span, based on deflection of $1/360$ of span, is 100 psf, except panels with a span rating of 48 on center are limited to 65 psf total uniform load at maximum span.

Wood structural panels are the class of panels manufactured with fully waterproof adhesive and include plywood, oriented strand board (OSB), and composite panels made up of a combination of wood veneers and reconstructed wood layers.

Topic: Structural Framing
Reference: IRC R505.2

Category: Floors
Subject: Steel Floor Framing

Code Text: *Load-bearing floor framing members shall comply with Figure R505.2(1) and with the dimensional and minimum thickness requirements specified in Tables R505.2(1) and R505.2(2). Screws for steel-to-steel connections shall be installed with a minimum edge distance and center-to-center spacing of 0.5 inch, shall be self-drilling tapping, and shall conform to SAE J78.*

Discussion and Commentary: The provisions of Section R505 apply to the construction of floor systems using cold-formed steel framing. Cold-formed steel structural members have profiles with rounded corners and slender flat elements with large width-to-thickness ratios. Cold-formed steel structural members are produced from carbon or low alloy steel sheet, strip, plate or bars not more than one inch thick. Forming of the steel shapes is done at or near room temperature using bending brakes, press brakes or roll forming machines.

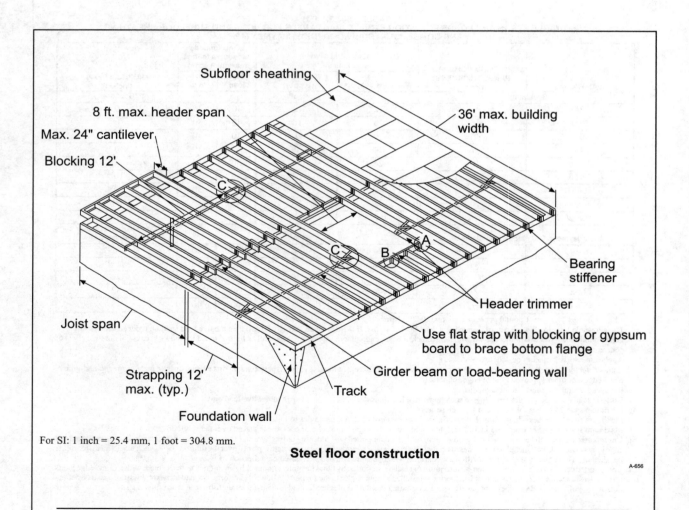

For SI: 1 inch = 25.4 mm, 1 foot = 304.8 mm.

Steel floor construction

A-656

The prescriptive provisions for cold-formed steel framing are only applicable to one- and two-story buildings with a maximum story height of 10 feet and footprint of 36 feet by 60 feet. The wind speed, Seismic Design Category and ground snow load are also limited.

Topic: Concrete Slab-on-ground Floors
Reference: IRC R506

Category: Floors
Subject: Concrete Floors on Ground

Code Text: *Concrete slab-on-ground floors shall be a minimum 3.5 inches (89 mm) thick (for expansive soils, see Section R403.1.8). The specified compressive strength of concrete shall be as set forth in Section R402.2. The area within the foundation walls shall have all vegetation, top soil and foreign material removed. A 4-inch-thick (102 mm) base course consisting of clean graded sand, gravel, crushed stone or crushed blast-furnace slag passing a 2-inch (51 mm) sieve shall be placed on the prepared subgrade when the slab is below grade. See exception for base course where slab is installed on well-drained or sand-gravel mixture soils.*

Discussion and Commentary: Removal of construction debris and foreign materials, such as lumber formwork, stakes, tree stumps and other vegetation, limits the attraction of termites, insects and vermin. Top soil and vegetation should also be removed in that it is generally loosely compacted to such an extent as to allow soil settlement. Differential settlement can result in cracking of the floor slab and the interior wall/ceiling finishes.

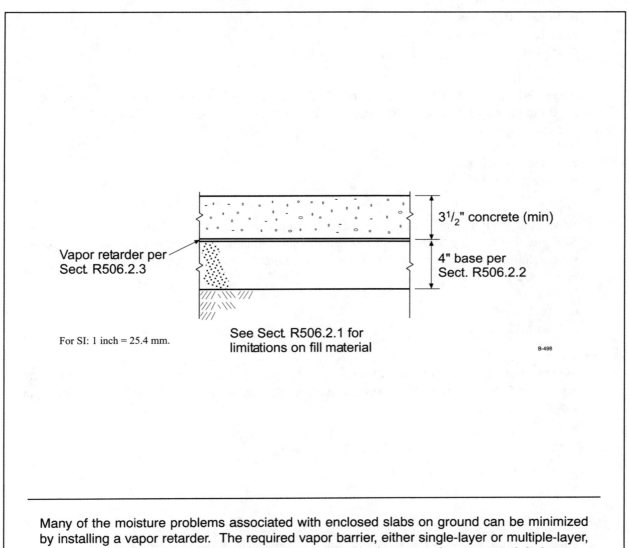

Vapor retarder per
Sect. R506.2.3

$3^1/_2$" concrete (min)

4" base per
Sect. R506.2.2

For SI: 1 inch = 25.4 mm.

See Sect. R506.2.1 for
limitations on fill material

B-498

Many of the moisture problems associated with enclosed slabs on ground can be minimized by installing a vapor retarder. The required vapor barrier, either single-layer or multiple-layer, must be properly installed with lapped joints, and the barrier cannot be punctured during construction.

QUIZ

Study Session 6

1. A minimum design live load of _____ shall be used for the determination of the maximum allowable floor joist spans in attics that are accessed by a fixed stairway.
    a. 10 psf
    b. 20 psf
    c. 30 psf
    d. 40 psf

2. Where 2-inch-by-10-inch floor joists of #2 Hem Fir are spaced at 16 inches on center, what is the maximum span where such joists support a sleeping room and a dead load of 10 psf?
    a. 16 feet, 0 inches
    b. 16 feet, 10 inches
    c. 17 feet, 8 inches
    d. 19 feet, 8 inches

3. Where 2-inch-by-10-inch floor joists of #1 SPF are spaced at 16 inches on center, what is the maximum span where such joists support a living room and a dead load of 20 psf?
    a. 14 feet, 1 inch
    b. 15 feet, 5 inches
    c. 16 feet, 0 inches
    d. 16 feet, 9 inches

4. Where the ground snow load does not exceed 30 psf, what is the maximum span of a built-up SPF girder consisting of three 2x12s when it is located at an exterior bearing wall and supports a roof, a ceiling and a center-bearing floor in a building having a width of 28 feet?
    a. 5 feet, 2 inches
    b. 7 feet, 8 inches
    c. 8 feet, 0 inches
    d. 8 feet, 11 inches

5. Where a built-up girder of three 2x10s is used for an exterior bearing wall, what is the minimum number of jack studs required to support a girder that carries a roof, ceiling and one clear span floor?
    a. 1
    b. 2
    c. 3
    d. 4

6.   What is the maximum girder span for an interior bearing wall using a built-up girder consisting of three 2x10s and supporting one floor in a building having a width of 36 feet?
a. 5 feet, 5 inches
b. 6 feet, 9 inches
c. 7 feet, 7 inches
d. 8 feet, 10 inches

7.   Doubled joists under parallel bearing partitions that are separated to permit the installation of piping or vents shall be blocked at maximum intervals of _____ on center.
a. 16 inches
b. 32 inches
c. 48 inches
d. 60 inches

8.   The ends of floor joists shall bear a minimum of _____ on wood or metal.
a. 1 inch
b. $1^{1}/_{2}$ inches
c. $2^{1}/_{2}$ inches
d. 3 inches

9.   Where floor joists frame from opposite sides across the top of a wood girder and are lapped, the minimum lap shall be _____.
a. 3 inches
b. 4 inches
c. 6 inches
d. 8 inches

10.   Bridging shall be provided to support floor joists laterally where the minimum 2x joist depth exceeds _____.
a. 8 inches
b. 10 inches
c. 12 inches
d. 14 inches

11.   What is the maximum permitted length of a notch in a floor joist?
a. 2 inches
b. twice the notch depth
c. $1/_{3}$ the depth of the joist
d. $1/_{6}$ the depth of the joist

12.   A hole bored through a floor joist shall have a maximum diameter of _____.
a. 2 inches
b. $1/_{6}$ the depth of the joist
c. $1/_{4}$ the depth of the joist
d. $1/_{3}$ the depth of the joist

13. Tail joists exceeding _____ in length shall be supported by framing anchors or on complying ledger strips.
   a. 4 feet
   b. 6 feet
   c. 8 feet
   d. 12 feet

14. Where a ceiling is suspended below the floor framing, draftstops shall be installed so that the maximum area of any concealed space is _____ square feet.
   a. 100
   b. 400
   c. 1,000
   d. 1,500

15. Where wood structural panels are used for required draftstops in concealed floor/ceiling assemblies, what is the minimum thickness mandated?
   a. $5/16$ inch
   b. $3/8$ inch
   c. $15/32$ inch
   d. $23/32$ inch

16. Wood structural panels $15/32$-inch thick are to be used as subfloor sheathing and are to be covered with $3/4$-inch wood finish flooring installed at right angles to the joists. What is the maximum allowable span of the wood structural panels if the span rating of the panels is $32/16$?
   a. 0 inches; the panels may not be used as subfloor sheathing
   b. 16 inches
   c. 20 inches
   d. 24 inches

17. Where Species Group 2 sanded plywood is used as combination subfloor underlayment, what is the minimum required plywood thickness where the joists are spaced at 16 inches on center?
   a. $1/2$ inch
   b. $5/8$ inch
   c. $3/4$ inch
   d. $7/8$ inch

18. Steel floor framing constructed in accordance with the IRC is limited to buildings a maximum of _____ stories with each story a maximum of _____ feet in height.
   a. 2, 10
   b. 2, 12
   c. 3, 10
   d. 3, 12

19. Where complying screws are used for the connection of steel floor framing members, screws shall extend through the steel a minimum of _____.
   a. $1/8$ inch
   b. $3/8$ inch
   c. 3 exposed threads
   d. 5 exposed threads

20. In a steel floor framing system, No. 8 screws spaced at a maximum of _____ on center on the edges and _____ on center at the intermediate supports shall be used to fasten the sub-floor to the floor joists.
   a. 6 inches, 10 inches
   b. 6 inches, 12 inches
   c. 8 inches, 12 inches
   d. 8 inches, 14 inches

21. Cold-formed steel joists are installed at 16 inches on center in the floor framing system for a dwelling. If the nominal joist size is 800S162-43, the maximum span for a 40 psf live load is _____.
   a. 10 feet, 5 inches
   b. 12 feet, 3 inches
   c. 14 feet, 1 inch
   d. 15 feet, 6 inches

22. Concrete slab-on-ground floors shall be a minimum of _____ thick.
   a. 3 inches
   b. $3^1/2$ inches
   c. 4 inches
   d. 5 inches

23. The maximum fill depth when preparing a site for construction of a concrete slab-on-ground floor is _____ for earth and _____ for clean sand or gravel.
   a. 6 inches, 12 inches
   b. 8 inches, 16 inches
   c. 8 inches, 24 inches
   d. 12 inches, 24 inches

24. A base course is not required for the prepared subgrade for a concrete floor slab below grade where the soil is classified as _____.
   a. GM
   b. SC
   c. CH
   d. OH

25. Where a vapor retarder is required between a concrete floor slab and the prepared subgrade, the joints of the vapor retarder shall be lapped a minimum of _____.
   a. 2 inches
   b. 4 inches
   c. 6 inches
   d. 8 inches

26. Where supporting only a light-frame exterior wall and roof, what is the maximum cantilever span for 2 x 10 floor joists spaced at 16 inches on center, provided the roof has a width of 32 feet and the ground snow load is 30 psf?
   a. 18 inches
   b. 21 inches
   c. 22 inches
   d. 26 inches

27. Where supporting an exterior balcony in an area having a ground snow load of 50 psf, what is the maximum cantilever span for 2 x 12 floor joists spaced at 16 inches on center?
   a. 49 inches
   b. 57 inches
   c. 67 inches
   d. 72 inches

28. Unless the joists are of sufficient size to carry the load, bearing partitions perpendicular to floor joists may be offset a maximum of _____ from supporting girders, walls or partitions.
   a. 6 inches
   b. 12 inches
   c. the depth of the floor sheathing
   d. the depth of the floor joists

29. Where an opening in floor framing is framed with a single header joist the same size as the floor joist, the maximum header joist span shall be _____ feet.
   a. 3
   b. 4
   c. 6
   d. 12

30. If a base course is required for a concrete floor slab installed below grade, the thickness of the base course shall be _____ inches.
   a. 3
   b. 4
   c. 6
   d. 8

INTERNATIONAL RESIDENTIAL CODE
Study Session 7
Chapters 6 and 7—Wall Construction and Wall Covering

OBJECTIVE: To obtain an understanding of the requirements dealing with wall construction and wall covering, including wood wall framing details; headers in both exterior and interior walls; fireblocking; braced wall panels and braced wall lines; cripple walls; steel wall framing; masonry wall construction; exterior windows and exterior doors; gypsum board plaster, ceramic tile and other interior wall finishes; and exterior wall finishes such as siding, plaster and veneer.

REFERENCE: Chapters 6 and 7, 2003 *International Residential Code*

KEY POINTS:

- How is the maximum allowable spacing of wood studs determined? How is a double top plate to be installed? A single top plate?
- What is the appropriate number and type of fasteners for connecting wood structural members? What is the maximum allowable spacing of such fasteners? What alternate fastening methods are acceptable?
- Where must a bored hole in a stud be located? What is the maximum permitted size of a bored hole located in a bearing wall? In a nonbearing wall?
- What is the maximum notch size allowed in a bearing stud wall? In a nonbearing wall? How are notches and bored holes in a top plate to be addressed?
- What is the method for determining the maximum header span in an interior bearing wall? In an exterior bearing wall? In a nonbearing wall?
- What is a wood structural panel box header? How is a box header to be constructed?
- Where is fireblocking required? What materials can be used as fireblocking?
- How are foundation cripple studs to be sized? When must cripple walls be sheathed?
- What methods are acceptable for the construction of braced wall panels? What percentage of a braced wall line must consist of braced wall panels? How close to the end of a braced wall line is a braced wall panel required? What is minimum required panel length?
- What alternate methods of construction are provided for braced wall panels? What are the code benefits of continuous structural panel sheathing?
- How do braced wall panels need to be connected to the floor framing and top plates? What special conditions apply in Seismic Design Categories D_1 and D_2?
- What are the general provisions for load-bearing steel wall framing? How are such walls to be fastened? How are connections to be made?
- What is the minimum required thickness of masonry walls? Parapet walls?
- How must masonry walls be laterally supported? What is the minimum spacing between lateral supports? How are walls to be anchored at floors and roofs?
- How shall glass masonry be constructed?
- What is the performance level required for exterior windows and doors? How must such windows and doors be tested and identified? How should they be attached to the structure?
- What limitations are placed on insulating concrete form wall construction?
- How must gypsum board be applied on an interior wall? How must it be fastened?
- Where is the use of water-resistant gypsum backing board prohibited?
- When is weather-resistant sheathing paper required on an exterior wall?
- What is the criteria for the installation of weep screeds in exterior plaster applications?
- How shall masonry veneer be supported? What method of anchorage is mandated?
- How must fiber cement siding be installed?

Topic: Fasteners for Structural Members   **Category:** Wall Construction
Reference: IRC R602.3   **Subject:** Wood Wall Construction

Code Text: *Components of exterior walls shall be fastened in accordance with Tables R602.3(1) through R602.3(4).*

Discussion and Commentary: The fastener schedules provide the minimum nailing requirements for the connection of structural wood framing members, wood structural panels, and other types of wall sheathing materials. For dimensional lumber, the number of nails and maximum nail spacing is specified. The use of staples is permitted in specific locations. Wood structural panels and other sheathing materials are regulated by thickness for nail size, number and spacing. Staples and other alternate attachments are also permitted when in compliance with the code. Although many of the listed building elements are to some degree related to wall construction, a number of the connection locations are specific to floor and roof framing.

TABLE R602.3(1)
FASTENER SCHEDULE FOR STRUCTURAL MEMBERS

DESCRIPTION OF BUILDING ELEMENTS	NUMBER AND TYPE OF FASTENER[a,b,c,d]	SPACING OF FASTENERS
Joist to sill or girder, toe nail	3-8d	—
1″ × 6″ subfloor or less to each joist, face nail	2-8d 2 staples, 1³/₄	— —
2″ subfloor to joist or girder, blind and face nail	2-16d	—
Sole plate to joist or blocking, face nail	16d	16″ o.c.
Top or sole plate to stud, end nail	2-16d	—
Stud to sole plate, toe nail	3-8d or 2-16d	—
Double studs, face nail	10d	24″ o.c.
Double top plates, face nail	10d	24″ o.c.
Sole plate to joist or blocking at braced wall panels	3-16d	16″ o.c.
Double top plates, minimum 48-inch offset of end joints, face nail in lapped area	8-16d	—
Blocking between joists or rafters to top plate, toe nail	3-8d	—
Rim joist to top plate, toe nail	8d	6″ o.c.
Top plates, laps at corners and intersections, face nail	2-10d	—
Built-up header, two pieces with ¹/₂″ spacer	16d	16″ o.c. along each edge
Continued header, two pieces	16d	16″ o.c. along each edge
Ceiling joists to plate, toe nail	3-8d	—
Continuous header to stud, toe nail	4-8d	—
Ceiling joist, laps over partitions, face nail	3-10d	—
Ceiling joist to parallel rafters, face nail	3-10d	—
Rafter to plate, toe nail	2-16d	—
1″ brace to each stud and plate, face nail	2-8d 2 staples, 1³/₄	— —

The code allows the use of built-up members as headers and girders in various applications. The fastening schedules require staggered nailing with minimum 10d nails on built-up members so they may act as a single member. Thus, the loads are properly transferred to all elements of the built-up member.

Topic: Design and Construction
Reference: IRC R602.3

Category: Wall Construction
Subject: Wood Wall Framing

Code Text: *The size, height and spacing of studs shall be in accordance with Table R602.3(5).* See exceptions for limits on use of utility grade studs and studs exceeding 10 feet in height. *Wood stud walls shall be capped with a double top plate installed to provide overlapping at corners and intersections with bearing partitions. End joints in top plates shall be offset at least 24 inches (610 mm). Plates shall be a nominal 2 inches (51 mm) in depth and have a width at least equal to the width of the studs.* See exception for use of a single top plate.

Discussion and Commentary: Unless the studs are supporting two floors and a roof, it is possible to use 2-inch by 4-inch studs for the framing of bearing walls up to 10 feet in height. The center-to-center spacing varies based on the loading conditions and the lumber grade. If the stud height exceeds 10 feet, it is likely that larger studs and closer spacing will be required.

TABLE R602.3.1
MAXIMUM ALLOWABLE LENGTH OF WOOD WALL STUDS EXPOSED TO WIND SPEEDS OF 100 MPH OR LESS
IN SEISMIC DESIGN CATEGORIES A, B, C and D[a,b,c]

HEIGHT (feet)	ON-CENTER SPACING (inches)			
	24	16	12	8
Supporting a roof only				
>10	2×4	2×4	2×4	2×4
12	2×6	2×4	2×4	2×4
14	2×6	2×6	2×6	2×4
16	2×6	2×6	2×6	2×4
18	NA[a]	2×6	2×6	2×6
20	NA[a]	NA[a]	2×6	2×6
24	NA[a]	NA[a]	NA[a]	2×6
Supporting one floor and a roof				
>10	2×6	2×4	2×4	2×4
12	2×6	2×6	2×6	2×4
14	2×6	2×6	2×6	2×6
16	NA[a]	2×6	2×6	2×6
18	NA[a]	2×6	2×6	2×6
20	NA[a]	NA[a]	2×6	2×6
24	NA[a]	NA[a]	NA[a]	2×6
Supporting two floors and a roof				
>10	2×6	2×6	2×4	2×4
12	2×6	2×6	2×6	2×6
14	2×6	2×6	2×6	2×6
16	NA[a]	NA[a]	2×6	2×6
18	NA[a]	NA[a]	2×6	2×6
20	NA[a]	NA[a]	NA[a]	2×6
22	NA[a]	NA[a]	NA[a]	NA[a]
24	NA[a]	NA[a]	NA[a]	NA[a]

For SI: 1 inch = 25.4 mm, 1 foot = 304.8 mm, 1 pound per square foot = 0.0479 kN/m^2, 1 pound per square inch = 6.895 kPa, 1 mile per hour = 1.609 km/h.

a. Design required.

b. Applicability of this table assumes the following: Snow load not exceeding 25 psf, but not less than 1310 psi determined by multiplying the AF&PA NDS tabular base design value by the repetitive use factor, and by the size factor for all species except southern pine, E not less than 1.6 by 10^6 psi, tributary dimensions for floors and roofs not exceeding 6 feet, maximum span for floors and roof not exceeding 12 feet, eaves not greater than 2 feet in dimension and exterior sheathing. Where the conditions are not within these parameters, design is required.

c. Utility, standard, stud and No. 3 grade lumber of any species are not permitted.

(continued)

The installation of double top plates is mandated for a variety of reasons. They tie the building together at corners and at wall intersections, serve as beams by supporting joists, rafters, and trusses that do not bear above studs, and work as chords for floor and roof diaphragms.

Topic: Drilling and Notching
Reference: IRC R602.6

Category: Wall Construction
Subject: Wood Wall Framing

Code Text: *Any stud in an exterior wall or bearing partition may be cut or notched to a depth not exceeding 25 percent of its width. Studs in nonbearing partitions may be notched to a depth not to exceed 40 percent of a single stud width. Any stud may be bored or drilled, provided that the diameter of the resulting hole is no greater than 40 percent of the stud width, the edge of the hole is no closer than $^5/_8$ inch (15.9 mm) to the edge of the stud, and the hole is not located in the same section as a cut or notch.* See exceptions for the use of approved stud shoes and conditions for larger bored holes.

Discussion and Commentary: In order to maintain the structural integrity of stud wall systems, limits are imposed on the notching or drilling of studs. Where a bored hole is oversized, it must be considered a notch for compliance purposes.

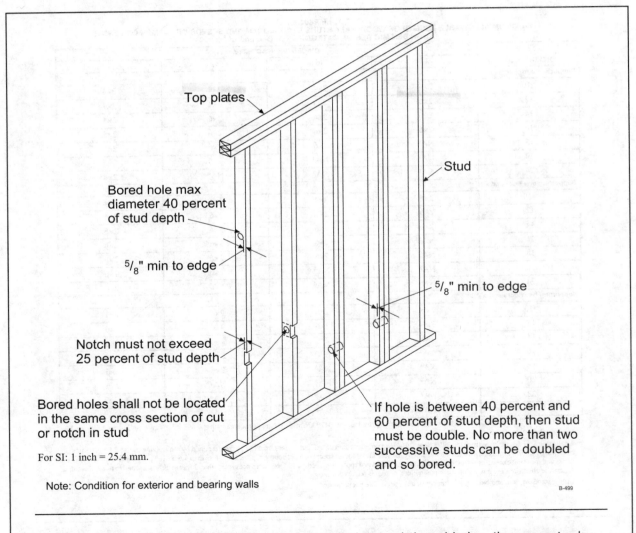

Top plates

Stud

Bored hole max
diameter 40 percent
of stud depth

$^5/_8$" min to edge

$^5/_8$" min to edge

Notch must not exceed
25 percent of stud depth

Bored holes shall not be located
in the same cross section of cut
or notch in stud

For SI: 1 inch = 25.4 mm.

If hole is between 40 percent and
60 percent of stud depth, then stud
must be double. No more than two
successive studs can be doubled
and so bored.

Note: Condition for exterior and bearing walls

B-499

Where nonmetallic sheathed (NM) cable is installed through bored holes, the nearest edge of the hole must be located at least 1$^1/_4$ inches from the edge of stud. Otherwise, a minimum $^1/_{16}$-inch steel plate, shoe or sleeve must be installed to provide the necessary physical protection.

Category: Wall Construction
Subject: Wood Wall Framing

Code Text: *For header spans see Tables R502.5(1) and R502.5(2). Load-bearing headers are not required in interior or exterior nonbearing walls. A single flat 2-inch-by-4-inch (51 mm by 102 mm) member may be used as a header in interior or exterior nonbearing walls for openings up to 8 feet (2438 mm) in width if the vertical distance to the parallel nailing surface above is not more than 24 inches (610 mm). For such nonbearing headers, no cripples or blocking are required above the header.*

Discussion and Commentary: Where it is necessary to transfer floor, ceiling, wall and roof loads to the foundation, headers must be installed over openings created for doors, windows and similar items. The maximum allowable spans are provided for various conditions, with the minimum permitted size of built-up header and supporting jack studs identified.

TABLE R502.5(1)
GIRDER SPANS[a] AND HEADER SPANS[a] FOR EXTERIOR BEARING WALLS
(Maximum spans for Douglas fir-larch, hem-fir, southern pine and spruce-pine-fir[b] and required number of jack studs)

GIRDERS AND HEADERS SUPPORTING	SIZE	GROUND SNOW LOAD (psf)[e]											
		30						50					
		Building width[c] (feet)											
		20		28		36		20		28		36	
		Span	NJ[d]	Span	NJ[d]	Span	NJ[d]	Span	NJ[d]	Span	NJ[d]	Span	NJ[d]
Roof and ceiling	2-2×4	3-6	1	3-2	1	2-10	1	3-2	1	2-9	1	2-6	1
	2-2×6	5-5	1	4-8	1	4-2	1	4-8	1	4-1	1	3-8	2
	2-2×8	6-10	1	5-11	2	5-4	2	5-11	2	5-2	2	4-7	2
	2-2×10	8-5	2	7-3	2	6-6	2	7-3	2	6-3	2	5-7	2
	2-2×12	9-9	2	8-5	2	7-6	2	8-5	2	7-3	2	6-6	2
	3-2×8	8-4	1	7-5	1	6-8	1	7-5	1	6-5	2	5-9	2
	3-2×10	10-6	1	9-1	2	8-2	2	9-1	2	7-10	2	7-0	2
	3-2×12	12-2	2	10-7	2	9-5	2	10-7	2	9-2	2	8-2	2
	4-2×8	9-2	1	8-4	1	7-8	1	8-4	1	7-5	1	6-8	1
	4-2×10	11-8	1	10-6	1	9-5	2	10-6	1	9-1	2	8-2	2
	4-2×12	14-1	1	12-2	2	10-11	2	12-2	2	10-7	2	9-5	2
Roof, ceiling and one center-bearing floor	2-2×4	3-1	1	2-9	1	2-5	1	2-9	1	2-5	1	2-2	1
	2-2×6	4-6	1	4-0	1	3-7	2	4-1	1	3-7	2	3-3	2
	2-2×8	5-9	2	5-0	2	4-6	2	5-2	2	4-6	2	4-1	2
	2-2×10	7-0	2	6-2	2	5-6	2	6-4	2	5-6	2	5-0	2
	2-2×12	8-1	2	7-1	2	6-5	2	7-4	2	6-5	2	5-9	3
	3-2×8	7-2	1	6-3	2	5-8	2	6-5	2	5-8	2	5-1	2
	3-2×10	8-9	2	7-8	2	6-11	2	7-11	2	6-11	2	6-3	2
	3-2×12	10-2	2	8-11	2	8-0	2	9-2	2	8-0	2	7-3	2
	4-2×8	8-1	1	7-3	1	6-7	1	7-5	1	6-6	1	5-11	2
	4-2×10	10-1	1	8-10	2	8-0	2	9-1	2	8-0	2	7-2	2
	4-2×12	11-9	2	10-3	2	9-3	2	10-7	2	9-3	2	8-4	2
Roof, ceiling and one clear span floor	2-2×4	2-8	1	2-4	1	2-1	1	2-7	1	2-3	1	2-0	1
	2-2×6	3-11	1	3-5	2	3-0	2	3-10	2	3-4	2	3-0	2
	2-2×8	5-0	2	4-4	2	3-10	2	4-10	2	4-2	2	3-9	2
	2-2×10	6-1	2	5-3	2	4-8	2	5-11	2	5-1	2	4-7	3
	2-2×12	7-1	2	6-1	3	5-5	3	6-10	2	5-11	3	5-4	3
	3-2×8	6-3	2	5-5	2	4-10	2	6-1	2	5-3	2	4-8	2
	3-2×10	7-7	2	6-7	2	5-11	2	7-5	2	6-5	2	5-9	2

Box headers are permitted when minimum $^{15}/_{32}$-inch wood structural panels are used. Header depths of both 9 inches and 15 inches are addressed, with panels attached to both sides or one side only. Adequate attachment to the framing is critical, with nails spaced at 3 inches on center.

Topic: Fireblocking
Reference: IRC R602.8

Category: Wall Construction
Subject: Wood Wall Framing

Code Text: *Fireblocking shall be provided to cut off all concealed draft openings (both vertical and horizontal) and to form an effective barrier between stories, and between a top story and the roof space. Fireblocking shall be provided in wood-frame construction in the following locations: 1) in concealed spaces of stud walls and partitions, including furred spaces, at the ceiling and floor level and at 10 foot (3048 mm) intervals both vertical and horizontal; 2) at all interconnections between concealed vertical and horizontal spaces; 3) in concealed spaces between stair stringers at the top and bottom of the run; 4) at openings around vents, pipes and ducts at ceiling and floor level; 5) at chimneys and fireplaces per Section R1001.16; and 6) at cornices of a two-family dwelling at the line of dwelling unit separation.*

Discussion and Commentary: Fireblocking is defined as the installation of building materials to resist the free passage of flame to other areas of the building through concealed spaces.

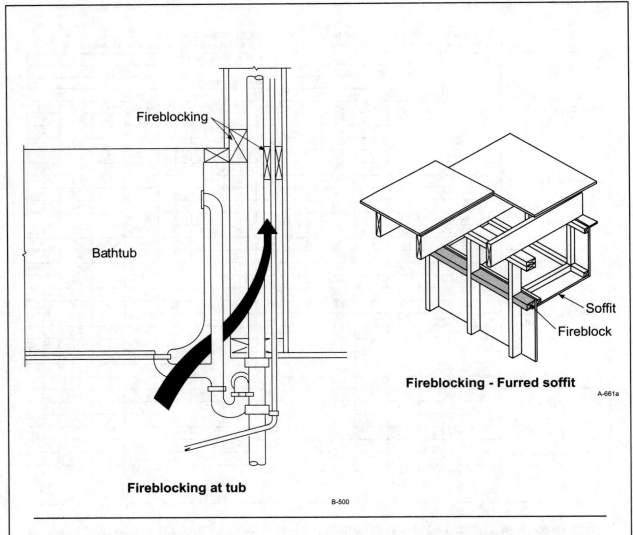

Fireblocking at tub

B-500

Fireblocking - Furred soffit

A-661a

The technique of wood-frame platform construction provides natural fireblocking in many locations to restrict fire movement between horizontal and vertical concealed spaces. Where platform framing and vertical openings occur, a more prescriptive approach is necessary.

Topic: Wall Bracing
Reference: IRC R602.10

Category: Wall Construction
Subject: Wood Wall Framing

Code Text: *Braced wall lines shall consist of braced wall panel construction methods in accordance with Section R602.10.3. The amount and location of bracing shall be in accordance with Table R602.10.1 and the amount of bracing shall be the greater of that required by the Seismic Design Category or the design wind speed.*

Discussion and Commentary: Braced wall lines are the lateral-resisting elements in conventional construction that are similar to shear walls in engineered structures. In order to properly transfer the lateral loads through the floor or roof diaphragms to the braced wall panels in the braced wall lines, the structure must be adequately connected. There are eight different types of braced wall panels that may be used.

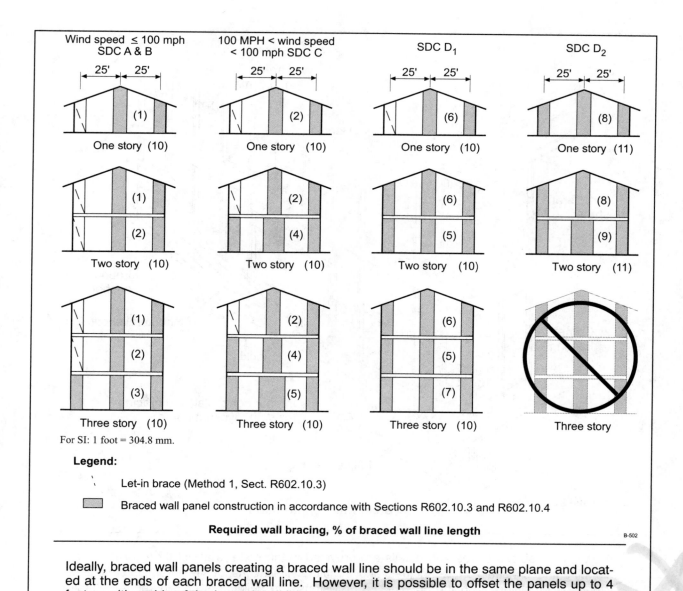

For SI: 1 foot = 304.8 mm.

Legend:

   Let-in brace (Method 1, Sect. R602.10.3)

   Braced wall panel construction in accordance with Sections R602.10.3 and R602.10.4

Required wall bracing, % of braced wall line length

B-502

Ideally, braced wall panels creating a braced wall line should be in the same plane and located at the ends of each braced wall line. However, it is possible to offset the panels up to 4 feet on either side of the braced wall line, with the first panel up to 12$\frac{1}{2}$ feet from the end of the wall.

Topic: Structural Framing
Reference: IRC R603.2, R603.3

Category: Wall Construction
Subject: Steel Wall Framing

Code Text: *Load-bearing steel wall framing members shall comply with Figure R603.2(1) and the dimensional and minimum thickness requirements specified in Tables R603.2(1) and R603.2(2). All exterior steel framed walls and interior load-bearing steel framed walls shall be constructed in accordance with the provisions of* Section R603.3 *and* Figure R603.3.1(1) or R603.3.1(2).

Discussion and Commentary: The use of load-bearing steel members is regulated prescriptively not only for wall framing but also for floor and roof system framing. In order to verify that the steel members are in compliance with appropriate materials standards, the framing members must have a legible label, stencil, stamp or embossment with the manufacturer's identification as well as the minimum steel thickness, coating designation and yield strength.

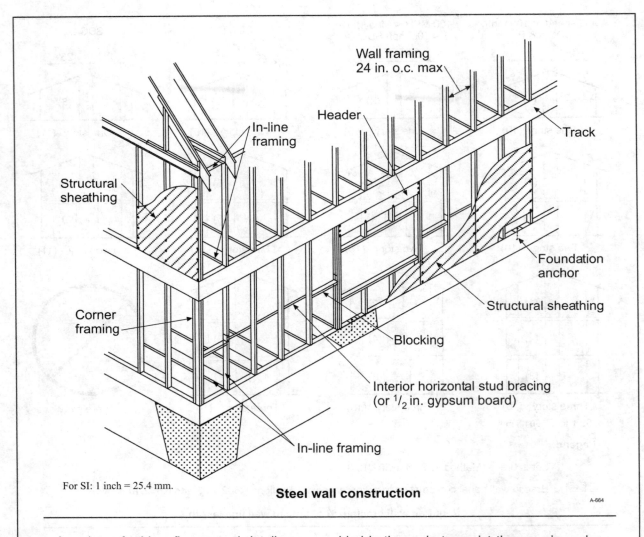

For SI: 1 inch = 25.4 mm.

Steel wall construction

A-664

A variety of tables, figures and details are provided in the code to assist the user in understanding the construction techniques and structural limitations of steel-framed structures. When in compliance with the prescriptive provisions, the building is considered structurally sound.

Topic: Masonry Thickness
Reference: IRC R606.2

Category: Wall Construction
Subject: General Masonry Construction

Code Text: *The minimum thickness of masonry bearing walls more than one story high shall be 8 inches (203 mm). Solid masonry walls of one-story dwellings and garages shall not be less than 6 inches (152 mm) in thickness when not greater than 9 feet (2743 mm) in height, provided that when gable construction is used, an additional 6 feet (1829 mm) is permitted to the peak of the gable. Masonry walls shall be laterally supported in either the horizontal or vertical direction at intervals as required by Section R606.8.*

Discussion and Commentary: The minimum required width of masonry units is based directly on the height of the wall, both in number of stories and vertical dimension. Lateral support may be provided horizontally at complying intervals along the length of masonry walls, determined by the walls' length-to-thickness ratio, or vertically at the floors and/or the roof.

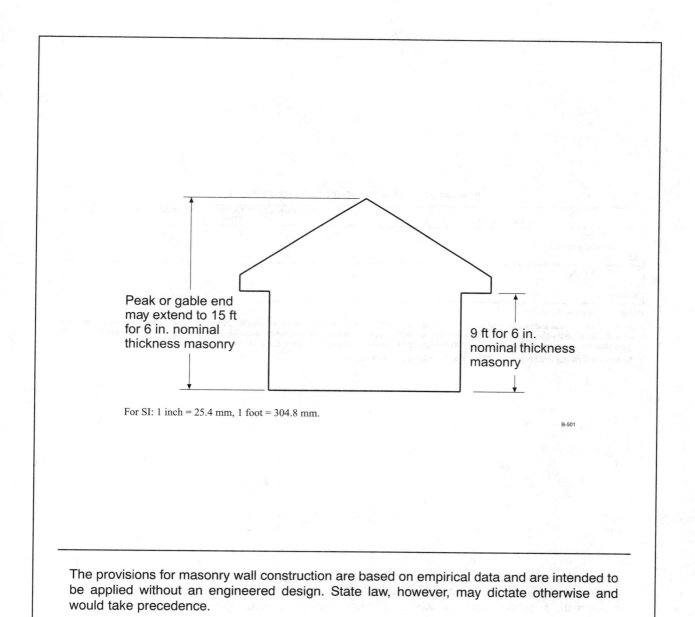

Peak or gable end may extend to 15 ft for 6 in. nominal thickness masonry

9 ft for 6 in. nominal thickness masonry

For SI: 1 inch = 25.4 mm, 1 foot = 304.8 mm.

B-501

The provisions for masonry wall construction are based on empirical data and are intended to be applied without an engineered design. State law, however, may dictate otherwise and would take precedence.

Topic: Lateral Support   **Category:** Wall Construction
Reference: IRC R606.8   **Subject:** General Masonry Construction

Code Text: *Masonry walls shall be laterally supported in either the horizontal or the vertical direction. The maximum spacing between lateral supports shall not exceed the distances in Table R606.8. Lateral support shall be provided by cross walls, pilasters, buttresses or structural frame members when the limiting distance is taken horizontally, or by floors or roofs when the limiting distance is taken vertically.*

Discussion and Commentary: The limitations on the maximum unsupported height or length of masonry walls specified in Table R606.8 provides reasonable performance. At the base of the wall, footings are a lateral support point. Thus, the unsupported height from the footing to the anchorage point at the floor or roof is the unsupported height, which must be limited to the values in Table R606.8.

TABLE R606.8
SPACING OF LATERAL SUPPORT FOR MASONRY WALLS

CONSTRUCTION	MAXIMUM WALL LENGTH TO THICKNESS OR WALL HEIGHT TO THICKNESS[a,b]
Bearing walls: Solid or solid grouted All other	 20 18
Nonbearing walls: Exterior Interior	 18 36

For SI: 1 foot = 304.8 mm.

a. Except for cavity walls and cantilevered walls, the thickness of a wall shall be its nominal thickness measured perpendicular to the face of the wall. For cavity walls, the thickness shall be determined as the sum of the nominal thicknesses of the individual wythes. For cantilever walls, except for parapets, the ratio of height to nominal thickness shall not exceed 6 for solid masonry, or 4 for hollow masonry. For parapets, see Section R606.2.4.

b. An additional unsupported height of 6 feet is permitted for gable end walls.

In lieu of unsupported dimensional limitations being measured vertically from footing to supporting floor or roof, the span limitations may be met with the use of pilasters, columns, piers, cross walls or similar elements whose relative stiffness is greater than the wall.

Topic: Vertical Lateral Support   **Category:** Wall Covering
Reference: IRC R606.8.2   **Subject:** General Masonry Construction

Code Text: *Masonry walls in Seismic Design Category A, B or C shall be anchored to roof structures with metal strap anchors spaced in accordance with the manufacturer's instructions, $^1/_2$-inch (12.7 mm) bolts spaced not more than 6 feet (1829 mm) on center, or other approved anchors. Anchors shall be embedded at least 16 inches (406 mm) into the masonry, or be hooked or welded to bond beam reinforcement placed not less than 6 inches (152 mm) from the top of the wall.*

Discussion and Commentary: All anticipated lateral forces must have a path of tranfer from the roof and floor diaphragms to the masonry walls for transfer to the building's foundation. It is necessary to provide adequate connections to maintain the required load path. Where the building is located in one of the higher Seismic Design Categories, additional conditions are mandated.

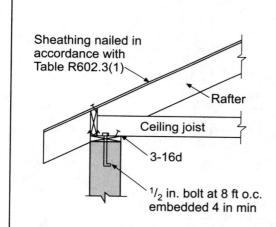

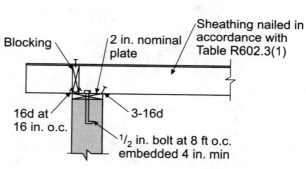

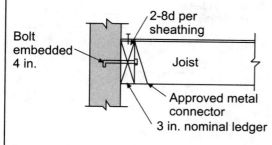

Bolt embedded 4 in.

See table for bolt size and spacing

For SI: 1 inch = 25.4 mm, 1 foot = 304.8 mm.

Ledger bolt size and spacing

Joist span	Bolt size and spacing	
	Roof	Floor
10'	$^1/_2$ at 2' 6" $^7/_8$ at 3' 6"	$^1/_2$ at 2' 0" $^7/_8$ at 2' 9"
10-15'	$^1/_2$ at 1' 9" $^7/_8$ at 2' 6"	$^1/_2$ at 1' 4" $^7/_8$ at 2' 0"
15-20'	$^1/_2$ at 1' 3" $^7/_8$ at 2' 0"	$^1/_2$ at 1' 0" $^7/_8$ at 1' 6"

A-665a

Where the floor diaphragm is connected to a masonry wall, a ledger is used. The ledger fasteners transfer shear forces when loading is in the plane of the wall and prevent separation of the wall and flooring system when the forces occur out of plane.

Topic: Performance and Anchorage
Reference: IRC R613

Category: Wall Construction
Subject: Exterior Windows and Glass Doors

Code Text: *Exterior windows and doors shall be designed to resist the design wind loads specified in Table R301.2(2) adjusted for height and exposure per Table R301.2(3). Window and glass door assemblies shall be anchored in accordance with the published manufacturer's recommendations to achieve the design pressure specified. Substitute anchoring systems used for substrates not specified by the fenestration manufacturer shall provide equal or greater anchoring performance as demonstrated by accepted engineering practice.*

Discussion and Commentary: Glass doors and windows are components of the exterior wall. As such, they are regulated through the specification of performance criteria for such components, as well as their supporting elements, in order to protect against glass breakage due to high wind pressure.

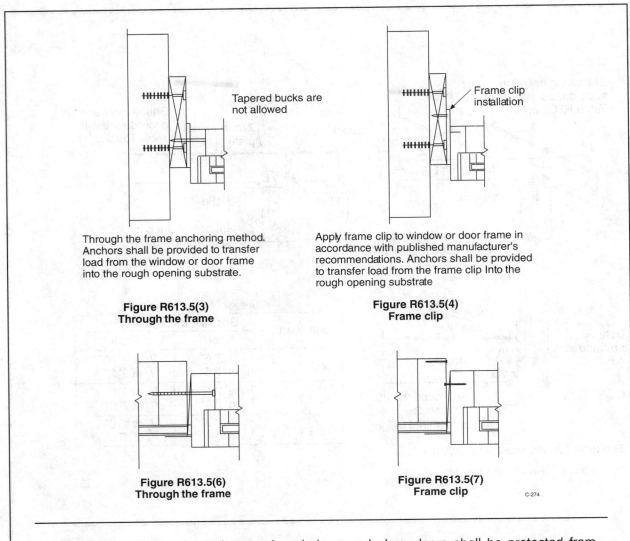

Tapered bucks are not allowed

Frame clip installation

Through the frame anchoring method. Anchors shall be provided to transfer load from the window or door frame into the rough opening substrate.

Figure R613.5(3)
Through the frame

Apply frame clip to window or door frame in accordance with published manufacturer's recommendations. Anchors shall be provided to transfer load from the frame clip Into the rough opening substrate

Figure R613.5(4)
Frame clip

Figure R613.5(6)
Through the frame

Figure R613.5(7)
Frame clip

C-274

In hurricane-prone regions, the exterior windows and glass doors shall be protected from windborne debris in accordance with Section R301.2.1.2. Impact-resistant glazing, precut wood structural panels or design as a partially enclosed building are options.

Code Text: *Maximum spacing of supports and the size and spacing of fasteners used to attach gypsum board shall comply with Table R702.3.5. Gypsum board shall be applied at right angles or parallel to framing members. All edges and ends of gypsum board shall occur on the framing members, except those edges and ends that are perpendicular to the framing members. Interior gypsum board shall not be installed where it is directly exposed to the weather or to water.*

Discussion and Commentary: Table R702.3.5 is a comprehensive table identifying the minimum thickness and fastening requirements for gypsum wallboard. The wallboard thickness ($3/8$-inch, $1/2$-inch, or $5/8$-inch, location of the wallboard (wall or ceiling), orientation of the wallboard to the framing members (parallel or perpendicular), spacing of framing members (16 inches or 24 inches on center), type of fasteners (nails or screws) and use of adhesive (with or without) are all set forth in the table.

TABLE R702.3.5
MINIMUM THICKNESS AND APPLICATION OF GYPSUM BOARD

THICKNESS OF GYPSUM BOARD (inches)	APPLICATION	ORIENTATION OF GYPSUM BOARD TO FRAMING	MAXIMUM SPACING OF FRAMING MEMBERS (inches o.c.)	MAXIMUM SPACING OF FASTENERS (inches) Nails[a]	MAXIMUM SPACING OF FASTENERS (inches) Screws[b]	SIZE OF NAILS FOR APPLICATION TO WOOD FRAMING[c]
Application without adhesive						
$3/8$	Ceiling[d]	Perpendicular	16	7	12	13 gage, $1^1/4''$ long, $19/64''$ head; 0.098 diameter, $1^1/4''$ long, annular-ringed; or 4d cooler nail, 0.080'' diameter, $1^3/8''$ long, $7/32''$ head.
	Wall	Either direction	16	8	16	
$1/2$	Ceiling	Either direction	16	7	12	13 gage, $1^3/8''$ long, $19/64''$ head; 0.098 diameter, $1^1/4''$ long, annular-ringed; 5d cooler nail, 0.086 diameter, $1^5/8''$ long, $15/64''$ head; or gypsum board nail, 0.086 diameter, $1^5/8''$ long, $9/32''$ head.
	Ceiling[d]	Perpendicular	24	7	12	
	Wall	Either direction	24	8	12	
	Wall	Either direction	16	8	16	
$5/8$	Ceiling	Either direction	16	7	12	13 gage, $1^5/8''$ long, $19/64''$ head; 0.098 diameter, $1^3/8''$ long, annular-ringed; 6d cooler nail, 0.092 diameter, $1^7/8''$ long, $1/4''$ head; or gypsum board nail, 0.0915 diameter, $1^7/8''$ long, $19/64''$ head.
	Ceiling[e]	Perpendicular	24	7	12	
	Wall	Either direction	24	8	12	
	Wall	Either direction	16	8	16	
Application with adhesive						
$3/8$	Ceiling[d]	Perpendicular	16	16	16	Same as above for $3/8''$ gypsum board
	Wall	Either direction	16	16	24	
$1/2$ or $5/8$	Ceiling	Either direction	16	16	16	Same as above for $1/2''$ and $5/8''$ gypsum board, respectively
	Ceiling[d]	Perpendicular	24	12	16	
	Wall	Either direction	24	16	24	
two $3/8$ layers	Ceiling	Perpendicular	16	16	16	Base ply nailed as above for $1/2''$ gypsum board; face ply installed with adhesive
	Wall	Either direction	24	24	24	

For SI: 1 inch = 25.4 mm.

a. For application without adhesive, a pair of nails spaced not less than 2 inches apart or more than $2^1/2$ inches apart may be used with the pair of nails spaced 12 inches on center.

b. Screws shall be Type S or W per ASTM C 1002 and shall be sufficiently long to penetrate wood framing not less than $5/8$ inch and metal framing not less than $3/8$ inch.

c. Where metal framing is used with a clinching design to receive nails by two edges of metal, the nails shall be not less than $5/8$ inch longer than the gypsum board thickness and shall have ringed shanks. Where the metal framing has a nailing groove formed to receive the nails, the nails shall have barbed shanks or be 5d, $13^1/2$ gage, $1^5/8$ inches long, $15/64$-inch head for $1/2$-inch gypsum board; and 6d, 13 gage, $1^7/8$ inches long, $15/64$-inch head for $5/8$-inch gypsum board.

d. Three-eighths-inch-thick single-ply gypsum board shall not be used on a ceiling where a water-based textured finish is to be applied, or where it will be required to support insulation above a ceiling. On ceiling applications to receive a water-based texture material, either hand or spray applied, the gypsum board shall be applied perpendicular to framing. When applying a water-based texture material, the minimum gypsum board thickness shall be increased from $3/8$ inch to $1/2$ inch for 16-inch on center framing, and from $1/2$ inch to $5/8$ inch for 24-inch on center framing or $1/2$-inch sag-resistant gypsum ceiling board shall be used.

e. Type X gypsum board for garage ceilings beneath habitable rooms shall be installed perpendicular to the ceiling framing and shall be fastened at maximum 6 inches o.c. by minimum $1^7/8$ inches 6d coated nails or equivalent drywall screws.

Where screws are used for attaching gypsum board to wood framing, they shall be either Type W or Type S and must penetrate the wood at least $5/8$ inch. Type S screws are to be used where attachment is made to light-gage steel framing, with a minimum penetration of $3/8$ inch.

Topic: Weather-resistant Sheathing Paper
Reference: IRC R703.2

Category: Wall Covering
Subject: Exterior Covering

Code Text: *Asphalt-saturated felt free from holes and breaks, weighing not less than 14 pounds per 100 square feet (0.683 kg/m²) and complying with ASTM D 226 or other approved weather-resistant material shall be applied over studs or sheathing of all exterior walls as required by Table R703.4. Such felt or material shall be applied horizontally, with the upper layer lapped over the lower layer not less than 2 inches (51 mm). Where joints occur, felt shall be lapped not less than 6 inches (152 mm).*

Discussion and Commentary: Structural members are adversely affected by moisture, particularly under cyclic conditions of wetting and drying. Due to the potential for water accumulation within the wall cavity, a weather-resistant membrane is mandated behind any exterior siding or veneer.

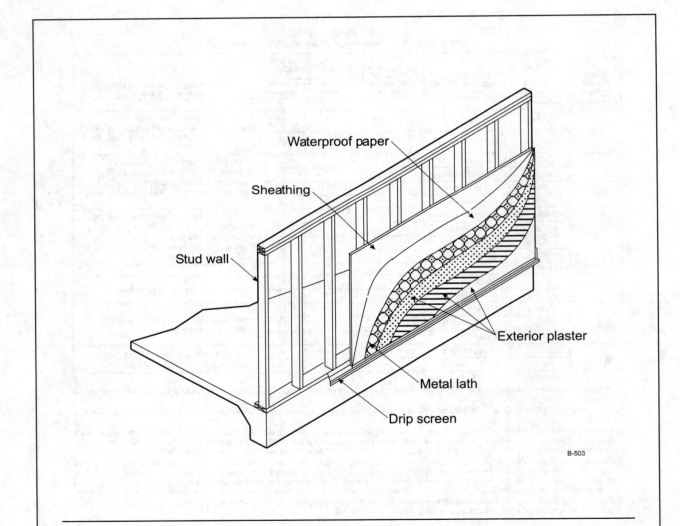

B-503

Under certain conditions, the value of asphalt felt or a similar weather-resistant barrier is questionable or its use unnecessary. It is permissible to eliminate the membrane in detached accessory buildings or where the siding, finish materials or lath provides the needed protection.

2003 IRC Study Companion

Topic: Masonry Veneer Anchorage
Reference: IRC R703.7.4

Category: Wall Covering
Subject: Exterior Covering

Code Text: *Masonry veneer shall be anchored to the supporting wall with corrosion-resistant metal ties. Where veneer is anchored to wood backings through the use of corrugated sheet metal ties, the distance separating the veneer from the sheathing material shall be a maximum of 1 inch (25.4 mm). Each tie shall be spaced not more than 24 inches (610 mm) on center horizontally and vertically and shall support not more than 2.67 square feet (0.248 m²) of wall area. See exception for closer spacing in high-wind or high-seismic areas.*

Discussion and Commentary: Two types of ties are approved for the attachment of masonry veneer to wood construction: corrugated sheet metal ties and metal strand wire ties. The clearance between the veneer and the wood backing is limited and varies based upon which type of tie is used.

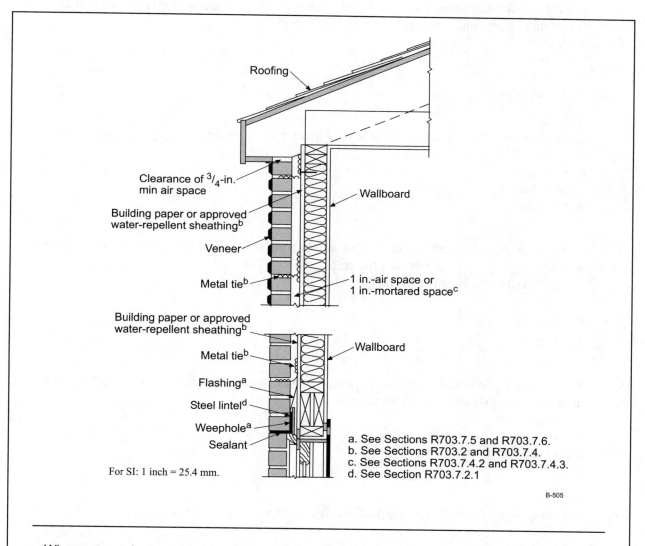

Roofing

Clearance of ³/₄-in. min air space

Building paper or approved water-repellent sheathing[b]

Veneer

Metal tie[b]

Wallboard

1 in.-air space or 1 in.-mortared space[c]

Building paper or approved water-repellent sheathing[b]

Metal tie[b]

Flashing[a]

Steel lintel[d]

Weephole[a]

Sealant

Wallboard

For SI: 1 inch = 25.4 mm.

a. See Sections R703.7.5 and R703.7.6.
b. See Sections R703.2 and R703.7.4.
c. See Sections R703.7.4.2 and R703.7.4.3.
d. See Section R703.7.2.1

B-505

Where veneer is attached to a wall as facing material, it is intended to provide ornamentation, protection, insulation or a combination of these features. It does not provide any structural strength to the wall, nor is it intended to carry any load other than its own weight.

QUIZ

Study Session 7

1. What is the maximum center-to-center stud spacing for a 2-inch-by-6-inch, 8-foot-high wood stud bearing wall supporting two floors and a roof?
    a. 8 inches
    b. 12 inches
    c. 16 inches Table R602.3.5
    d. 24 inches

2. An exterior bearing wall consists of 2-inch-by-4-inch wood studs at 12 inches on center. If located in Seismic Design Category C and exposed to a wind speed of 90 mph, what is the maximum stud height when the wall supports a roof only?
    a. 10 feet
    b. 12 feet Table 602.3.1
    c. 14 feet
    d. 16 feet

3. The minimum offset for end joints in a double top plate shall be _____.
    a. 24 inches R602.3.2
    b. 36 inches
    c. 48 inches
    d. 60 inches

4. A single top plate is permitted in a wood stud bearing wall where the rafters or joists are located within _____of the center of the studs.
    a. 0 inches; no tolerance is permitted
    b. 1 inch
    c. $1^1/_2$ inch
    d. 5 inches

5. Where nailed to the studs, what is the maximum stud spacing permitted for $^3/_8$-inch wood structural panel wall sheathing with a span rating of $^{12}/_0$?
    a. 12 inches
    b. 16 inches
    c. 20 inches
    d. 24 inches

6. In a nonbearing partition, a wood stud may be notched a maximum of _____.
    a. $^5/_8$ inch
    b. $1^3/_8$ inch
    c. 40 percent of the stud depth
    d. 60 percent of the stud depth

2003 IRC Study Companion

7. A bored hole shall not be located with _____ of the edge of a wood stud.
   a. $^3/_8$ inch
   b. $^1/_2$ inch
   c. $^5/_8$ inch
   d. 1 inch

8. In a nonbearing interior partition, what is the maximum diameter permitted for a bored hole in a wood stud?
   a. $1^1/_2$ inches
   b. 25 percent of the stud depth
   c. 40 percent of the stud depth
   d. 60 percent of the stud depth

9. A 15-inch-deep box header in an exterior wall is constructed with wood structural panels on both sides. What is the maximum allowable header span for a condition where the header supports a clear-span roof truss with a span of 26 feet?
   a. 4 feet
   b. 5 feet
   c. 7 feet
   d. 8 feet

10. Which of the following materials is not specifically identified by the IRC as a fireblocking material?
    a. $^1/_4$-inch cement-based millboard
    b. $^1/_2$-inch gypsum board
    c. $^{15}/_{32}$-inch wood structural panel 602.8.1
    d. $^3/_4$-inch particleboard

11. Where unfaced fiberglass batt insulation is used as a fireblocking material in the wall cavity of a wood stud wall system, the insulation shall be installed with a minimum vertical height of _____.
    a. 16 inches 602.8.1.1
    b. 3 feet
    c. 4 feet
    d. the entire stud space

12. A foundation cripple wall shall be considered an additional story for stud sizing requirements where the wall height exceeds _____.
    a. 14 inches
    b. 30 inches
    c. 48 inches
    d. 72 inches

13. For one-story buildings located in Seismic Design Category C with a wind speed of 90 mph, particleboard wall sheathing panels used as wall bracing shall be provided for a minimum of _____ of the braced wall line.
    a. 16 percent
    b. 25 percent
    c. 30 percent
    d. 45 percent

14. Braced wall panels shall begin a maximum of _____ from the end of a braced wall line.
    a. 2 feet, 8 inches
    b. 4 feet
    c. 8 feet
    d. 12 feet, 6 inches

15. In steel-framed wall construction, rafters may be offset a maximum of _____ from the centerline of the load-bearing steel studs.
    a. 0 inches; no tolerance is permitted
    b. $^3/_4$ inch
    c. 1 inch
    d. 5 inches

16. Unless appropriately patched, holes in load-bearing steel wall framing members shall be located a minimum of _____ from the end of the member.
    a. $1^1/_2$ inches
    b. 3 inches
    c. 4 inches
    d. 10 inches

17. In unreinforced masonry construction, what is the maximum vertical spacing of longitudinal reinforcement where units are laid in stack bond?
    a. 12 inches
    b. 16 inches
    c. 20 inches
    d. 24 inches

18. Where ledgers are used at the connection between a masonry wall and a wood floor system having floor joists spanning 16 feet, $^1/_2$-inch ledger bolts shall be located at a maximum of _____ on center, provided the building is in Seismic Design Category A, B or C and the wind loads are less than 30 psf.
    a. 1 foot, 0 inches
    b. 1 foot, 3 inches
    c. 1 foot, 9 inches
    d. 2 feet, 0 inches

19. When $1/2$-inch gypsum board is used as an interior wall covering and installed perpendicular to framing members at 16 inches on center, the maximum spacing of nails is _____ on center where adhesive is used.
   a. 7 inches
   b. 8 inches
   c. 12 inches
   d. 16 inches

20. Screws for attaching gypsum board to light-gage steel framing shall penetrate the steel a minimum of _____.
   a. $1/4$ inch
   b. $3/8$ inch
   c. $1/2$ inch
   d. $5/8$ inch

21. One-half-inch thick water-resistant gypsum backing board is permitted to be used on a ceiling where the framing members are spaced a maximum of _____ on center.
   a. 12 inches
   b. 16 inches
   c. 20 inches
   d. 24 inches

22. Where $3/8$-inch particleboard is used as an exterior wall covering, what type of fasteners are required if the particleboard is attached directly to the studs?
   a. 0.120 nail, 2 inches long
   b. 6d box nail
   c. 8d box nail
   d. direct attachment to the studs is prohibited

23. In Seismic Design Category D_1, metal ties for anchoring masonry veneer to a supporting wall shall support a maximum of _____ of wall area.
   a. 2 square feet
   b. $2^2/3$ square feet
   c. $3^1/2$ square feet
   d. $4^1/2$ square feet

24. An Exterior Insulation Finish System (EIFS) shall terminate a minimum of _____ above the finished ground level.
   a. 1 inch
   b. 2 inches
   c. 6 inches
   d. 8 inches

25. What is the minimum required size of a steel angle spanning 8 feet used as a lintel supporting one story of masonry veneer above?
   a. 3 x 3 x $^1/_4$
   b. 4 x 3 x $^1/_4$
   c. 5 x $3^1/_2$ x $^5/_{16}$
   d. 6 x $3^1/_2$ x $^5/_{16}$

26. Utility grade studs, where used in loadbearing walls supporting only a roof and a ceiling, shall have a maximum height of _____ feet.
   a. 8
   b. 10
   c. 12
   d. 14

27. Double top plates shall be face-nailed together in the required lapped area with _____ fasteners.
   a. 4 - 16d
   b. 6 - 10d
   c. 6 - 16d
   d. 8 - 16d

28. $^{15}/_{32}$-inch wood structural panels attached to wall framing with 15 ga. staples shall be fastened at a maximum of _____ inches at the panel edges and _____ inches at the intermediate supports.
   a. 3, 6
   b. 4, 8
   c. 5, 10
   d. 6, 12

29. Where the top plate of an interior load-bearing wall is notched by more than _____ of its width to accommodate piping, a complying metal tie shall be installed.
   a. 25 percent
   b. 40 percent
   c. 50 percent
   d. 60 percent

30. A weep screed installed on exterior stud walls in an exterior plaster application shall be placed a minimum of _____ inch(es) above paved areas.
   a. 1
   b. 2
   c. 4
   d. 6

INTERNATIONAL RESIDENTIAL CODE
Study Session 8
Chapters 8 and 9—Roof/Ceiling Construction and Roof Assemblies

OBJECTIVE: To gain an understanding of the requirements for ceiling construction, roof construction and roof coverings, including ceiling joist and rafter sizing; cutting and notching of framing members; framing at openings; purlins; steel roof framing; roof ventilation and attic access; and roof covering materials such as asphalt shingles, concrete or clay tiles, wood shingles and wood shakes.

REFERENCE: Chapters 8 and 9, 2003 *International Residential Code*

KEY POINTS:
- How must roof drainage water be discharged?
- How must wood framing members be identified? What special concerns must be addressed when using fire-retardant-treated wood?
- What are the acceptable framing methods at the roof ridge? At valleys and hips? Under what condition are ridges, valleys and hips required to be designed as beams?
- How are rafters to be tied together? At what maximum intervals must rafter ties be located?
- How is the maximum allowable ceiling joist span determined? Roof rafter span? How does the location of rafter ties affect the maximum allowable rafter span?
- What methods are prescribed for framing around ceiling and roof openings? When is lateral support mandated?
- What is the purpose of a purlin system? How are purlins sized? What is the maximum spacing of purlin braces? The maximum unbraced length?
- How is the maximum allowable span for roof sheathing determined? How must roof sheathing be installed?
- How are trusses to be connected to the wall plates to address uplift pressures?
- What are the general limitations for the prescriptive use of steel roof framing systems? What are the general requirements?
- When is roof ventilation required? What is minimum required ventilating area? How can this area be reduced?
- When is attic access required? What is the minimum size needed for an attic access opening? How much headroom clearance must be provided over the access opening?
- At what distance must combustible insulation be separated from recessed lighting fixtures, fan motors, and other heat-producing devices?
- What is Class A, Class B or Class C roofing? When is such roofing required? What specific materials are considered Class A roof coverings?
- Where is flashing required? How is roof drainage to be addressed?
- How must roofing materials be identified?
- What are the installation requirements for asphalt shingles? Clay and concrete tile? Wood shingles? Wood shakes?
- What other types of roof coverings are regulated by the code?
- What level of work is considered reroofing? What is a roof repair? Roof replacement? When must the existing roof coverings be removed prior to any reroofing work?

Topic: Framing Details
Reference: IRC R802.3

Category: Roof-Ceiling Construction
Subject: Wood Roof Framing

Code Text: *Rafters shall be framed to ridge board or to each other with a gusset plate as a tie. Ridge board shall be at least 1-inch (25.4 mm) nominal thickness and not less in depth than the cut end of the rafter. At all valleys and hips there shall be a valley or hip rafter not less than 2-inch (51 mm) nominal thickness and not less in depth than the cut end of the rafter.*

Discussion and Commentary: For the roof loads to be carried to the exterior walls, a continuous tie must be created at the roof ridge. This is typically accomplished by installing the rafters in direct opposition to each other and nailing them directly to a ridge board having a depth at least that of the rafter ends. Another option is the use of a gusset plate or similar element that ties the roof system together at the ridge line. Continuity from rafter to rafter is the key requirement.

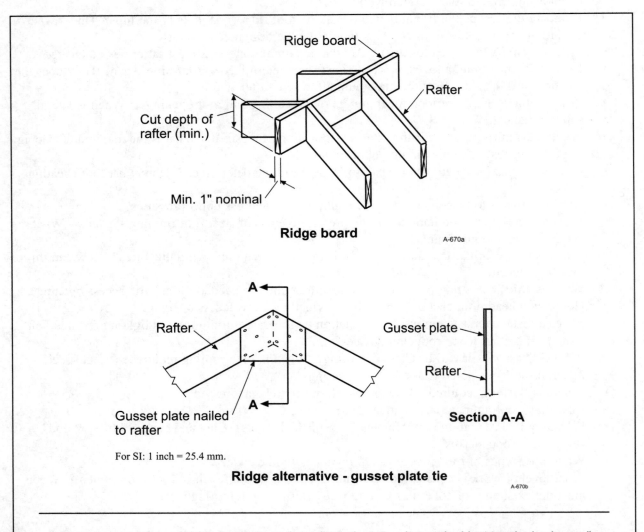

For SI: 1 inch = 25.4 mm.

Ridge alternative - gusset plate tie

A-670b

Where the roof has a relatively shallow slope (less than 3 units vertical in 12 units horizontal), the ridge must be designed as a beam in order to carry the necessary tributary loading. Positive bearing for the rafters must be provided unless an approved alternative method is used.

Topic: Ceiling Joist and Rafter Connections
Reference: IRC R802.3.1

Category: Roof-Ceiling Construction
Subject: Wood Roof Framing

Code Text: *Ceiling joists and rafters shall be nailed to each other in accordance with Tables R602.3(1) and R802.5.1(9), and the assembly shall be nailed to the top wall plate in accordance with Table R602.3(1). Ceiling joists shall be continuous or securely joined where they meet over interior partitions and nailed to adjacent rafters to provide a continuous tie across the building when such joists are parallel to the rafters.*

Discussion and Commentary: Where ceiling joists are located parallel to roof rafters, they must be tied together. Adequate attachment is mandated to transfer the thrust from the rafters to the joists. Where the ceiling joists do not run parallel to the rafters, it is necessary to use other means, such as rafter ties spaced a maximum of 4 feet on center and located near the top plate.

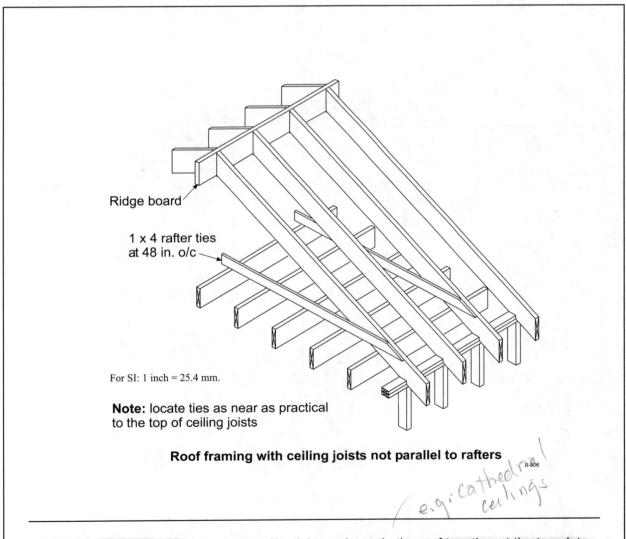

Ridge board

1 x 4 rafter ties
at 48 in. o/c

For SI: 1 inch = 25.4 mm.

Note: locate ties as near as practical
to the top of ceiling joists

Roof framing with ceiling joists not parallel to rafters

e.g. cathedral ceilings

B-506

There are many locations where no ceiling joists exist to tie the roof together at the top plate line. Under such conditions, the ridge must be designed to act as a girder and support the necessary tributary roof load. Both the girder and its supporting elements must be designed.

Topic: Allowable Ceiling Joist Spans

Reference: IRC R802.4

Category: Roof-Ceiling Construction

Subject: Wood Roof Framing

Code Text: *Spans for ceiling joists shall be in accordance with Tables R802.4(1) and R802.4(2).*

Discussion and Commentary: Two tables are provided for the more common ceiling joist conditions. The maximum allowable spans are provided for attics with no storage (10 psf live load) and limited storage (20 psf). Table R301.4 indicates that the 10 psf criterion is to be used only where the roof slopes at 3 in 12 or less. It is assumed that higher roof slopes create conditions that allow for some degree of storage, thus mandating a higher design load. Similar to the listings for floor joists and rafter spans, four common lumber species are listed, along with a variety of lumber grades. Four center-to-center spacing conditions are included, as are four sizes of dimension lumber. The combination of these criteria provides all the information necessary to determine the maximum allowable ceiling joist spans.

TABLE R802.4(1)—continued
CEILING JOIST SPANS FOR COMMON LUMBER SPECIES
(Uninhabitable attics without storage, live load = 20 psf, L/Δ = 240)

CEILING JOIST SPACING (inches)	SPECIES AND GRADE		DEAD LOAD = 5 psf			
			2x4	2x6	2x8	2x10
			Maximum ceiling joist spans			
			(feet - inches)	(feet - inches)	(feet - inches)	(feet - inches)
24	Douglas fir-larch	SS	10-5	16-4	21-7	Note a
	Douglas fir-larch	#1	10-0	15-9	20-1	24-6
	Douglas fir-larch	#2	9-10	14-10	18-9	22-11
	Douglas fir-larch	#3	7-8	11-2	14-2	17-4
	Hem-fir	SS	9-10	15-6	20-5	Note a
	Hem-fir	#1	9-8	15-2	19-7	23-11
	Hem-fir	#2	9-2	14-5	18-6	22-7
	Hem-fir	#3	7-8	11-2	14-2	17-4
	Southern pine	SS	10-3	16-1	21-2	Note a
	Southern pine	#1	10-0	15-9	20-10	Note a
	Southern pine	#2	9-10	15-6	20-1	23-11
	Southern pine	#3	8-2	12-0	15-4	18-1
	Spruce-pine-fir	SS	9-8	15-2	19-11	25-5
	Spruce-pine-fir	#1	9-5	14-9	18-9	22-11
	Spruce-pine-fir	#2	9-5	14-9	18-9	22-11
	Spruce-pine-fir	#3	7-8	11-2	14-2	17-4

Check sources for availability of lumber in lengths greater than 20 feet.

For SI: 1 inch = 25.4 mm, 1 foot = 304.8 mm, 1 pound per square foot = 0.0479 kN/m^2.

a. Span exceeds 26 feet in length

In a limited number of applications, the span tables may not address the loading conditions under consideration. Therefore, it may be necessary to refer to another source to determine the maximum allowable joist spans. AF&PA provides span tables for such situations.

Topic: Allowable Rafter Spans
Reference: IRC R802.5

Category: Roof-Ceiling Construction
Subject: Wood Roof Framing

Code Text: *Spans for rafters shall be in accordance with Tables R802.5.1(1) through R802.5.1(8). The span of each rafter shall be measured along the horizontal projection of the rafter.*

Discussion and Commentary: Eight span tables are available in the code to determine the maximum allowable span of rafters in light-frame wood construction. The tables are specific for snow loads of 30 psf, 50 psf and 70 psf, as well as a live load of 20 psf. Each of these loading conditions includes two different deflection criteria. Where a ceiling is attached to the rafters, such as a cathedral ceiling, the deflection is limited to L/240. Where there is no ceiling material supported by the rafters, a maximum deflection of L/180 is permitted. The tables' format is similar to that for floor joists and ceiling joists.

TABLE R802.5.1(1)—continued
RAFTER SPANS FOR COMMON LUMBER SPECIES
(Roof live load=20 psf, ceiling not attached to rafters, L/Δ=180)

RAFTER SPACING (Inches)	SPECIES AND GRADE		DEAD LOAD = 10 psf					DEAD LOAD = 20 psf				
			2x4	2x6	2x8	2x10	2x12	2x4	2x6	2x8	2x10	2x12
			Maximum rafter spans[a]									
			(feet - inches)	(feet - inches)	(feet - inches)	(feet - inches)	(feet - inches)	(feet - inches)	(feet - inches)	(feet - inches)	(feet - inches)	(feet - inches)
24	Douglas fir-larch	SS	9-1	14-4	18-10	23-4	Note b	8-11	13-1	16-7	20-3	23-5
	Douglas fir-larch	#1	8-7	12-6	15-10	19-5	22-6	7-5	10-10	13-9	16-9	19-6
	Douglas fir-larch	#2	8-0	11-9	14-10	18-2	21-0	6-11	10-2	12-10	15-8	18-3
	Douglas fir-larch	#3	6-1	8-10	11-3	13-8	15-11	5-3	7-8	9-9	11-10	13-9
	Hem-fir	SS	8-7	13-6	17-10	22-9	Note b	8-7	12-10	16-3	19-10	23-0
	Hem-fir	#1	8-4	12-3	15-6	18-11	21-11	7-3	10-7	13-5	16-4	19-0
	Hem-fir	#2	7-11	11-7	14-8	17-10	20-9	6-10	10-0	12-8	15-6	17-11
	Hem-fir	#3	6-1	8-10	11-3	13-8	15-11	5-3	7-8	9-9	11-10	13-9
	Southern pine	SS	8-11	14-1	18-6	23-8	Note b	8-11	14-1	18-6	22-11	Note b
	Southern pine	#1	8-9	13-9	17-9	21-1	25-2	8-3	12-3	15-4	18-3	21-9
	Southern pine	#2	8-7	12-3	15-10	18-11	22-2	7-5	10-8	13-9	16-5	19-3
	Southern pine	#3	6-5	9-6	12-1	14-4	17-1	5-7	8-3	10-6	12-5	14-9
	Spruce-pine-fir	SS	8-5	13-3	17-5	21-8	25-2	8-4	12-2	15-4	18-9	21-9
	Spruce-pine-fir	#1	8-0	11-9	14-10	18-2	21-0	6-11	10-2	12-10	15-8	18-3
	Spruce-pine-fir	#2	8-0	11-9	14-10	18-2	21-0	6-11	10-2	12-10	15-8	18-3
	Spruce-pine-fir	#3	6-1	8-10	11-3	13-8	15-11	5-3	7-8	9-9	11-10	13-9

Check sources for availability of lumber in lengths greater than 20 feet.

For SI: 1 inch = 25.4 mm, 1 foot = 304.8 mm, 1 pound per square foot = 0.0479 kN/m².

a. The tabulated rafter spans assume that ceiling joists are located at the bottom of the attic space or that some other method of resisting the outward push of the rafters on the bearing walls, such as rafter ties, is provided at that location. When ceiling joists or rafter ties are located higher in the attic space, the rafter spans shall be multiplied by the factors given below:

H_C/H_R	Rafter Span Adjustment Factor
2/3 or greater	0.50
1/2	0.58
1/3	0.67
1/4	0.76
1/5	0.83
1/6	0.90
1/7.5 and less	1.00

where: H_C = Height of ceiling joists or rafter ties measured vertically above the top of the rafter support walls.

H_R = Height of roof ridge measured vertically above the top of the rafter support walls.

b. Span exceeds 26 feet in length.

It is important that the horizontal thrust created by roof loads be resisted by the connection of the rafters to ceiling joists or rafter ties. This connection must occur very near the top plate line. Where the horizontal ties are higher in the space, a reduction in the rafter span is required.

Topic: Purlins
Reference: IRC R802.5.1

Category: Roof-Ceiling Construction
Subject: Wood Roof Framing

Code Text: *Purlins are permitted to be installed to reduce the span of rafters as shown in Figure R802.5.1. Purlins shall be sized no less than the required size of the rafters that they support. Purlins shall be continuous and shall be supported by 2-inch by 4-inch (51 mm by 102 mm) braces installed to bearing walls at a slope not less than 45 degrees from the horizontal. The braces shall be spaced not more than 4 feet (1219 mm) on center and the unbraced length of braces shall not exceed 8 feet (2438 mm).*

Discussion and Commentary: Purlins provide an effective method for reducing rafter size by reducing the rafter span. The span is measured horizontally from the exterior wall to the purlin and from the purlin to the ridge, with the controlling span determined by the greater length.

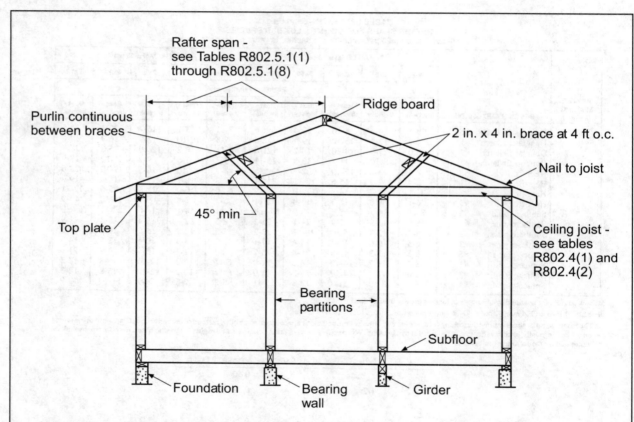

Note: Where ceiling joists run perpendicular to the rafters, rafter ties shall be nailed to the rafter near the plate line and spaced not more than 4 ft on center.

For SI: 1 inch = 25.4 mm, 1 foot = 304.8 mm, 1 degree = 0.01745 rad.

Braced rafter construction

A-669

It is critical that purlin braces be supported by bearing walls. The use of 2-inch by 4-inch braces is based on a maximum span of 4 feet between such purlin supports. Although there is no specific distance limit between bearing points at the bearing wall, the 45-degree criteria must be met.

Topic: Roof Tie-Down
Reference: IRC R802.11.1

Category: Roof-Ceiling Construction
Subject: Wood Roof Framing

Code Text: *Roof assemblies which are subject to wind uplift pressures of 20 pounds per square foot (0.958 kN/m²) or greater shall have roof rafters or trusses attached to their supporting wall assemblies by connections capable of providing the resistance required in Table R802.11. Wind uplift pressures shall be determined using an effective wind area of 100 square feet (9.3 m²) and Zone 1 in Table R301.2(2), as adjusted for height and exposure per Table R301.2(3).*

Discussion and Commentary: Because roof uplift caused by high winds can be a significant factor in roof-system damage, the code requires roof-to-wall connections capable of resisting this force. In addition, it is necessary to maintain the integrity of the continuous load path by providing a complying means to transmit the uplift forces from the rafter or truss ties to the foundation.

TABLE R802.11
REQUIRED STRENGTH OF TRUSS OR RAFTER CONNECTIONS TO RESIST WIND UPLIFT FORCES[a,b,c,e,f]
(Pounds per connection)

BASIC WIND SPEED (3-second gust)	ROOF SPAN (feet)							OVERHANGS[d] (pounds/feet)
	12	20	24	28	32	36	40	
85	-72	-120	-145	-169	-193	-217	-241	-38.55
90	-91	-151	-181	-212	-242	-272	-302	-43.22
100	-131	-218	-262	-305	-349	-393	-436	-53.36
110	-175	-292	-351	-409	-467	-526	-584	-64.56

For SI: 1 inch = 25.4 mm, 1 foot = 305 mm, 1 mph = 1.61 km/hr, 1 pound/foot = 14.5939 N/m, 1 pound = 0.454 kg.

a. The uplift connection requirements are based on a 30 foot mean roof height located in Exposure B. For Exposures C and D and for other mean roof heights, multiply the above loads by the Adjustment Coefficients in Table R-301.2(3).

b. The uplift connection requirements are based on the framing being spaced 24 inches on center. Multiply by 0.67 for framing spaced 16 inches on center and multiply by 0.5 for framing spaced 12 inches on center.

c. The uplift connection requirements include an allowance for 10 pounds of dead load.

d. The uplift connection requirements do not account for the effects of overhangs. The magnitude of the above loads shall be increased by adding the overhang loads found in the table. The overhang loads are also based on framing spaced 24 inches on center. The overhang loads given shall be multiplied by the overhang projection and added to the roof uplift value in the table.

e. The uplift connection requirements are based upon wind loading on end zones as defined in Section M1609.6 of the *International Building Code*. Connection loads for connections located a distance of 20% of the least horizontal dimension of the building from the corner of the building are permitted to be reduced by multiplying the table connection value by 0.7 and multiplying the overhang load by 0.8.

f. For wall-to-wall and wall-to-foundation connections, the capacity of the uplift connector is permitted to be reduced by 100 pounds for each full wall above. (For example, if a 600-pound rated connector is used on the roof framing, a 500-pound rated connector is permitted at the next floor level down.)

In utilizing Table R301.2(2), the roof slope must be known. The determination of the component pressure is only necessary for the negative values shown in the table, as these values represent uplift pressures.

Topic: Structural Framing
Reference: IRC R804.2, R804.3

Category: Roof-Ceiling Construction
Subject: Steel Roof Framing

Code Text: *Load-bearing steel roof framing members shall comply with Figure R804.2(1) and the dimensional and minimum thickness requirements specified in Tables R804.2(1) and R804.2(2). The clear span of cold-formed steel ceiling joists shall not exceed the limits set forth in Table R804.3.1(1) or R804.3.1(2). The horizontal projection of the rafter span, as shown in Figure R804.3, shall not exceed the limits set forth in Table R804.3.3(1).*

Discussion and Commentary: The conventional use of steel roof framing members is regulated in much the same manner as steel floor and wall framing systems. It is important that roof rafters and ceiling joists bear directly above load-bearing studs, with a maximum tolerance of $^3/_4$ inch between the centerline of the stud and the roof joist or rafter.

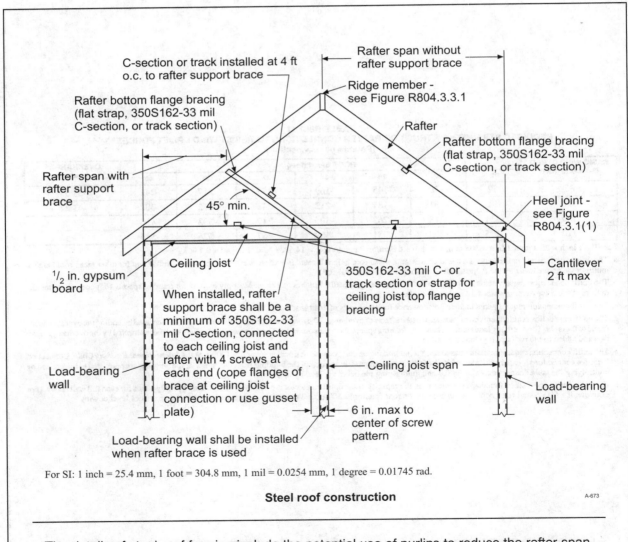

For SI: 1 inch = 25.4 mm, 1 foot = 304.8 mm, 1 mil = 0.0254 mm, 1 degree = 0.01745 rad.

Steel roof construction

A-673

The details of steel roof framing include the potential use of purlins to reduce the rafter span. The unsupported length of the purlin brace is limited to 4 feet. Otherwise, the criteria do not differ from those for light-frame wood construction.

Topic: Minimum Ventilation Required
Reference: IRC R806.1, R806.2

Category: Roof-Ceiling Construction
Subject: Roof Ventilation

Code Text: *Enclosed attics and enclosed rafter spaces formed where ceilings are applied directly to the underside of roof rafters shall have cross ventilation for each separate space by ventilating openings protected against the entrance of rain or snow. The total net free ventilating area shall not be less than 1 to 150 of the area of the space ventilated.* See exceptions for decrease in required ventilating area.

Discussion and Commentary: Large amounts of water vapor can migrate into the attic and condense on wood roof components. As the wetting and drying cycle continues, rotting and decay are possible. If the attic or enclosed rafter spaces are properly ventilated, water will not accumulate on the building components.

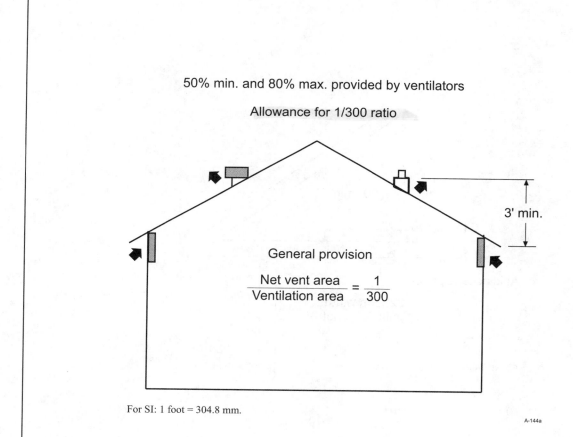

50% min. and 80% max. provided by ventilators

Allowance for 1/300 ratio

3' min.

General provision

$$\frac{\text{Net vent area}}{\text{Ventilation area}} = \frac{1}{300}$$

For SI: 1 foot = 304.8 mm.

A-144a

A vapor barrier restricts the passage of moisture to the attic such that a reduction in the required venting area is possible by installing a barrier on the warm side of the ceiling. Under this method, it is critical that exhaust fans be directed to the outdoors and not terminate in the attic.

Topic: Access Location and Size

Reference: IRC R807

Category: Roof-Ceiling Construction

Subject: Attic Access

Code Text: *In buildings with combustible ceiling or roof construction, an attic access opening shall be provided to attic areas that exceed 30 square feet (2.8 m²) and have a vertical height of 30 inches (762 mm) or greater. The rough-framed opening shall not be less than 22 inches by 30 inches (559 mm by 762 mm) and shall be located in a hallway or other readily accessible location. A 30-inch (762 mm) minimum unobstructed headroom in the attic space shall be provided at some point above the access opening.*

Discussion and Commentary: During the life of a dwelling structure, it is quite probable that the attic will need to be accessed for a variety of reasons. Therefore, an adequately sized access opening must be provided with adequate headroom above the opening.

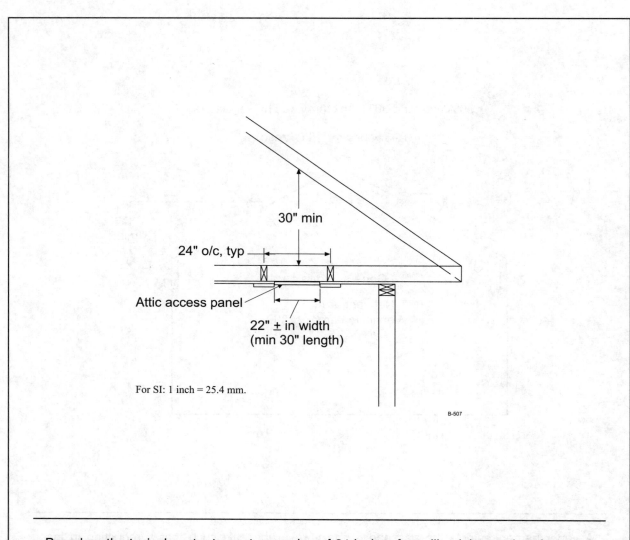

For SI: 1 inch = 25.4 mm.

Based on the typical center-to-center spacing of 24 inches for ceiling joists and roof trusses, the access opening is designed to occur between the members without any special modifications. A lip or frame supporting the access panel may encroach into the required 22-inch width.

Topic: Roof Covering Materials
Reference: IRC R902.1

Category: Roof Assemblies
Subject: Roof Classification

Code Text: *Roofs shall be covered with materials as set forth in Sections R904 and R905. Class A, B or C roofing shall be installed in areas designated by law as requiring their use or when the edge of the roof is less than 3 feet (914 mm) from a property line. Classes A, B and C roofing required to be listed by Section R902.1 shall be tested in accordance with UL 790 or ASTM E 108.*

Discussion and Commentary: The general scope of Chapter 9 includes the use of roof coverings as a weather-resistant material for protection of the building's interior. To a limited degree, the roof covering materials must also be resistant to external fire conditions. Because the primary consideration is that of fire spread from building to building, the requirements are limited to those structures located in very close proximity to adjoining property lines.

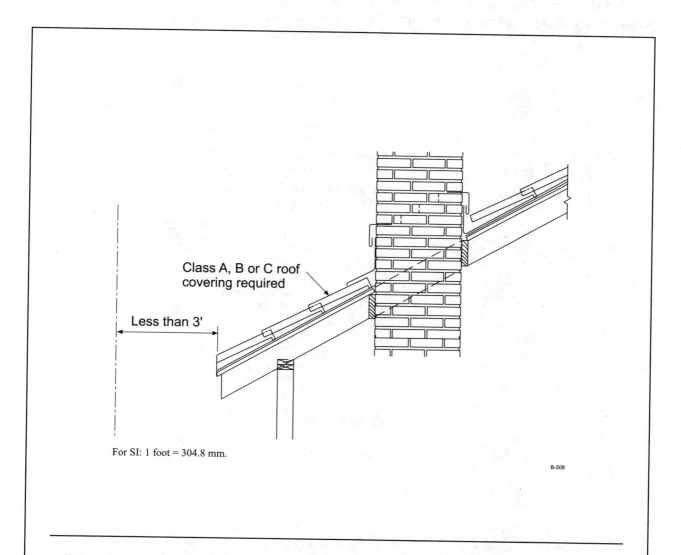

Class A, B or C roof covering required

Less than 3'

For SI: 1 foot = 304.8 mm.

B-508

Although not specifically listed, slate, clay and concrete roof tiles, exposed concrete roof decks, and metal sheets and shingles are equivalent to Class A roofing. Otherwise, a listed Class A, B or C roof covering must be used when the roof is located within 3 feet of the lot line.

Topic: Roof Drainage
Reference: IRC R903.4

Category: Roof Assemblies
Subject: Weather Protection

Code Text: *Unless roofs are sloped to drain over roof edges, roof drains shall be installed at each low point of the roof. Where required for roof drainage, scuppers shall be placed level with the roof surface in a wall or parapet. The scupper shall be located as determined by the roof slope and contributing roof area.*

Discussion and Commentary: Although roofs on most buildings regulated by the IRC are distinctly pitched and simply drain roof water over the edge (often to a gutter and downspout system), occasionally a flat roof system is provided. In such situations, it is necessary to install roof drains as well as overflow drains. The overflow drains shall be installed with the inlet flow line located 2 inches above the low point of the roof. Overflow drains shall not be connected to roof drain lines and must discharge to a location approved by the building official. The *International Plumbing Code®* (IPC®), Section 1105, requires that the inlet area of roof drain strainers to be one and one-half times the area of the conductor or leader and, when they are the flat-surface type, two times the area of the conductor or leader.

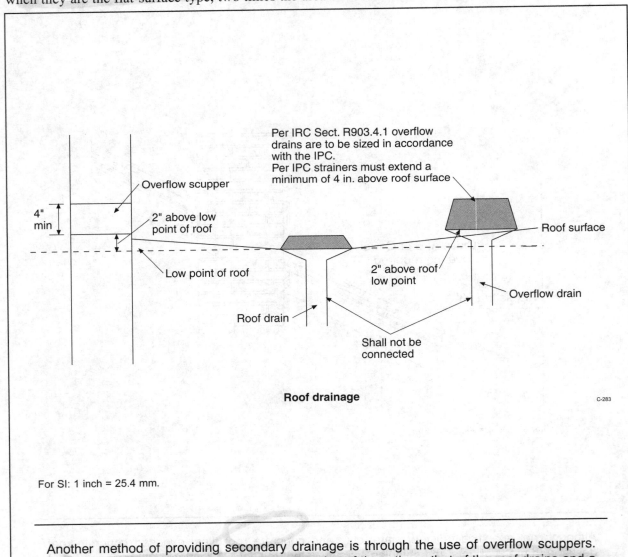

Per IRC Sect. R903.4.1 overflow drains are to be sized in accordance with the IPC.
Per IPC strainers must extend a minimum of 4 in. above roof surface

Overflow scupper

4" min

2" above low point of roof

Low point of roof

Roof drain

Roof surface

2" above roof low point

Overflow drain

Shall not be connected

Roof drainage

C-283

For SI: 1 inch = 25.4 mm.

Another method of providing secondary drainage is through the use of overflow scuppers. The scuppers shall have a minimum opening size of three times that of the roof drains and a minimum opening height of 4 inches.

Topic: Asphalt Shingles
Reference: IRC R905.2

Category: Roof Assemblies
Subject: Roof Covering Requirements

Code Text: *Asphalt shingles shall only be used on roof slopes of two units vertical in 12 units horizontal (2:12) or greater. For roof slopes from . . . 2:12 up to . . . 4:12, double underlayment application is required in accordance with Section R905.2.7. Fasteners for asphalt shingles shall be . . . of a length to penetrate through the roofing materials and a minimum of $^3/_4$ inch (19.1 mm) into the roof sheathing. Where the roof sheathing is less than $^3/_4$ inch (19.1 mm) thick, the fasteners shall penetrate through the sheathing. For normal application, asphalt shingles shall be secured to the roof with not less than four fasteners per strip shingle or two fasteners per individual shingle.*

Discussion and Commentary: A very common roofing material, asphalt shingles are composed of organic or glass felt coated with mineral granules.

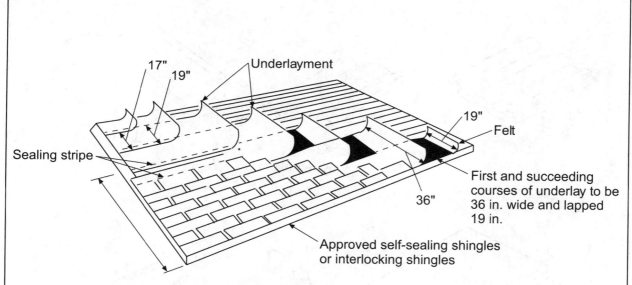

Note: In areas where average daily temperature in January is 25°F or less, special methods, rather than normal underlayment, shall extend up from eaves for enough to overlie a point 24 in. inside the wall line of the building.

For SI: 1 inch = 25.4 mm, °C = [(°F)-32/1.8].

source NRCA

Application of asphalt shingle on slopes between 2:12 and 4:12

A-674a

Where asphalt shingles are installed, they are regulated for sheathing, roof slope, underlayment, fasteners and attachment. As with other types of roof covering materials, they must also be installed in conformance with the manufacturer's installation instructions.

Topic: Clay and Concrete Tile
Reference: IRC R905.3

Category: Roof Assemblies
Subject: Roof Covering Requirements

Code Text: *Clay and concrete tile shall be installed on roof slopes of two and one-half units vertical in 12 units horizontal ($2^1/_2$:12) or greater. For roof slopes from . . . $2^1/_2$:12 to . . . 4:12, double underlayment application is required in accordance with Section R905.3.3. Nails shall be . . . of sufficient length to penetrate the deck a minimum of $^3/_4$ inch (19.1 mm) or through the thickness of the deck, whichever is less.*

Discussion and Commentary: Tile has been used as a roofing material for many centuries. It not only provides for weather protection but is also resistant to exterior fire conditions. Both flat tile and roll tile, the two common configurations, can be interlocked through the use of ribs located along the tile edges.

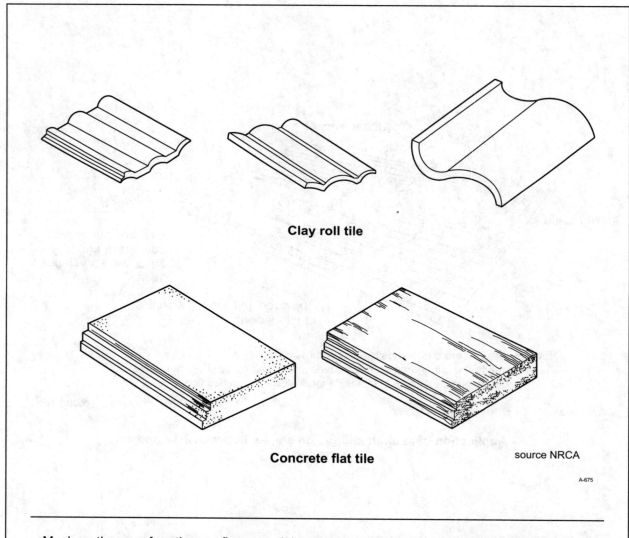

Clay roll tile

Concrete flat tile

source NRCA

A-675

Much as they are for other roofing materials, special underlayment requirements are mandated for tile roofs having low slopes. In addition, the underlayment application is more highly regulated for structures located in areas subject to high winds.

Topic: Wood Shingles
Reference: IRC R905.7

Category: Roof Assemblies
Subject: Roof Covering Requirements

Code Text: *Wood shingles shall be installed on slopes of three units vertical in 12 units horizontal (25-percent slope) or greater. Weather exposure for wood shingles shall not exceed those set in Table R905.7.5. Fasteners . . . shall be corrosion-resistant with a minimum penetration of $^1/_2$ inch (12.7 mm) into the sheathing. Wood shingles shall be attached to the roof with two fasteners per shingle, positioned no more than . . . 1 inch (25.4 mm) above the exposure line.*

Discussion and Commentary: Wood shingles are wood roofing materials that are sawed in a manner to produce a uniform butt thickness. There are three different grades of wood shingles, with each grade available in three different lengths. The application method is specifically described so that following it will eliminate the potential for water intrusion.

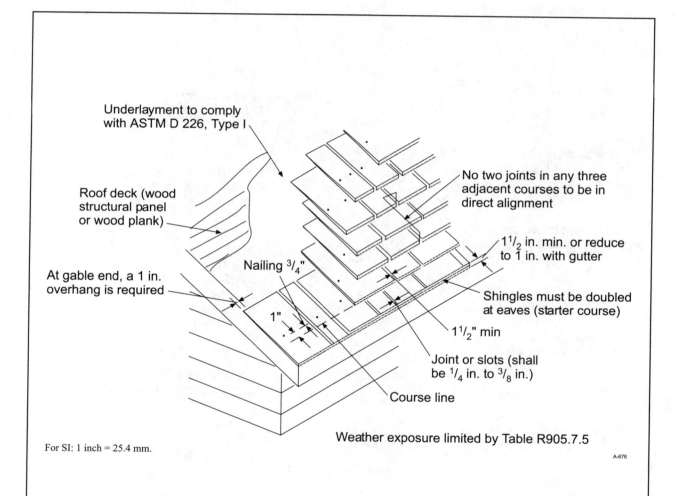

Underlayment to comply with ASTM D 226, Type I

Roof deck (wood structural panel or wood plank)

At gable end, a 1 in. overhang is required

Nailing $^3/_4$"

1"

No two joints in any three adjacent courses to be in direct alignment

$1^1/_2$ in. min. or reduce to 1 in. with gutter

Shingles must be doubled at eaves (starter course)

$1^1/_2$" min

Joint or slots (shall be $^1/_4$ in. to $^3/_8$ in.)

Course line

Weather exposure limited by Table R905.7.5

For SI: 1 inch = 25.4 mm.

A-676

When fastening wood shingles to either spaced or solid sheathing, it is critical that care be taken so that the shingles are not split. Unless the sheathing is less than $^1/_2$-inch thick, the fasteners must penetrate into the sheathing a minimum of $^1/_2$ inch.

Topic: Wood Shakes
Reference: IRC R905.8

Category: Roof Assemblies
Subject: Roof Covering Requirements

Code Text: *Wood shakes shall only be used on slopes of three units vertical in 12 units horizontal (25-percent slope) or greater. Fasteners for wood shakes shall be corrosion-resistant, with a minimum penetration of $^1/_2$ inch (12.7 mm) into the sheathing. Wood shakes shall be attached to the roof with two fasteners per shake, positioned no more than 1 inch (25.4 mm) from each edge and no more than 2 inches (51 mm) above the exposure line. Shakes shall be interlaid with 18-inch-wide (457 mm) strips of not less than No. 30 felt shingled between each course in such a manner that no felt is exposed to the weather.*

Discussion and Commentary: Wood shakes are roofing materials that are split from logs, creating a more inconsistent surface finish than that of wood shingles. Shakes are either 18 inches or 24 inches long and graded as No. 1 or No. 2.

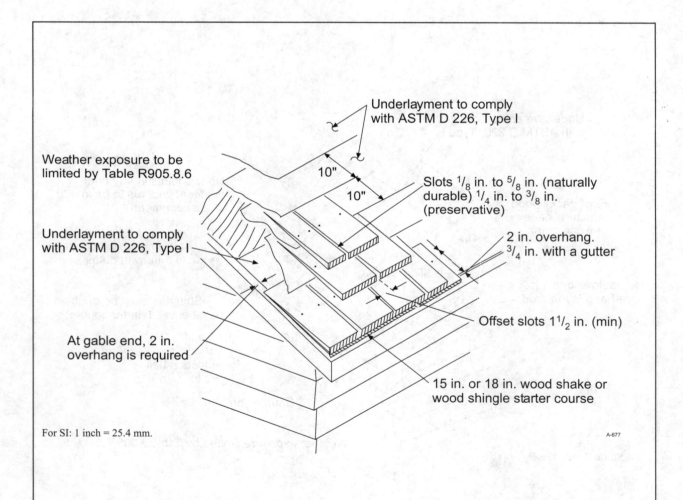

A key element of a wood shake roof covering system is the application of interlayment installed shingle fashion between each course of shakes. By positioning the lower edge of the minimum No. 30 felt above the butt end of the shake, the felt extends twice the required weather exposure.

QUIZ

Study Session 8

1. On a site having expansive soil, roof drainage water shall discharge a minimum of _____ from the foundation walls or to an approved drainage system.
   a. 3 feet
   b. 5 feet
   c. 6 feet
   d. 10 feet

2. Where the roof slopes are less than _____, the structural members that support rafters and ceiling joists (such as ridge beams, hips, and valleys) shall be designed as beams.
   a. 2:12
   b. 3:12
   c. 4:12
   d. 5:12

3. The maximum spacing of rafter ties shall be _____.
   a. 16 inches
   b. 4 feet
   c. 5 feet
   d. 8 feet

4. Ends of ceiling joists shall be lapped a minimum of _____ unless butted and toenailed to the supporting member.
   a. $1^1/_2$ inches
   b. 3 inches
   c. 4 inches
   d. 6 inches

5. Where SPF #2 ceiling joists create an uninhabitable attic without storage and are spaced 24 inches on center, what is the maximum allowable span when 2-inch-by-6-inch members are used?
   a. 11 feet, 2 inches
   b. 14 feet, 5 inches
   c. 14 feet, 9 inches
   d. 15 feet, 11 inches

6. A roof system is subjected to a 50 psf ground snow load and creates a dead load of 20 psf. The ceiling is not attached to the rafters. Assuming that rafter ties are provided at the top plate line, what is the maximum span of 2-inch-by-8-inch Hem-Fir #1 rafters spaced at 24 inches on center?
    a. 9 feet, 9 inches
    b. 10 feet, 6 inches
    c. 11 feet, 6 inches
    d. 11 feet, 10 inches

7. Purlins shall be supported by braces having a maximum unbraced length of _____ and spaced at a maximum of _____ on center.
    a. 8 feet, 4 feet
    b. 8 feet, 6 feet
    c. 12 feet, 4 feet
    d. 12 feet, 6 feet

8. The ends of a ceiling joist shall have a minimum of _____ bearing on wood and a minimum of _____ on metal.
    a. $1^1/_2$ inches, $1^1/_2$ inches
    b. $1^1/_2$ inches, 3 inches
    c. 3 inches, $1^1/_2$ inches
    d. 3 inches, 3 inches

9. Notches in solid sawn lumber ceiling joists shall not be located in the middle _____ of the span and are limited in depth to _____ the depth of the joist.
    a. one-fourth, one-sixth
    b. one-fourth, one-third
    c. one-third, one-sixth
    d. one-third, one-third

10. An opening in a roof system may be framed with a single header and single trimmer joists provided the header joist is located a maximum of _____ from the trimmer joist bearing.
    a. 3 feet
    b. 4 feet
    c. 6 feet
    d. 12 feet

11. Where $^{15}/_{32}$-inch wood structural panels having a span rating of $^{32}/_{16}$ are used as roof sheathing, what is the maximum span without edge support?
    a. 16 inches
    b. 24 inches
    c. 28 inches
    d. 32 inches

12. A dwelling is located in an area having a 110 mph wind speed and Exposure B. The ground snow load is 20 psf. What is the maximum allowable rafter span for 800S162-54 steel rafters installed at 24 inches on center with a roof slope of 12:12?
    a. 12 feet, 10 inches
    b. 14 feet, 6 inches
    c. 18 feet, 5 inches
    d. 22 feet, 7 inches

13. In a steel-framed roof system, roof cantilevers shall not exceed _____ in horizontal projection.
    a. 12 inches
    b. 18 inches
    c. 24 inches
    d. 30 inches

14. Openings provided for roof ventilation shall be covered with corrosion-resistant wire mesh with openings a minimum of _____ and a maximum of _____.
    a. $1/8$ inch, $1/4$ inch
    b. $1/8$ inch, $3/8$ inch
    c. $1/4$ inch, $3/8$ inch
    d. $1/4$ inch, $1/2$ inch

15. A required attic access opening shall have a minimum rough opening size of _____.
    a. 18 inches by 24 inches
    b. 18 inches by 30 inches
    c. 22 inches by 30 inches
    d. 24 inches by 36 inches

16. Unless listed for lesser clearances, combustible insulation shall be separated a minimum of _____ from recessed lighting fixtures, fan motors and other heat-producing devices.
    a. 1 inch
    b. 2 inches
    c. 3 inches
    d. 6 inches

17. Class A, B or C roofing shall be installed unless the edge of the roof is a minimum of _____ from the property line.
    a. 12 inches
    b. 2 feet
    c. 3 feet
    d. 5 feet

18. In the attachment of asphalt shingles, special methods of fastening are required where the minimum roof slope of _____ is exceeded.
a. 12 units vertical in 12 units horizontal
b. 16 units vertical in 12 units horizontal
c. 18 units vertical in 12 units horizontal
d. 20 units vertical in 12 units horizontal

19. Double underlayment is required for an asphalt shingle application where the roof has a maximum slope of _____.
a. 2:12
b. $2^1/_2$:12
c. 3:12
d. 4:12

20. In an asphalt shingle roof application, a cricket shall be installed on the ridge side of any chimney with a minimum width of _____.
a. 30 inches
b. 32 inches
c. 36 inches
d. 42 inches

21. Nails used to attach concrete roof tiles to the roof deck shall penetrate the deck a minimum of _____ or through the thickness of the deck, whichever is less.
a. $^1/_2$ inch
b. $^5/_8$ inch
c. $^3/_4$ inch
d. $^7/_8$ inch

22. Wood roof shingles shall be laid with a minimum side lap of _____ between joints in courses.
a. $^3/_8$ inch
b. $^1/_2$ inch
c. 1 inch
d. $1^1/_2$ inch

23. Where No. 2 wood shingles of naturally durable wood are installed on a roof having a 5:12 pitch, the maximum weather exposure for 16-inch shingles shall be _____.
a. $3^1/_2$ inches
b. 4 inches
c. $5^1/_2$ inches
d. $6^1/_2$ inches

24.   For preservative tapersawn shakes installed on a roof, the spacing between wood shakes shall be
       a minimum of _____ and a maximum of _____.
       a. $1/8$ inch, $3/8$ inch
       b. $1/8$ inch, $5/8$ inch
       c. $1/4$ inch, $3/8$ inch
       d. $3/8$ inch, $5/8$ inch

25.   For a wood shake roof system, sheet metal roof valley flashing shall extend a minimum of
       _____ from the centerline of the valley in each direction.
       a. 4 inches
       b. 7 inches
       c. 11 inches
       d. 12 inches

26.   A notch located in the end of a solid lumber ceiling joist shall have a maximum depth of _____.
       a. 1 inch
       b. 2 inches
       c. one-third the joist depth
       d. one-fourth the joist depth

27.   In the framing of an opening in ceiling construction, approved hangers shall be used for the
       header joist to trimmer joist connections where the header joist span exceeds _____ feet.
       a. 4
       b. 6
       c. 8
       d. 12

28.   Where eave or cornice vents are installed, a minimum _____ air space shall be provided
       between the insulation and the roof sheathing at the vent location.
       a. $1/2$-inch
       b. 1-inch
       c. 2-inch
       d. 3-inch

29.   An access opening is required in attics of combustible construction where the attic space
       exceeds _____ square feet in area.
       a. 30
       b. 50
       c. 100
       d. 120

30.   Overflow scuppers utilized for secondary roof drainage shall have a minimum size _____ of the roof drain.
        a. equal to that
        b. twice the size
        c. three times the size
        d. four times the size

INTERNATIONAL RESIDENTIAL CODE
Study Session 9
Chapters 10 and 11—Chimneys/Fireplaces and Energy Effiency

OBJECTIVE: To obtain an understanding of the provisions for chimneys and fireplaces, as well as the requirements for energy-efficient design and construction.

REFERENCE: Chapters 10 and 11, 2003 *International Residential Code*

KEY POINTS:

- When is reinforcement mandated in the construction of a masonry chimney? How must such chimneys be supported?
- Where are masonry chimneys prohibited from changing size or shape? How shall corbeling be done? Offsets?
- At what minimum height is a masonry chimney required to terminate?
- What are the requirements for spark arrestors installed on masonry chimneys?
- What are the various methods and materials used as flue linings? How are multiple flues to be constructed? What methods are provided for determining the minimum required flue area?
- When is a chimney cleanout opening required? Where must it be located?
- How far must combustible material be located from a masonry chimney located in the interior of a building? Located entirely outside a building?
- How is a factory-built chimney regulated?
- How must a masonry fireplace be supported and reinforced?
- What is the minimum thickness of masonry fireplace walls? What is the maximum width of joints between firebricks?
- Of what material must a lintel be constructed? What is the minimum required bearing length at each end of the lintel? Where is the fireplace throat to be located?
- What construction materials are permitted for hearths and hearth extensions? What is the minimum thickness of a hearth? A hearth extension?
- What are the minimum dimensions of a hearth extension? What is the minimum depth of a firebox?
- What is a climate zone and how is it determined? What is an alternative to the climate zones set forth in the code? How does the climate zone relate to the energy efficiency requirements?
- How must insulation be labeled? When is a certificate of compliance permitted? How shall the insulation information be permanently identified in an attic area?
- How is the performance of fenestration regulated? Replacement fenestration? Recessed light fixtures?
- When is the *International Energy Conservation Code*® applicable?
- How can energy compliance be accomplished for detached one- and two-family dwellings? For townhouses?
- How is the minimum required *R*-value for exterior walls regulated? Ceilings? Floors? Basement walls? Slab perimeter? How is the *R*-value determined?
- In what manner is the performance of mechanical systems regulated? When is duct insulation required? Piping insulation?

Topic: Construction
Reference: IRC R1001.6, R1001.7, R1001.8

Category: Chimneys and Fireplaces
Subject: Masonry Chimneys

Code Text: *Chimneys shall extend at least 2 feet (610 mm) higher than any portion of a building within 10 feet (3048 mm), but shall not be less than 3 feet (914 mm) above the point where the chimney passes through the roof. Masonry chimney walls shall be constructed of solid masonry units or hollow masonry units grouted solid with not less than a 4-inch (102 mm) nominal thickness. Masonry chimneys shall be lined.*

Discussion and Commentary: A chimney must terminate some distance above a roof or other surface in order to provide for the necessary upward draft in the chimney. The minimum extensions mandated by the code, based on experience, were determined considering the possibility of downward drafts creating eddying, which neutralizes the required upward draft in the chimney as well as flue gas temperature.

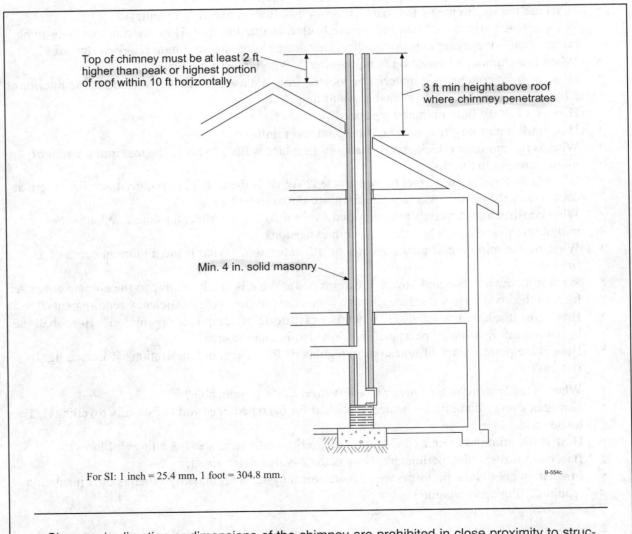

Top of chimney must be at least 2 ft higher than peak or highest portion of roof within 10 ft horizontally

3 ft min height above roof where chimney penetrates

Min. 4 in. solid masonry

For SI: 1 inch = 25.4 mm, 1 foot = 304.8 mm.

B-554c

Changes in direction or dimensions of the chimney are prohibited in close proximity to structural components. Such changes create the potential for leaks or cracks in the chimney wall or flue lining, providing a point for ignition of combustible members located within any concealed areas.

Topic: Cleanout Openings
Reference: IRC R1001.14

Category: Chimneys and Fireplaces
Subject: Masonry Chimneys

Code Text: *Cleanout openings shall be provided within 6 inches (152 mm) of the base of each flue within every masonry chimney.* See exception for chimney flues where cleaning is possible through the fireplace opening. *The upper edge of the cleanout shall be located at least 6 inches (152 mm) below the lowest chimney inlet opening. The height of the opening shall be at least 6 inches (152 mm). The cleanout shall be provided with a noncombustible cover.*

Discussion and Commentary: The chimney interior can become coated with carbonaceous material that attacks the masonry and mortar, resulting in deposits that fall to the bottom of the chimney. Chimney cleaning also results in the accumulation of deposits at the chimney base. The deposits can be removed as necessary through the cleanout opening at the base of the chimney.

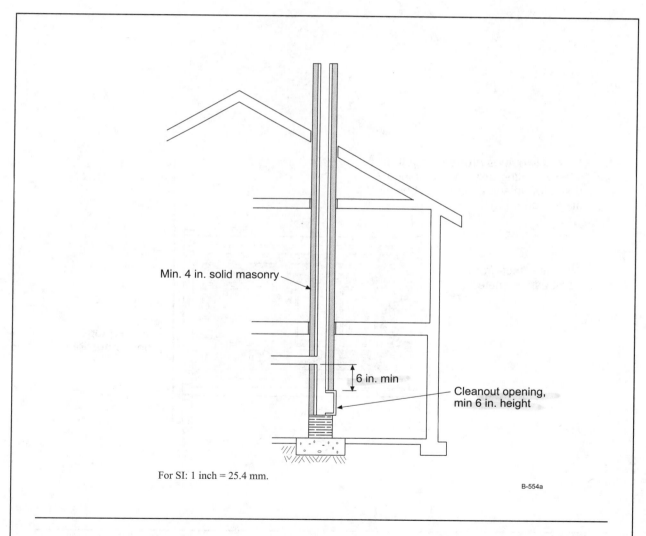

Min. 4 in. solid masonry

6 in. min

Cleanout opening, min 6 in. height

For SI: 1 inch = 25.4 mm.

B-554a

The opening for the chimney cleanout must be provided with a tight-fitting cover of cast iron or a similar material. Any space or gap created by a poor-fitting cover can create a hazardous condition or possibly result in a troublesome reduction in the intended draft effect.

Topic: Chimney Clearances
Reference: IRC R1001.15

Category: Chimneys and Fireplaces
Subject: Masonry Chimneys

Code Text: *Any portion of a masonry chimney located in the interior of the building or within the exterior wall of the building shall have a minimum air space clearance to combustibles of 2 inches (51 mm). Chimneys located entirely outside the exterior walls of the building, including chimneys that pass through the soffit or cornice, shall have a minimum air space clearance of 1 inch (25.4 mm).* See exceptions for alternate chimney designs or systems.

Discussion and Commentary: On account of the hot gases and products of combustion, the chimney walls become quite hot. As the heat is transferred through the masonry material, any combustible material in close proximity to the heated walls may reach the point of ignition. Thus, an air space of limited depth must be created around the chimney to help in the dissipation of heat.

Chimney inside wall 2" clearance
outside 1" "

Max of two flues grouped together where venting only one appliance (joints of adjacent flues must be staggered 4 in.)

Air space

Min clearance from combustible material

2"

2 in. fire stop (noncombustible)

4" min

For SI: 1 inch = 25.4 mm.

B-509

The depth of the air space at a masonry chimney need not be as great where the chimney is located entirely outside of the exterior walls of the dwelling. The exterior surface will remain cooler because of its direct exposure to the ambient outdoor temperature.

Topic: Chimney Crickets
Reference: IRC R1001.17

Category: Chimneys and Fireplaces
Subject: Masonry Chimneys

Code Text: *Chimneys shall be provided with crickets when the dimension parallel to the ridgeline is greater than 30 inches (762 mm) and does not intersect the ridgeline. The intersection of the cricket and the chimney shall be flashed and counterflashed in the same manner as normal roof-chimney intersections. Crickets shall be constructed in compliance with Figure R1001.17 and Table R1001.17.*

Discussion and Commentary: The potential for water intrusion at chimney openings in roofs is of concern, particularly where the length of the chimney wall that runs perpendicular to the roof slope is sizable. By providing a cricket, positive water flow away from the roof opening can be ensured.

TABLE R1001.17
CRICKET DIMENSIONS

ROOF SLOPE	H
12 - 12	$1/2$ of W
8 - 12	$1/3$ of W
6 - 12	$1/4$ of W
4 - 12	$1/6$ of W
3 - 12	$1/8$ of W

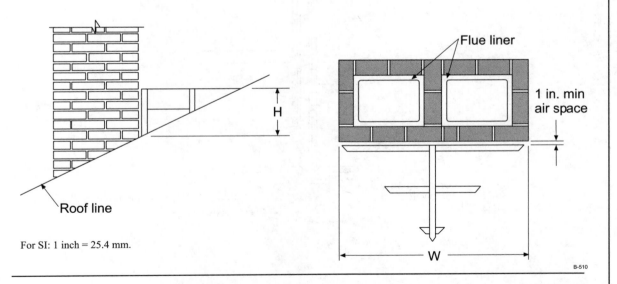

For SI: 1 inch = 25.4 mm.

The use of flashing and counterflashing at a roof opening such as a chimney is based on their ability to prevent moisture from breaching the weather protective membrane. Where corrosion-resistant metal is used as flashing, it must be minimum No. 26 galvanized sheet steel.

Topic: Support

Reference: IRC R1003.2

Category: Chimneys and Fireplaces

Subject: Masonry Fireplaces

Code Text: *Footings for masonry fireplaces and their chimneys shall be constructed of concrete or solid masonry at least 12 inches (305 mm) thick and shall extend at least 6 inches (152 mm) beyond the face of the fireplace or foundation wall on all sides. Footings shall be founded on natural, undisturbed earth or engineered fill below frost depth. In areas not subjected to freezing, footings shall be at least 12 inches (305 mm) below finished grade.*

Discussion and Commentary: Because of the extensive weight of a masonry chimney, the chimney foundation is likely to support a greater load than the foundation for the surrounding construction. The minimum size mandated is expected to carry such a load. The relationship between the building foundation and the chimney foundation must be considered in order to avoid detrimental settling.

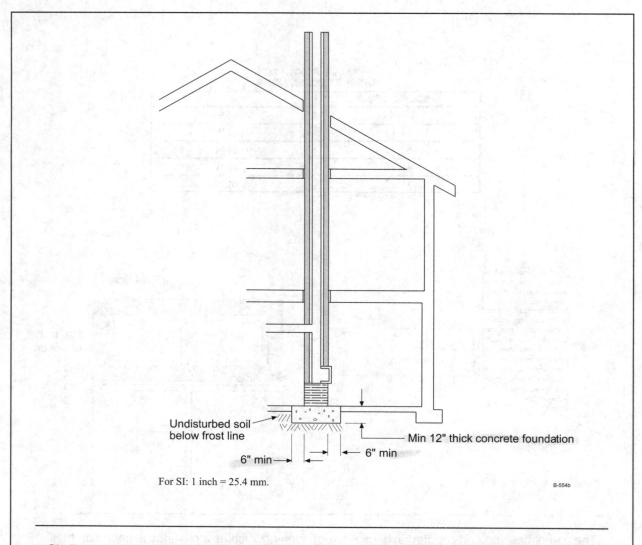

Undisturbed soil below frost line

Min 12" thick concrete foundation

6" min

6" min

For SI: 1 inch = 25.4 mm.

B-554b

Similar to other conventional foundation systems, the bottom of the footing must extend to below the frost line to eliminate the potential for movement due to freeze/thaw conditions. It is critical that the footing and foundation provide a solid base to avoid any possible settlement.

Topic: Firebox Walls
Reference: IRC R1003.5

Category: Chimneys and Fireplaces
Subject: Masonry Fireplaces

Code Text: *Masonry fireboxes shall be constructed of solid masonry units, hollow masonry units grouted solid, stone or concrete. When a lining of firebrick at least 2 inches (51 mm) in thickness or other approved lining is provided, the total minimum thickness of back and side walls shall each be 8 inches (203 mm) of solid masonry, including the lining. The width of joints between firebricks shall not be greater than $^1/_4$ inch (6.4 mm). When no lining is provided, the total minimum thickness of back and side walls shall be 10 inches (254 mm) of solid masonry.*

Discussion and Commentary: A variety of noncombustible materials is acceptable for the construction of masonry fireplaces. Where an approved lining is provided, the total masonry thickness is reduced from that needed where no lining is installed. Care should be taken to limit the joint width between firebricks in order to reduce the potential for joint deterioration.

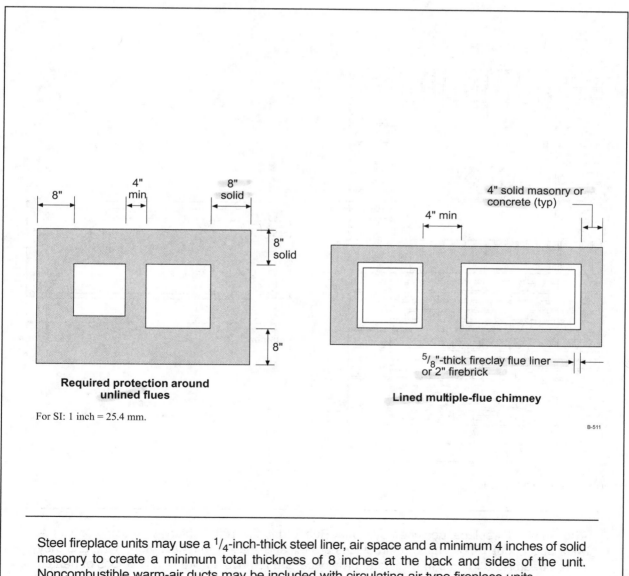

Required protection around unlined flues

For SI: 1 inch = 25.4 mm.

Lined multiple-flue chimney

B-511

Steel fireplace units may use a $^1/_4$-inch-thick steel liner, air space and a minimum 4 inches of solid masonry to create a minimum total thickness of 8 inches at the back and sides of the unit. Noncombustible warm-air ducts may be included with circulating-air-type fireplace units.

Topic: Firebox Dimensions
Reference: IRC R1003.6

Category: Chimneys and Fireplaces
Subject: Masonry Fireplaces

Code Text: *The firebox of a concrete or masonry fireplace shall have a minimum depth of 20 inches (508 mm). The throat shall not be less than 8 inches (203 mm) above the fireplace opening. The throat opening shall not be less than 4 inches (102 mm) in depth. The cross-sectional area of the passageway above the firebox, including the throat, damper and smoke chamber, shall not be less than the cross-sectional area of the flue. See exception for Rumford fireplaces.*

Discussion and Commentary: To provide for positive drafting, the size of the firebox is regulated, as is the location and depth of the throat opening. These criteria are based on many years of trial and error in the construction of fireplaces and chimneys.

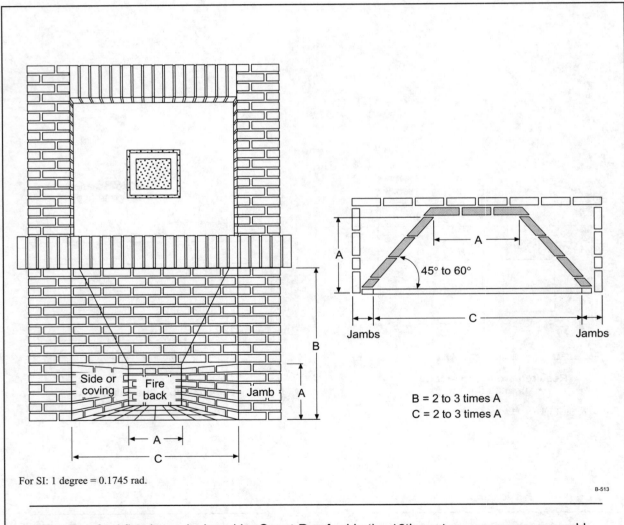

For SI: 1 degree = 0.1745 rad.

B-513

The Rumford fireplace, designed by Count Rumford in the 18th century, was common world-wide until about 1850. Rumford fireplaces are tall and shallow to reflect more heat, and they have streamlined throats to eliminate turbulence and remove smoke with little loss of heated room air.

2003 IRC Study Companion

Topic: Hearth and Hearth Extension

Category: Chimneys and Fireplaces

Reference: IRC R1003.9, R1003.10

Subject: Masonry Fireplaces

Code Text: *The minimum thickness of fireplace hearths shall be 4 inches (102 mm). The minimum thickness of hearth extensions shall be 2 inches (51 mm).* See exception for permitted reduction in thickness. *Hearth extensions shall extend at least 16 inches (406 mm) in front of and at least 8 inches (203 mm) beyond each side of the fireplace opening. Where the fireplace opening is 6 square feet (0.557 m²) or larger, the hearth extension shall extend at least 20 inches (508 mm) in front of and at least 12 inches (305 mm) beyond each side of the fireplace opening.*

Discussion and Commentary: The hearth is the floor of the firebox, whereas the hearth extension is the projection of the hearth beyond the firebox opening. The purpose of the hearth extension is to provide a noncombustible area adjacent to the opening where sparks, embers and ashes from the firebox may fall without the risk of igniting of combustible flooring materials.

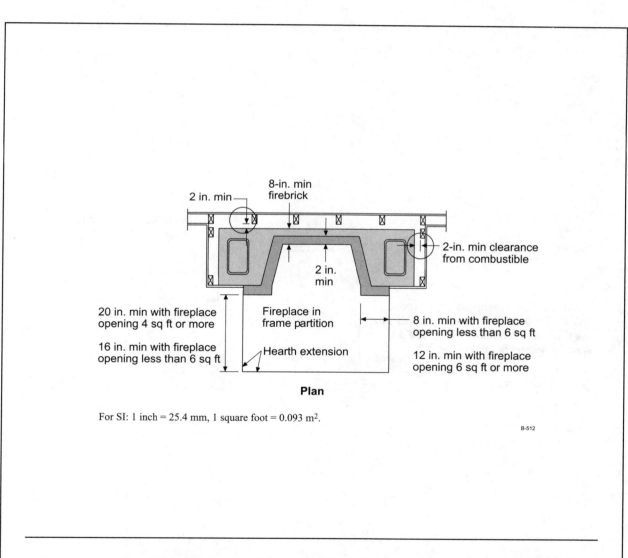

Plan

For SI: 1 inch = 25.4 mm, 1 square foot = 0.093 m².

B-512

Always noncombustible, a hearth extension must be a minimum of 2 inches thick unless the bottom of the firebox opening is located at least 8 inches above the top of the hearth extension. Under such conditions, the hearth extension need be only 3/8-inch-thick material.

Topic: Fireplace Clearance
Reference: IRC R1003.11, R1003.12

Category: Chimneys and Fireplaces
Subject: Masonry Fireplaces

Code Text: *All wood beams, joists, studs and other combustible material shall have a clearance of not less than 2 inches (51 mm) from the front faces and sides of masonry fireplaces and not less than 4 inches (102 mm) from the back faces of masonry fireplaces. See four exceptions for permitted reductions in clearance. Woodwork or other combustible materials shall not be placed within 6 inches (152 mm) of a fireplace opening. Combustible material within 12 inches (305 mm) of the fireplace opening shall not project more than $^1/_8$ inch (3.2 mm) for each 1-inch (25.4 mm) distance from such opening.*

Discussion and Commentary: Radiant heat transfer through the materials used to construct a masonry fireplace and/or chimney necessitates a minimum separation between the masonry and combustible materials, such as wood floor, wall or roof framing.

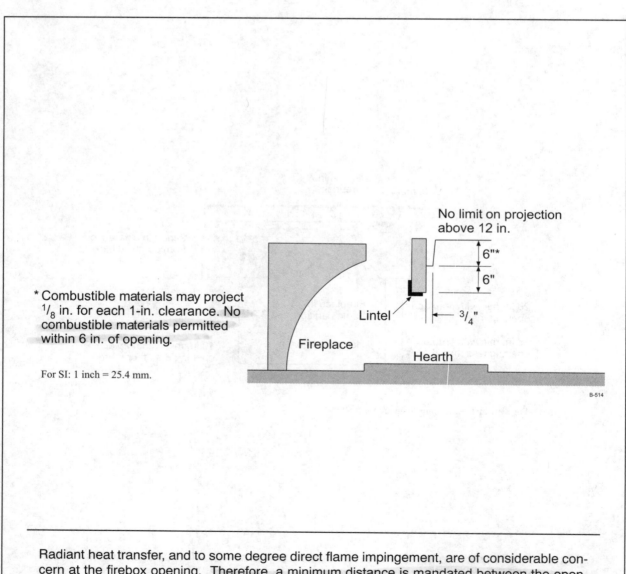

* Combustible materials may project $^1/_8$ in. for each 1-in. clearance. No combustible materials permitted within 6 in. of opening.

For SI: 1 inch = 25.4 mm.

No limit on projection above 12 in.

6"*

6"

Lintel

$^3/_4$"

Fireplace

Hearth

B-514

Radiant heat transfer, and to some degree direct flame impingement, are of considerable concern at the firebox opening. Therefore, a minimum distance is mandated between the opening and any adjacent combustible material.

Topic: Exterior Air Intake
Reference: IRC R1005.2

Category: Chimneys and Fireplaces
Subject: Exterior Air Supply

Code Text: *The exterior air intake shall be capable of providing all combustion air from the exterior of the dwelling or from spaces within the dwelling ventilated with outside air such as nonmechanically ventilated crawl or attic spaces. The exterior air intake shall not be located within the garage or basement of the dwelling nor shall the air intake be located at an elevation higher than the firebox. The exterior air intake shall be covered with a corrosion-resistant screen of $^1/_4$-inch (6.4 mm) mesh.*

Discussion and Commentary: Unless it is possible to mechanically ventilate and control the indoor pressure of the room in which a fireplace is located so that the pressure remains neutral or positive, both factory-built and masonry fireplaces must have an exterior air supply. The presence of a complying exterior air intake allows for an adequate supply of combustion air to the fireplace.

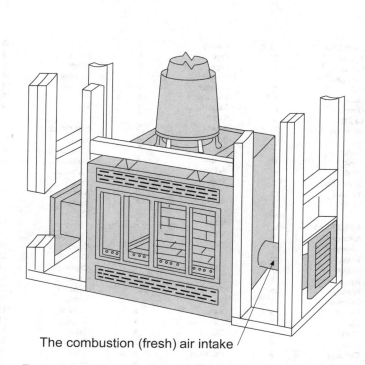

The combustion (fresh) air intake

Factory-built fireplace installed in accordance with listing and manufacturer's instructions

B-515

Combustion air ducts are regulated for clearance, size and location. The outlet may be located in the back or sides of the firebox chamber, or as an alternative, located on or near the floor within a distance of 24 inches to the firebox opening.

Topic: Compliance
Reference: IRC N1101.2

Category: Energy Efficiency
Subject: General Provisions

Code Text: *Where applicable, provisions are based on the climate zone where the building is located. The climate zone where the building is located shall be based on zone assignments in Table N1101.2 for the county and state in which the building is being constructed. Alternatively, the climate zone shall be permitted to be determined by the heating degree days assigned by the building official.*

Discussion and Commentary: Where the building is a one- or two-family dwelling, there are two acceptable methods for obtaining compliance for energy efficiency. The provisions of Chapter 11 may be used provided the glazing area does not exceed 15 percent of the gross area of exterior walls; or, as an option, the *International Energy Conservation Code*® (IECC)® may be utilized. For townhouse construction, a limit of 25 percent glazed area of the gross area of exterior walls is mandated for use of Chapter 11. Otherwise, use of the IECC is required.

Houston[a]	4	**ARIZONA**		Poinsett	8	Tehama	6	**CONNECTICUT**	
Jackson	8	Apache	13	Polk	8	Trinity	9	Fairfield	
Jefferson[a]	6	Cochise	6	Pope	8	Tulare	6	Hartford	
Lamar	7	Coconino	14	Prairie[a]	7	Tuolume	8	Litchfield	
Lauderdale	8	Gila	8	Pulaski[a]	7	Ventura	4	Middlesex	
Lawrence	8	Graham	6	Randolph	8	Yolo	6	New Haven	
Lee[a]	6	Greenlee	6	Saline[a]	7	Yuba	6	New London	
Limestone	8	La Paz	3	Scott[a]	7			Tolland	
Lowndes[a]	5	Maricopa	3	Searcy	9	**COLORADO**		Windham	
Macon[a]	6	Mohave	7	Sebastian	8	Adams	13		
Madison	8	Navajo	10	Sevier[a]	7	Alamosa	16	**DELAWARE**	
Marengo	5	Pima	4	Sharp	8	Arapahoe	13	Kent	
Marion	7	Pinal	4	St Francis[a]	7	Archuleta	16	New Castle	
Marshall	8	Santa Cruz	6	Stone	9	Baca	11	Sussex	
Mobile[a]	4	Yavapai	10	Union[a]	6	Bent	11		
Monroe[a]	5	Yuma	3	Van Buren	8	Boulder	13	**DC**	
Montgomery[a]	6			Washington	9	Chaffee	16	Washington	
Morgan	8	**ARKANSAS**		White[a]	7	Cheyenne	13		
Perry[a]	5	Arkansas[a]	6	Woodruff[a]	7	Clear Creek	17	**FLORIDA**	
Pickens[a]	6	Ashley[a]	6	Yell[a]	7	Conejos	16	Alachua[a]	
Pike[a]	5	Baxter	9			Costilla	16	Baker[a]	
Randolph	7	Benton	9			Crowley	11	Bay[a]	
Russell[a]	5	Boone	9	**CALIFORNIA**		Custer	16	Bradford[a]	
Shelby[a]	6	Bradley[a]	6	Alameda	6	Delta	13	Brevard[a]	
St Clair[a]	6	Calhoun[a]	6	Alpine	15	Denver	13	Broward[a]	
Sumter[a]	5	Carroll	9	Amador	8	Dolores	15	Calhoun[a]	
Talladega[a]	6	Chicot[a]	6	Butte	6	Douglas	13	Charlotte[a]	
Tallapoosa[a]	6	Clark[a]	6	Calaveras	8	Eagle	15	Citrus[a]	
Tuscaloosa[a]	6	Clay	8	Colusa	6	El Paso	13	Clay[a]	
Walker[a]	6	Cleburne	8	Contra Costa	6	Elbert	13	Collier[a]	
Washington[a]	5	Cleveland[a]	6	Del Norte	9	Fremont	11	Columbia[a]	
Wilcox[a]	5	Columbia[a]	6	El Dorado	8	Garfield	15	Dade[a]	
Winston	7	Conway[a]	7	Fresno	6	Gilpin	13	De Soto[a]	
		Craighead	8	Glenn	6	Grand	17	Dixie[a]	
				Humboldt	9			Duval[a]	

In addition to the requirements based on climate, such as insulation levels and maximum glazing efficiency, several provisions are applicable in all climate zones. Examples include sealing the building envelope and meeting the minimum heating and cooling equipment efficiency levels.

Topic: Insulation

Category: Energy Efficiency

Reference: IRC N1101.3.1

Subject: General Provisions

Code Text: *The thermal resistance (R-value) shall be indicated on all insulation and the insulation installed such that the R-value can be verified during inspection, or a certification of the installed R-value shall be provided at the job site by the insulation installer.*

Discussion and Commentary: The IRC requires certification of the installed density and *R*-value of blown-in or sprayed insulation in the wall cavities. Where installed in the attic, the certification must include both the initial installed thickness and the settled thickness, the coverage area and the number of bags of insulating material installed. Markers that indicate in 1-inch-high numbers the installed thickness of the insulation must be provided for every 300 square feet of attic area, attached to the trusses, rafters or joists.

TABLE N1101.3.2.1(1)
U-FACTOR DEFAULT TABLE FOR WINDOWS, GLAZED DOORS AND SKYLIGHTS

FRAME MATERIAL AND PRODUCT TYPE[a]	SINGLE GLAZED	DOUBLE GLAZED
Metal without thermal break		
Operable (including sliding and swinging glass doors)	1.27	0.87
Fixed	1.13	0.69
Garden window	2.60	1.81
Curtain wall	1.22	0.79
Skylight	1.98	1.31
Site-assembled sloped/overhead glazing	1.36	0.82
Metal with thermal break		
Operable (including sliding and swinging glass doors)	1.08	0.65
Fixed	1.07	0.63
Curtain wall	1.11	0.68
Skylight	1.89	1.11
Site-assembled sloped/overhead glazing	1.25	0.70
Reinforced vinyl/metal clad wood		
Operable (including sliding and swinging glass doors)	0.90	0.57
Fixed	0.98	0.56
Skylight	1.75	1.05
Wood/vinyl/fiberglass		
Operable (including sliding and swinging glass doors)	0.89	0.55
Fixed	0.98	0.56
Garden window	2.31	1.61
Skylight	1.47	0.84

a. Glass block assemblies with mortar but without reinforcing or framing shall have a *U*-factor of 0.60.

NOTE: This is IECC Table 102.5.2(1).

TABLE N1101.3.2.1(2)
U-FACTOR DEFAULT TABLE FOR NONGLAZED DOORS

DOOR TYPE	WITH FOAM CORE	WITHOUT FOAM CORE
Steel doors (1.75 inches thick)	0.35	0.60

	WITHOUT STORM DOOR	WITH STORM DOOR
Wood doors (1.75 inches thick)		
Panel with 0.438-inch panels	0.54	0.36
Hollow core flush	0.46	0.32
Panel with 1.125-inch panels	0.39	0.28
Solid core flush	0.40	0.26

For SI: 1 inch = 25.4 mm.

NOTE: This is IECC Table 102.5.2(2).

Fenestration includes skylights, roof windows, vertical windows, opaque doors, glazed doors, glass block and combination opaque/glazed doors. The U-factor for fenestration products shall be established in accordance with National Fenestration Rating Council (NFRC) 100.

Topic: Thermal Performance Criteria **Category:** Energy Efficiency
Reference: IRC N1102.1 **Subject:** Building Envelope

Code Text: *The minimum required insulation R-value or the area-weighted average maximum required fenestration U-factor for each element in the building thermal envelope (fenestration, roof/ceiling, opaque wall, floor, slab edge, crawl space wall and basement wall) shall be in accordance with the criteria in Table N1102.1.*

Discussion and Commentary: The required energy efficiency values of the various building elements are completely climate dependent and regulate minimum insulation levels (*R*-values) and maximum glazing efficiency (*U*-factors). These requirements are presented in Table N1102.1. The table is based on the Heating Degree Day (HDD) values for each jurisdiction, which are shown as climate zones in Table N1102.1.1.1(1). The climate zones for each county in the United States are indicated in Table N1101.2.

TABLE N1102.1
SIMPLIFIED PRESCRIPTIVE BUILDING ENVELOPE THERMAL COMPONENT CRITERIA
MINIMUM REQUIRED THERMAL PERFORMANCE (*U*-FACTOR AND *R*-VALUE)

BUILDING LOCATION		MAXIMUM GLAZING *U*-FACTOR [Btu / (hr·ft²·°F)]	MINIMUM INSULATION *R*-VALUE [(hr·ft²·°F) / Btu]					
Climate Zone	HDD		Ceilings	Walls	Floors	Basement walls	Slab perimeter *R*-value and depth	Crawl space walls
1	0-499	Any	R-13	R-11	R-11	R-0	R-0	R-0
2	500-999	0.90	R-19	R-11	R-11	R-0	R-0	R-4
3	1,000-1,499	0.75	R-19	R-11	R-11	R-0	R-0	R-5
4	1,500-1,999	0.75	R-26	R-13	R-11	R-5	R-0	R-5
5	2,000-2,499	0.65	R-30	R-13	R-11	R-5	R-0	R-6
6	2,500-2,999	0.60	R-30	R-13	R-19	R-6	R-4, 2 ft.	R-7
7	3,000-3,499	0.55	R-30	R-13	R-19	R-7	R-4, 2 ft.	R-8
8	3,500-3,999	0.50	R-30	R-13	R-19	R-8	R-5, 2 ft.	R-10
9	4,000-4,499	0.45	R-38	R-13	R-19	R-8	R-5, 2 ft.	R-11
10	4,500-4,999	0.45	R-38	R-16	R-19	R-9	R-6, 2 ft.	R-17
11	5,000-5,499	0.45	R-38	R-18	R-19	R-9	R-6, 2 ft.	R-17
12	5,500-5,999	0.40	R-38	R-18	R-21	R-10	R-9, 4 ft.	R-19
13	6,000-6,499	0.35	R-38	R-18	R-21	R-10	R-9, 4 ft.	R-20
14	6,500-6,999	0.35	R-49	R-21	R-21	R-11	R-11, 4 ft.	R-20
15	7,000-8,499	0.35	R-49	R-21	R-21	R-11	R-13, 4 ft.	R-20
16	8,500-8,999	0.35	R-49	R-21	R-21	R-18	R-14, 4 ft.	R-20
17	9,000-12,999	0.35	R-49	R-21	R-21	R-19	R-18, 4 ft.	R-20

For SI: 1 Btu/(hr·ft²·°F) = 5.68W/m²·K, 1 (hr·ft²·°F)/Btu = 0.176m²·K/W.

The specific criteria for determination of *R*-value and *U*-factor ratings are provided for each of the elements under consideration. For exterior walls, the rating is based solely on the insulating materials and not by the framing, drywall, structural sheathing or exterior siding materials.

2003 IRC Study Companion

Topic: Air Leakage
Reference: IRC N1102.1.10

Category: Energy Efficiency
Subject: Building Envelope

Code Text: *All joints, seams, penetrations; site-built windows, doors, and skylights; openings between window and door assemblies and their respective jambs and framing; and other sources of air leakage (infiltration and exfiltration) through the building thermal envelope shall be caulked, gasketed, weather-stripped, wrapped, or otherwise sealed to limit uncontrolled air movement.*

Discussion and Commentary: Sealing the building envelope is critical to good thermal performance for the building. It will prevent warm, conditioned air from leaking out around doors, windows and other cracks during the heating season, thereby reducing the cost of heating the residence. During the hot summer months proper sealing will stop hot air from entering the residence, helping to reduce the air conditioning load.

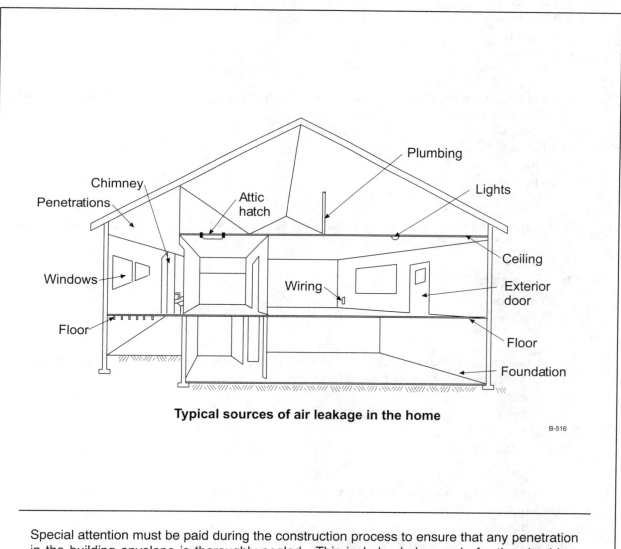

Typical sources of air leakage in the home

B-516

Special attention must be paid during the construction process to ensure that any penetration in the building envelope is thoroughly sealed. This includes holes made for the plumbing, electrical and HVAC systems. In addition, all joints in airflow retarders must be taped for continuity.

QUIZ

Study Session 9

1. The foundation for a masonry chimney shall be a minimum of _____ thick.
    a. 6 inches
    b. 8 inches
    c. 10 inches
    d. 12 inches

2. A masonry chimney wall shall not change size or shape for a minimum of _____ above or below where the chimney passes through floor, ceiling or roof components.
    a. 6 inches
    b. 12 inches
    c. 24 inches
    d. 30 inches

3. In all cases, a masonry chimney shall extend a minimum of _____ above the point where the chimney passes through the roof.
    a. 12 inches
    b. 18 inches
    c. 24 inches
    d. 36 inches

4. A flue liner for a masonry chimney shall be carried vertically, with a maximum permitted slope of _____ from the vertical.
    a. 15 degrees
    b. $22^1/_2$ degrees
    c. 30 degrees
    d. 45 degrees

5. When two or more flues are located in the same masonry chimney, they shall be separated by masonry wythes having a minimum thickness of _____.
    a. 2 inches
    b. 4 inches
    c. 6 inches
    d. 8 inches

6. A cleanout opening shall be provided within _____ of the base of a flue in a masonry chimney.
    a. 6 inches
    b. 8 inches
    c. 12 inches
    d. 18 inches

7. Where a masonry chimney is constructed totally outside the exterior walls of a dwelling, a minimum air space clearance to combustibles of _____ is required.
    a. 0 inches; no clearance is required
    b. $1/2$ inch
    c. 1 inch
    d. 2 inches

8. Foundations for masonry fireplaces and their chimneys shall extend a minimum of _____ beyond the face of the fireplace or support walls on all sides.
    a. 4 inches
    b. 6 inches
    c. 8 inches
    d. 12 inches

9. Where a 60-inch-wide masonry chimney and fireplace are located in Seismic Design Category D_1, the chimney shall be reinforced with _____ No. 4 continuous vertical reinforcing bars.
    a. 0; no reinforcement is required
    b. 2
    c. 4
    d. 6

10. What is the minimum thickness of the back and side walls of a masonry fireplace constructed without a firebrick lining?
    a. 6 inches
    b. 8 inches
    c. 10 inches
    d. 12 inches

11. Where firebricks are used as lining for a masonry fireplace, the maximum joint spacing shall be _____ between firebricks.
    a. $1/8$ inch
    b. $1/4$ inch
    c. $3/8$ inch
    d. $1/2$ inch

12. The damper or throat for a masonry fireplace shall be located a minimum of _____ above the lintel supporting the fireplace opening.
    a. 4 inches
    b. 6 inches
    c. 8 inches
    d. 12 inches

13. What is the minimum required thickness for the hearth of a masonry fireplace?
    a. $^3/_8$ inch
    b. $^3/_4$ inch
    c. 2 inches
    d. 4 inches

14. What is the minimum size of a hearth extension for a masonry fireplace having a 30-inch-high-by-42-inch-wide fireplace opening?
    a. 16 inches by 42 inches
    b. 16 inches by 58 inches
    c. 20 inches by 58 inches
    d. 20 inches by 66 inches

15. A minimum $^3/_8$-inch-thick noncombustible hearth extension is permitted where the bottom of the firebox opening is a minimum of _____ inches above the top of the hearth extension.
    a. 6
    b. 8
    c. 12
    d. 15

16. Wood studs shall have a minimum clearance of _____ from the back face of a masonry fireplace.
    a. 0 inches; no clearance is required
    b. 1 inch
    c. 2 inches
    d. 4 inches

17. Woodwork shall be located a minimum of _____ from the fireplace opening of a masonry fireplace.
    a. 1 inch
    b. 2 inches
    c. 4 inches
    d. 6 inches

18. Combustible trim projecting a distance of 1 inch beyond the fireplace opening of a masonry fireplace shall be located a minimum of _____ from the opening.
    a. 2 inches
    b. 6 inches
    c. 8 inches
    d. 12 inches

19. Factory-built fireplaces shall be tested in accordance with _____.
    a. ASTM C 199
    b. ASTM C 315
    c. UL 103
    d. UL 127

20. Unless the room is mechanically ventilated and controlled so that the indoor pressure is neutral or positive, a combustion air passageway having a minimum size of _____ square inches and a maximum size of _____ square inches shall be provided as an exterior air supply for a factory-built or masonry fireplace.
   a. 6, 25
   b. 6, 55   R1005.4
   c. 25, 55
   d. 25, 75

21. For the energy-efficiency requirements of the IRC, what range of Heating Degree Days (HDD) is designated for Sedgwick County, Kansas?
   a. 4,000 to 4,499
   b. 4,500 to 4,999
   c. 5,500 to 5,999
   d. 6,000 to 6,499

22. When a dwelling is located in a climate with an HDD of 3,200, the perimeter walls of a conditioned basement shall be insulated with insulation having a minimum R-value of _____ where using the simplified prescriptive building envelope method.
   a. 5
   b. 6
   c. 7
   d. 8

23. Opaque doors separating conditioned space from unconditioned space shall have a maximum U-factor of _____.
   a. 0.35
   b. 0.45
   c. 0.55
   d. 0.90

24. The exposed earth in all crawl space foundations shall be covered with a continuous vapor retarder having a maximum permeance rating of _____ perm.
   a. 1.0
   b. 0.5
   c. 0.1
   d. 0.01

25. Exposed insulating materials applied to the exterior of foundation walls shall be adequately protected to a depth of _____ below finished grade level.
   a. 6 inches
   b. 8 inches
   c. 12 inches
   d. 18 inches

26. Concrete foundations for masonry chimneys shall extend a minimum of _____ inches beyond each side of the exterior dimensions of the chimney.
    a. 3
    b. 6
    c. 8
    d. 12

27. The net free area of a spark arrestor for a masonry chimney shall be a minimum of _____ time(s) the net free area of the outlet of the chimney flue it serves.
    a. one
    b. two
    c. three
    d. four

28. Where a 54-inch-wide masonry chimney is constructed above a roof having a slope of 8:12, a required cricket shall have a minimum height of _____ inches.
    a. 9
    b. 13 ½
    c. 18
    d. 27

29. Except for Rumford fireplaces, the firebox of a masonry fireplace shall have a minimum depth of _____ inches.
    a. 12
    b. 18
    c. 20
    d. 24

30. A minimum clearance of _____ inches is required between the body of a masonry heater and adjacent framing.
    a. 3
    b. 4
    c. 6
    d. 12

INTERNATIONAL RESIDENTIAL CODE
Study Session 10
Chapters 13 and 14—General Mechanical System Requirements and Heating and Cooling Equipment

OBJECTIVE: To gain an understanding of the general provisions for the installation of mechanical appliances, equipment and systems, including requirements regulating heating and cooling equipment.

REFERENCE: Chapters 13 and 14, 2003 *International Residential Code*

KEY POINTS:

- What information must appear on the factory-applied nameplate affixed to an appliance?
- How much working space is needed in front of the control side of an appliance in order to service the appliance? How much of a reduction is allowed in front of a room heater?
- al furnace is installed in an alcove or compartment, what is the minimum total enclosed area? What is the minimum working clearance required along the sides of ? Along the back? On the top? How much front clearance is mandated where the open to the atmosphere?
- a passageway must be provided to an appliance installed in a compartment, alcove, or similar space? How must access to an attic appliance be addressed? For appliances n under-floor spaces?
- imum clearance is required between the ground and the appliance where the appliance ed from the floor? How far must the excavation extend beyond the appliance ?
- What is the minimum clearance needed between an appliance and unprotected combustible construction? How may such clearances be reduced? What is the minimum required clearance for solid fuel appliances?
- In what Seismic Design Categories must water heaters be anchored or strapped to resist horizontal displacement due to earthquake forces? What prescriptive method is set forth?
- How high above a floor in a garage is an appliance with an ignition source required to be located?
- What special provisions apply to hydrogen generating and refueling operations?
- How shall piping used as a portion of a mechanical system be protected where installed in concealed locations through holes and notches in framing members?
- How are radiant heating systems required to be installed? Duct heaters?
- What minimum clearance from a wall is mandated for a floor furnace? What is the minimum clearance from doors, draperies or similar combustible objects? What is the minimum required clearance from the furnace register to a combustible projecting object above?
- Where are vented wall furnaces prohibited from being located? What clearances are mandated between the furnace air inlet or outlet and the swing of a door?
- How must vented room heaters be installed? What minimum clearance is required on all sides of the appliance?
- How is condensate from refrigeration cooling equipment to be addressed? What methods are described for an auxiliary or secondary drain system? What is the minimum drain pipe size?
- What limitations are mandated for the use of fireplace stoves?

Topic: Central Furnaces
Reference: IRC M1305.1.1

Category: Mechanical Systems
Subject: Appliance Access

Code Text: *Central furnaces within compartments or alcoves shall have a minimum working space clearance of 3 inches (76 mm) along the sides, back and top with a total width of the enclosing space being at least 12 inches (305 mm) wider than the furnace. Furnaces having a firebox open to the atmosphere shall have at least a 6-inch (152 mm) working space along the front combustion chamber side.* See exception for replacement appliances.

Discussion and Commentary: A minimum amount of clearance is desired around a central furnace to allow for access to the unit, provide space to facilitate the removal of the furnace, create an air space for ventilating and cooling the appliance, and protect any surrounding combustible materials. The increased clearance at the front protects against flame roll-out and provides free movement of combustion air.

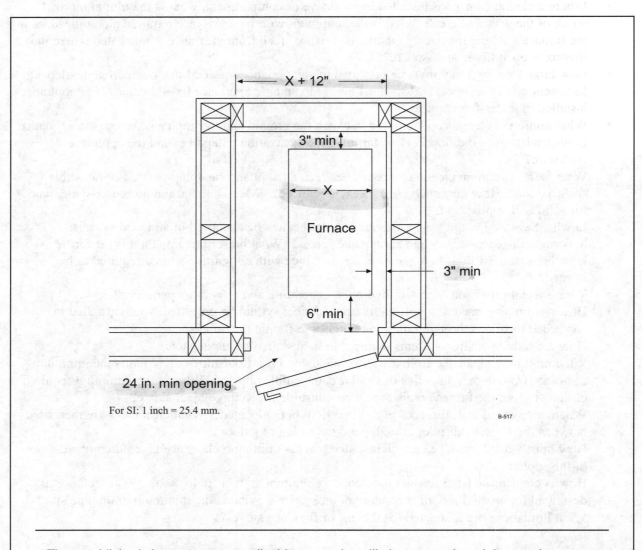

For SI: 1 inch = 25.4 mm.

B-517

The established clearances are applicable to new installllations even though lesser clearances are permitted by the manufacturer. Replacement appliances installed in existing compartments may have lesser clearances if in conformance with the manufacturer's instructions.

Topic: Appliances in Attics
Reference: IRC M1305.1.3

Category: Mechanical Systems
Subject: Appliance Access

Code Text: *Attics containing appliances requiring access shall be provided with an opening and a clear and unobstructed passageway large enough to allow removal of the largest appliance, but not less than 30 inches (762 mm) high and 22 inches (559 mm) wide and not more than 20 feet (6096 mm) in length when measured along the centerline of the passageway from the opening to the appliance.*

Discussion and Commentary: By placing the mechanical equipment in the attic, it is possible to better utilize the building's floor area. However, it is important that such equipment be accessible for service or removal. In addition to a properly sized access opening, a continuous passageway of solid flooring must be provided where needed to reach the appliance. A minimum 30-inch by 30-inch service area is required in front of any appliance access.

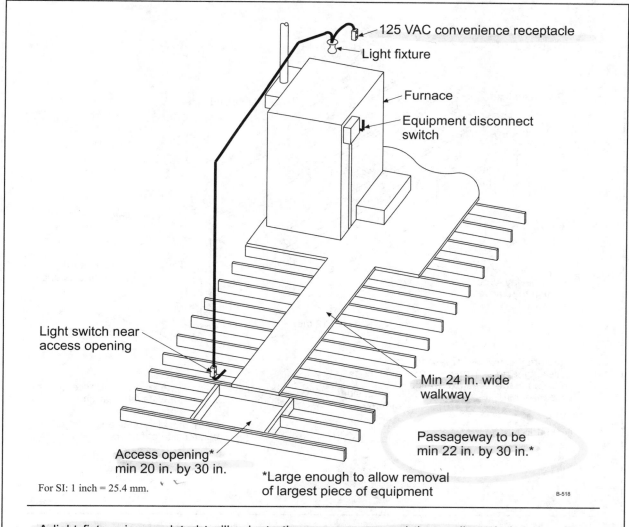

125 VAC convenience receptacle
Light fixture
Furnace
Equipment disconnect switch
Light switch near access opening
Min 24 in. wide walkway
Passageway to be min 22 in. by 30 in.*
Access opening* min 20 in. by 30 in.
*Large enough to allow removal of largest piece of equipment
For SI: 1 inch = 25.4 mm.

B-518

A light fixture is mandated to illuminate the passageway and the appliance's location. The switch controlling the fixture must be located at the entry to the passageway. In addition, a receptacle outlet shall be located at or near the appliance for use during maintenance or repair.

Topic: Appliances Under Floors
Reference: IRC M1305.1.4

Category: Mechanical Systems
Subject: Appliance Access

Code Text: *Underfloor spaces containing appliances requiring access shall be provided with an unobstructed passageway large enough to remove the largest appliance, but not less than 30 inches (762 mm) high and 22 inches (559 mm) wide, nor more than 20 feet (6096 mm) in length when measured along the centerline of the passageway from the opening to the appliance. If the depth of the passageway or the service space exceeds 12 inches (305 mm) below the adjoining grade, the walls shall be lined with concrete or masonry extending 4 inches (102 mm) above the adjoining grade in accordance with Chapter 4 (Foundations).*

Discussion and Commentary: The requirements for access to an underfloor furnace are similar to those for furnaces located in attics and other concealed areas. A switched light fixture and receptacle outlet are mandated to allow for safe and convenient appliance repair and maintenance.

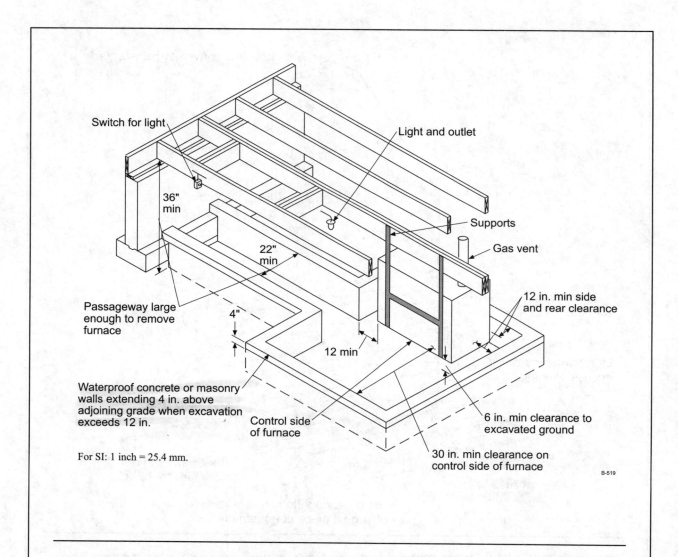

Switch for light

Light and outlet

36" min

Supports

22" min

Gas vent

Passageway large enough to remove furnace

4"

12 in. min side and rear clearance

12 min

Waterproof concrete or masonry walls extending 4 in. above adjoining grade when excavation exceeds 12 in.

Control side of furnace

6 in. min clearance to excavated ground

For SI: 1 inch = 25.4 mm.

30 in. min clearance on control side of furnace

B-519

Adequate clearance must be provided for an underfloor furnace. If the appliance is suspended from the floor, it must be at least 6 inches above the ground. Where the underfloor space is excavated, clearances of 12 inches on the sides and 30 inches on the control side are required.

Topic: Appliance Clearance
Reference: IRC M1306.1

Category: Mechanical Systems
Subject: Clearances from Combustible Construction

Code Text: *Appliances shall be installed with the clearances from unprotected combustible materials as indicated on the appliance label and in the manufacturer's installation instructions.*

Discussion and Commentary: The clearances established by the manufacturer's installation instructions are assumed to be from unprotected combustible construction. These clearances create the necessary airspace for reducing the conduction of heat that might ignite adjacent combustible materials. In addition, convection cooling provided by the airspace assists in the proper operation of the appliance. It is important to note that the provisions and guidelines are not intended to apply to the installation of unlisted appliances.

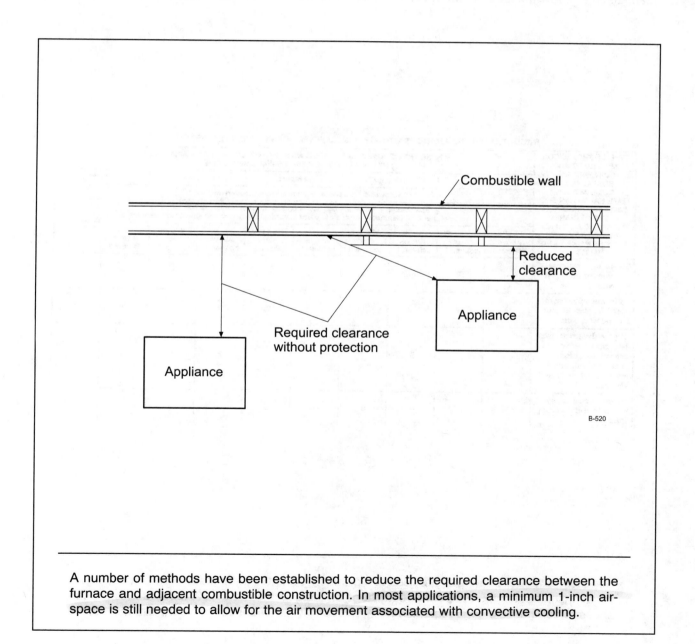

A number of methods have been established to reduce the required clearance between the furnace and adjacent combustible construction. In most applications, a minimum 1-inch airspace is still needed to allow for the air movement associated with convective cooling.

Topic: Clearance Reduction
Reference: IRC M1306.2

Category: Mechanical Systems
Subject: Clearances from Combustible Construction

Code Text: *Reduction of clearances shall be in accordance with the appliance manufacturer's instructions and Table M1306.2. Forms of protection with ventilated air space shall conform to the following requirements: 1) not less than 1-inch (25.4 mm) air space shall be provided between the protection and combustible wall surface; 2) air circulation shall be provided by having edges of the wall protection open at least 1 inch (25.4 mm); 3) if the wall protection is mounted on a single flat wall away from corners, air circulation shall be provided by having the bottom and top edges, or the side and top edges open at least 1 inch (25.4 mm); and 4) wall protection covering two walls in a corner shall be open at the bottom and top edges at least 1 inch (25.4 mm).*

Discussion and Commentary: The allowances for reduced clearances require that a 1-inch air space be maintained to allow the air circulation needed to keep the temperature rise within acceptable limits.

TABLE M1306.2
REDUCTION OF CLEARANCES WITH SPECIFIED FORMS OF PROTECTION[a,b,c,d,e,f,g,h,i,j,k]

TYPE OF PROTECTION APPLIED TO AND COVERING ALL SURFACES OF COMBUSTIBLE MATERIAL WITHIN THE DISTANCE SPECIFIED AS THE REQUIRED CLEARANCE WITH NO PROTECTION [SEE FIGURES M1306.1 AND M1306.2]	WHERE THE REQUIRED CLEARANCE WITH NO PROTECTION FROM APPLIANCE, VENT CONNECTOR, OR SINGLE WALL METAL PIPE IS:									
	36 inches		18 inches		12 inches		9 inches		6 inches	
	Allowable clearances with specified protection (inches)[b]									
	Use column 1 for clearances above an appliance or horizontal connector. Use column 2 for clearances from an appliance, vertical connector and single-wall metal pipe.									
	Above column 1	Sides and rear column 2	Above column 1	Sides and rear column 2	Above column 1	Sides and rear column 2	Above column 1	Sides and rear column 2	Above column 1	Sides and rear column 2
3½-in. thick masonry wall without ventilated air space	—	24	—	12	—	9	—	6	—	5
½-in. insulation board over 1-in. glass fiber or mineral wool batts	24	18	12	9	9	6	6	5	4	3
24 gage sheet metal over 1-in. glass fiber or mineral wool batts reinforced with wire on rear face with ventilated air space	18	12	9	6	6	4	5	3	3	3
3½-in. thick masonry wall with ventilated air space	—	12	—	6	—	6	—	6	—	6

Because solid-fuel-burning appliances can produce radiant high-intensity heat and have wide variations in heat output, there are restrictions on the use of Table M1306.2 for reducing clearances to combustible construction. A minimum of 12 inches must always be maintained.

Topic: Anchorage
Reference: IRC M1307.2

Category: Mechanical Systems
Subject: Appliance Installation

Code Text: *Appliances designed to be fixed in position shall be fastened or anchored in an approved manner. In Seismic Design Categories D_1 and D_2, water heaters shall be anchored or strapped to resist horizontal displacement due to earthquake motion. Strapping shall be at points within the upper one-third and lower one-third of the appliance's vertical dimensions. At the lower point, the strapping shall maintain a minimum distance of 4 inches (102 mm) above the controls.*

Discussion and Commentary: In those areas determined to have a high earthquake risk, it is important that any water heater be fastened in place to avoid damage. The strapping should be adequately attached to the studs or other framing members. It must be located an adequate distance from the control panel to allow access and use.

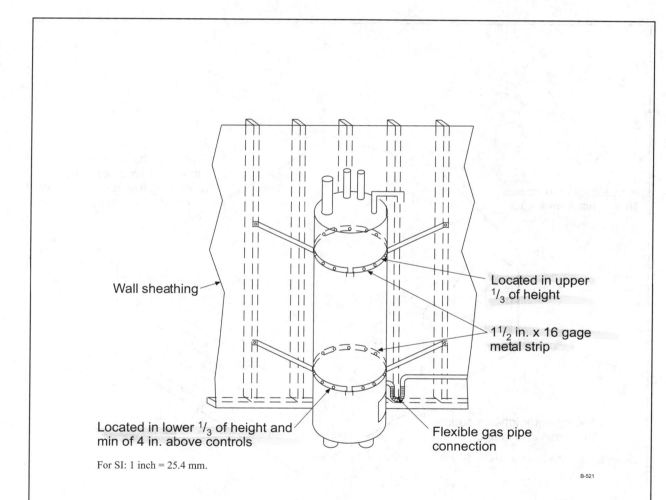

Wall sheathing

Located in upper $^1/_3$ of height

$1^1/_2$ in. x 16 gage metal strip

Located in lower $^1/_3$ of height and min of 4 in. above controls

Flexible gas pipe connection

For SI: 1 inch = 25.4 mm.

B-521

Large water heaters weigh several hundred pounds; if they are not secured, the water, gas and electrical connections may be damaged or severed in an earthquake, creating a hazardous situation. In addition to strapping, approved flexible connectors should be used.

Topic: Elevation of Ignition Source
Reference: IRC M1307.3

Category: Mechanical Systems
Subject: Appliance Installation

Code Text: *Appliances having an ignition source shall be elevated such that the source of ignition is not less than 18 inches (457 mm) above the floor in garages. For the purpose of Section M1307.3, rooms or spaces that are not part of the living space of a dwelling unit and that communicate with a private garage through openings shall be considered to be part of the garage. Appliances located in a garage or carport shall be protected from impact by automoviles*

Discussion and Commentary: Gasoline leakage or spillage in a garage is a possible danger. Gasoline fumes will evaporate from liquid puddles at the floor level. Any potential ignition source must be elevated to keep open flame, spark-producing elements and heating elements above the gasoline fume level.

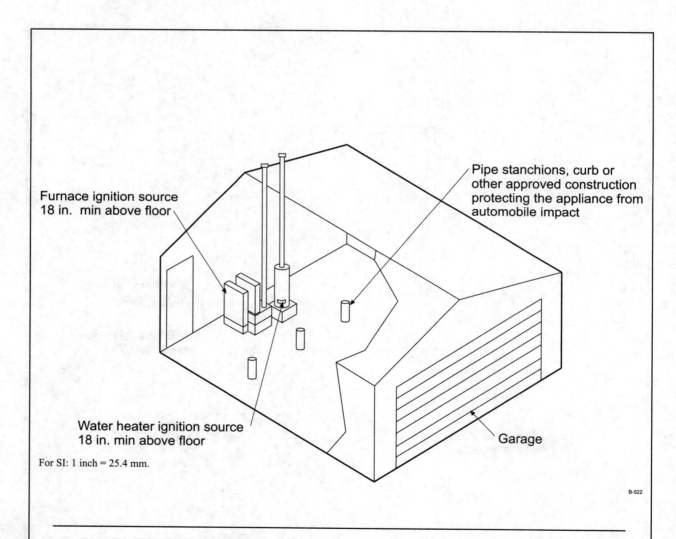

Furnace ignition source 18 in. min above floor

Pipe stanchions, curb or other approved construction protecting the appliance from automobile impact

Water heater ignition source 18 in. min above floor

Garage

For SI: 1 inch = 25.4 mm.

B-522

Appliances or equipment exposed to impact by vehicles could create a hazardous situation if their fuel or power connections were loosened or broken. Unless located out of the path of potential impact, the appliances must be protected by a barrier such as curb wheel stops or pipe stanchions.

Topic: Protection Against Damage
Reference: IRC M1308.2

Category: Mechanical Systems
Subject: Installations

Code Text: *In concealed locations where piping, other than cast-iron or galvanized steel, is installed through holes or notches in studs, joists, rafters or similar members less than 1.5 inches (38 mm) from the nearest edge of the member, the pipe shall be protected by shield plates. Protective shield plates shall be a minimum of 0.062-inch-thick (1.6 mm) steel, shall cover the area of the pipe where the member is notched or bored, and shall extend a minimum of 2 inches (51 mm) above sole plates and below top plates.*

Discussion and Commentary: It is critical that piping passing through joists, studs, rafters and other structural members be protected from damage, typically caused by wallboard fasteners and/or wood structural panels. Therefore, steel shield plates are to be utilized where the distance from the member surface to the notch or bored hole exceeds a specified distance.

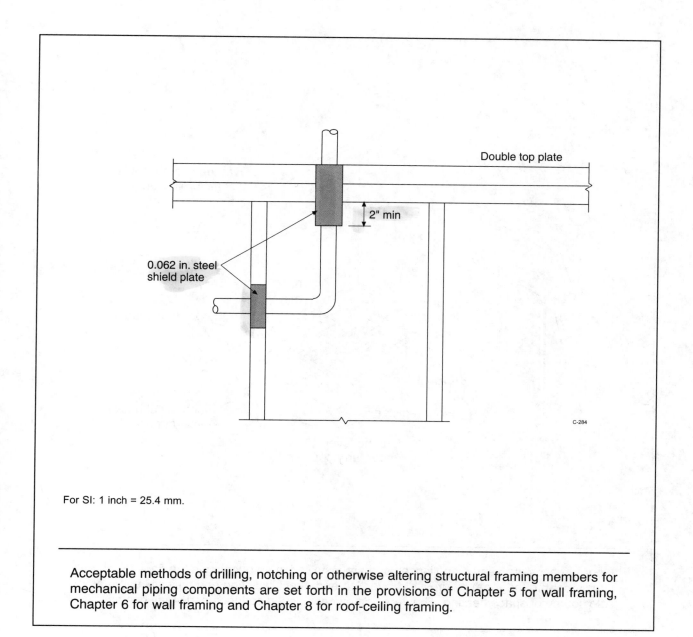

For SI: 1 inch = 25.4 mm.

Acceptable methods of drilling, notching or otherwise altering structural framing members for mechanical piping components are set forth in the provisions of Chapter 5 for wall framing, Chapter 6 for wall framing and Chapter 8 for roof-ceiling framing.

Topic: Location

Reference: IRC M1408.3

Category: Heating and Cooling Equipment

Subject: Vented Floor Furnaces

Code Text: *Floor registers of floor furnaces shall be installed not less than 6 inches (152 mm) from a wall. Wall registers of floor furnaces shall be installed not less than 6 inches (152 mm) from the adjoining wall at inside corners. The furnace register shall be located not less than 12 inches (305 mm) from doors in any position, draperies or similar combustible objects. The furnace register shall be located at least 5 feet (1524 mm) below any projecting combustible materials. The floor furnace shall not be installed where a door can swing within 12 inches (305 mm) of the grill opening.*

Discussion and Commentary: Flat floor furnaces and wall register floor furnaces are designed to be built directly into a wood-framed floor. However, there are clearances necessary because of the heat generated in the combustion chamber and heat coming from the register.

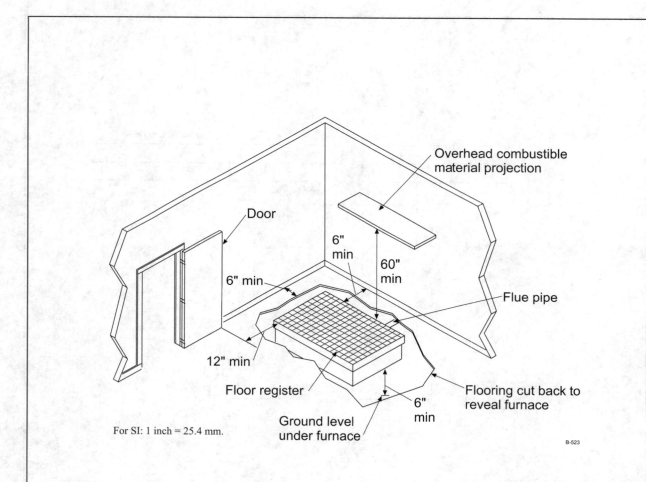

For SI: 1 inch = 25.4 mm.

B-523

Floor furnaces are vented appliances and are installled in an opening in the floor. Such units supply heat to the room by gravity convection and direct radiation and typically serve as the sole source of space heating. They are sometimes used in cottages and small homes.

2003 IRC Study Companion

Topic: Installation
Reference: IRC M1408.5

Category: Heating and Cooling Equipment
Subject: Vented Floor Furnaces

Code Text: *Floor furnaces shall be installed not closer than 6 inches (152 mm) to the ground. Clearance may be reduced to 2 inches (51 mm), provided that the lower 6 inches (152 mm) of the furnace is sealed to prevent water entry. Where excavation is required for a floor furnace installation, the excavation shall extend 30 inches (762 mm) beyond the control side of the floor furnace and 12 inches (305 mm) beyond the remaining sides.*

Discussion and Commentary: Floor furnaces must have supports independent of the floor register. A 6-inch clearance must also be maintained from the ground to protect against moisture, rust and corrosion of the furnace casing. This may be reduced to 2 inches if the unit is sealed against flooding. Floor furnaces are not permitted to be supported on the ground.

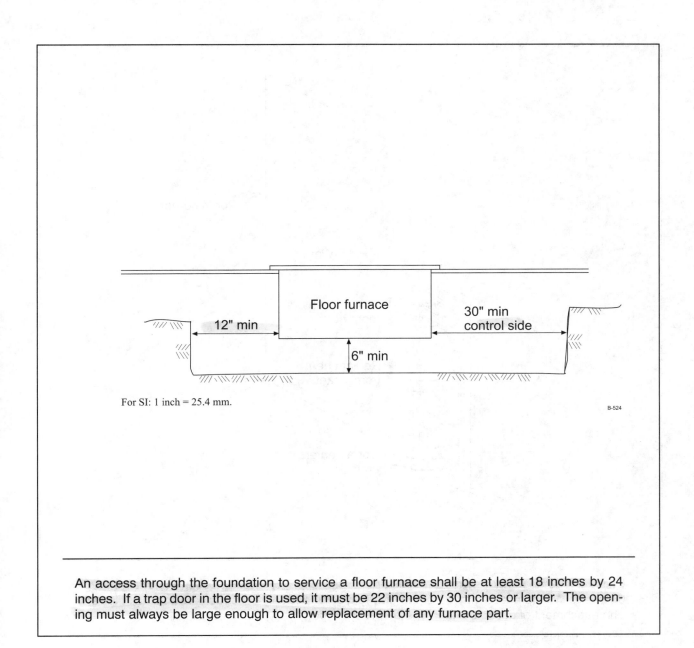

For SI: 1 inch = 25.4 mm.

B-524

An access through the foundation to service a floor furnace shall be at least 18 inches by 24 inches. If a trap door in the floor is used, it must be 22 inches by 30 inches or larger. The opening must always be large enough to allow replacement of any furnace part.

Topic: Location

Reference: IRC M1409.2

Category: Heating and Cooling Equipment

Subject: Vented Wall Furnaces

Code Text: *Vented wall furnaces shall be located so as not to cause a fire hazard to walls, floors, combustible furnishings or doors. Vented wall furnaces installed between bathrooms and adjoining rooms shall not circulate air from bathrooms to other parts of the building. Vented wall furnaces shall not be located where a door can swing within 12 inches (305 mm) of the furnace air inlet or outlet measured at right angles to the opening. Doorstops or door closers shall not be installed to obtain this clearance.*

Discussion and Commentary: Provided with the necessary internal clearances, wall furnaces are designed for installation in combustible stud cavities. To allow for necessary air movement and reduce the potential for the ignition of combustible materials, adequate clearances are mandated.

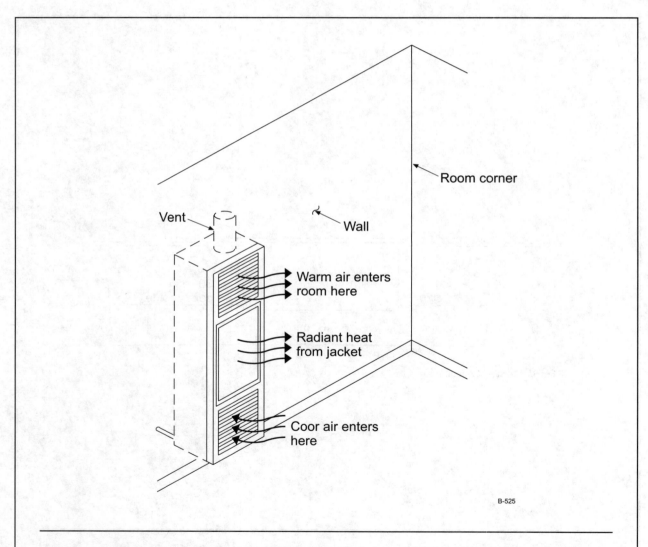

Unlike forced-air furnaces, wall furnaces are not designed for the distribution of hot air through ductwork. Wall furnaces may be of the gravity-type or possibly constructed with small blowers. The attachment of ducts to either type of unit could potentially cause overheating.

Topic: Condensate Disposal
Reference: IRC M1411.3

Category: Heating and Cooling Equipment
Subject: Refrigeration Cooling Equipment

Code Text: *Condensate from all cooling coils or evaporators shall be conveyed from the drain pan outlet to an approved place of disposal. Condensate shall not discharge into a street, alley or other areas so as to cause a nuisance. In addition to the requirements of Section M1411.3, a secondary drain or auxiliary drain pan shall be required for each cooling or evaporator coil where damage to any building components will occur as a result of overflow from the equipment drain pan or stoppage in the condensate drain piping. Drain piping shall be a minimum of $^3/_4$-inch (19.1 mm) nominal pipe size.*

Discussion and Commentary: The condensate created by cooling coils and evaporators can be a nuisance at best, a hazard at worst. Condensation can damage or ruin the appearance of building finish material, and over time undetected water damage can lead to possible structural failure.

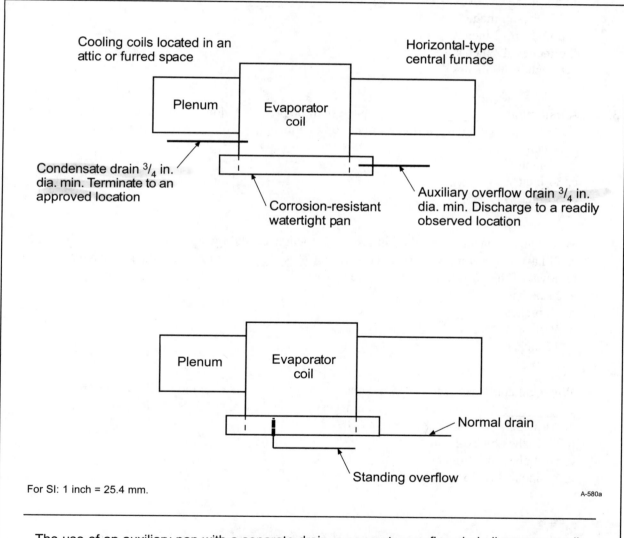

Cooling coils located in an attic or furred space

Horizontal-type central furnace

Plenum

Evaporator coil

Condensate drain $^3/_4$ in. dia. min. Terminate to an approved location

Corrosion-resistant watertight pan

Auxiliary overflow drain $^3/_4$ in. dia. min. Discharge to a readily observed location

Plenum

Evaporator coil

Normal drain

Standing overflow

For SI: 1 inch = 25.4 mm.

A-580a

The use of an auxiliary pan with a separate drain, a separate overflow drain line or an auxiliary pan without a separate drain line are three methods to address the concern created where the main condensate drain piping is blocked or the equipment drain pan overflows.

QUIZ

Study Session 10

1. In order for someone to service an appliance, a minimum of _____ of working space shall be provided in front of the control side of the appliance.
    a. 24 inches
    b. 30 inches
    c. 36 inches
    d. 48 inches

2. A central furnace located within a closet or similar compartment shall be provided with a minimum of _____ clearance along the sides, back and top of the furnace, with the total width of the enclosing space a minimum of _____ wider than the furnace.
    a. 3 inches, 6 inches
    b. 3 inches, 12 inches
    c. 6 inches, 12 inches
    d. 6 inches, 30 inches

3. A minimum _____ working space shall be provided along the front combustion chamber side of a furnace having a firebox open to the atmosphere.
    a. 6-inch
    b. 12-inch
    c. 24-inch
    d. 30-inch

4. Where an appliance is installed in a compartment accessed by a passageway, the passageway shall have a minimum unobstructed width of _____, but not less than necessary to allow removal of the largest appliance in the space.
    a. 24 inches
    b. 30 inches
    c. 36 inches
    d. 42 inches

5. Where an appliance is installed in an attic, the clear access opening shall be a minimum of _____, but in no case less than necessary to allow removal of the largest appliance.
    a. 20-inches-by-30-inches   M 1305.1.3
    b. 22-inches-by-30-inches
    c. 30-inches-by-30-inches
    d. 36-inches-by-36-inches

6. Appliances suspended from the floor shall have a minimum clearance of _____ from the ground.
   a. 3 inches
   b. 4 inches
   c. 6 inches
   d. 8 inches

7. Excavations for appliance installations shall extend a minimum depth of _____ below the appliance with a minimum clearance of _____ on all sides except the control side.
   a. 3 inches, 6 inches
   b. 6 inches, 6 inches
   c. 6 inches, 12 inches   M 1305.1.4.2.
   d. 12 inches, 12 inches

8. Where ventilated air space is used to reduce the required clearance between an appliance and unprotected combustible materials, what is the minimum air space that is typically required?
   a. $1/2$ inch
   b. 1 inch
   c. 2 inches
   d. 3 inches

9. An appliance requires a minimum clearance without protection of 12 inches above the appliance to combustible material. If $1/2$-inch-thick insulation board with a ventilated air space is used as protection, the minimum clearance may be reduced to _____.
   a. 4 inches
   b. 6 inches   Table M 1306.2
   c. 9 inches
   d. 12 inches

10. A solid-fuel-burning appliance requires a minimum clearance without protection of 18 inches from the sides and back to combustible material. If 0.024 sheet metal with a ventilated air space is used as protection, the minimum clearance may be reduced to _____.
    a. 4 inches
    b. 6 inches
    c. 9 inches
    d. 12 inches   M 1306.2.1

11. Where an appliance with an ignition source is located in the garage, the source of ignition shall be located a minimum of _____ above the garage floor.
    a. 12 inches
    b. 18 inches
    c. 36 inches
    d. 48 inches

12. A combustion air opening for a central furnace shall be unobstructed for a minimum distance of _____ in front of the opening.
   a. 3 inches
   b. 6 inches
   c. 12 inches
   d. 30 inches

13. The unobstructed total area of the outside and return air ducts or openings to a heat pump shall be a minimum of _____ square inches per 1,000 Btu/h output rating or as indicated by the conditions of the listing of the heat pump.
   a. 6
   b. 10
   c. 12.5
   d. 20

14. In order to permit the free drainage of defrost water, the supports or foundation for the outdoor unit of a heat pump shall be raised a minimum of _____ above the ground.
   a. 2 inches
   b. 3 inches
   c. 4 inches
   d. 6 inches

15. What is the maximum permitted operating temperature for a radiant heating system used on a gypsum assembly?
   a. 90°F
   b. 110°F
   c. 125°F
   d. 140°F

16. Unless listed and labeled for closer installation, a duct heater shall be located a minimum of _____ from a heat pump or air conditioner.
   a. 18 inches
   b. 24 inches
   c. 36 inches
   d. 48 inches

17. The furnace register of a vented floor furnace shall be located a minimum of _____ from doors in any position, draperies and other similar combustible objects.
   a. 3 inches
   b. 6 inches
   c. 12 inches
   d. 60 inches

18. The access opening through the foundation wall to a vented floor furnace shall have a minimum size of _____.
    a. 18-inches-by-24-inches
    b. 20-inches-by-30-inches
    c. 22-inches-by-30-inches
    d. 30-inches-by-30-inches

19. Where the lower 6 inches of a vented floor furnace is sealed to prevent water entry, a minimum _____ clearance shall be provided between the furnace and the ground.
    a. 2 inches
    b. 3 inches
    c. 4 inches
    d. 6 inches

20. A minimum clearance of _____ shall be provided between the inlet or outlet of a vented wall furnace and a door at any point in its swing, measured at a right angle to the opening.
    a. 6 inches
    b. 12 inches
    c. 18 inches
    d. 36 inches

21. Unless listed and labeled for such use, cooling coils of refrigeration cooling equipment shall not be located _____ of heat exchangers.
    a. within 12 inches
    b. within 36 inches
    c. upstream   M1411.2
    d. downstream

22. Drain piping for refrigeration cooling equipment shall be a minimum of _____ nominal pipe size.
    a. $1/2$ inch
    b. $3/4$ inch
    c. 1 inch
    d. $1^1/_2$ inch

23. Where an auxiliary drain pan with a separate drain is used for refrigeration cooling equipment, the minimum depth of the pan shall be _____.
    a. 1 inch
    b. $1^1/_2$ inches
    c. 2 inches
    d. 3 inches

24. Evaporative coolers shall be installed on a level platform or base located a minimum of _____ above the adjoining ground.
   a. 3 inches
   b. 4 inches
   c. 6 inches
   d. 12 inches

25. The supporting structure for a hearth extension for a fireplace stove shall be located _____ the supporting structure for the fireplace unit.
   a. at least 6 inches below
   b. at least 3 inches below
   c. at the same level as   M1414.2?
   d. at least 3 inches above

26. Where an appliance is located in an underfloor space, a minimum _____ level service space shall be provided at the front or service side of the appliance.
   a. 30-inch by 30-inch
   b. 36-inch by 30-inch
   c. 36-inch by 36-inch
   d. 42-inch by 36-inch

27. The required lower strapping for a water heater in Seismic Design Categories D_1 and D_2 shall be located a minimum of _____ inches above the controls.
   a. 3
   b. 4
   c. 6
   d. 8

28. A private garage containing a hydrogen generating appliance shall be limited to a maximum of _____ square feet in area.
   a. 720
   b. 800
   c. 850   M1307.4
   d. 1,000

29. Where piping other than cast-iron or galvanized steel is installed through bored holes of studs within a concealed wall space, steel shield plates are not required where the holes are a minimum of _____ inch(es) from the nearest edge of the stud.
   a. $3/4$
   b. $1 1/4$
   c. $1 3/8$
   d. $1 1/2$

30.   Unless listed for use on combustible floors without floor protection, a floor-mounted room
        heater shall be located so that the noncombustible flooring extends a minimum of _____ inches
        beyond the appliance on all sides.
        a. 12
        b. 18   m   1410.2
        c. 24
        d. 36

INTERNATIONAL RESIDENTIAL CODE
Study Session 11
Chapters 15, 16 and 19—Exhaust Systems, Duct Systems and Special Fuel-Burning Equipment

OBJECTIVE: To gain an understanding of the general provisions for the installation of exhaust systems, duct systems and special fuel-burning equipment, such as ranges, ovens and sauna heaters.

REFERENCE: Chapters 15, 16 and 19, 2003 *International Residential Code*

KEY POINTS:

- What are the installation requirements for clothes dryer exhaust systems? What is the maximum length of a flexible transition duct? Where is such a duct permitted?
- What is the maximum permitted length of a clothes dryer exhaust duct? How does the presence of one or more bends affect the allowable length?
- How is a range hood required to be ducted? What are the physical characteristics of the duct? What materials are permitted for duct construction?
- How must a domestic open-top broiler unit be ducted? What is the required clearance between the overhead exhaust hood and combustible construction? The minimum clearance between the cooking surface and the hood?
- What are the minimum exhaust rates for ventilating kitchens? Bathrooms?
- What is the maximum discharge air temperature permitted for equipment connected to an above-ground duct system? The maximum temperature when gypsum products are used? The maximum flame spread?
- Under what conditions can stud cavities be used as return air ducts?
- What are the limitations on the installation of an underground duct system? What is the maximum duct temperature permitted for underground ducts? What is the minimum size of the access opening to an under-floor plenum?
- What is the maximum permitted flame spread for duct insulation coverings and linings? The maximum permitted smoke-developed index?
- How should external duct insulation be identified? At what intervals must the identification be provided?
- What is the maximum permitted length of vibration isolators?
- What means of support are addressed for metal ducts? Nonmetallic ducts?
- What is the minimum area of return air ducts serving a warm-air furnace? Serving a central air conditioning unit and/or heat pump?
- Which sources of outside or return air are prohibited?
- What is the minimum area of supply air ducts serving a warm-air furnace? Serving a central air conditioning unit and/or heat pump?
- What is the minimum vertical clearance required between the cook-top of a range and any unprotected combustible material? Under what conditions may the clearance be reduced?
- How must sauna heaters be installed and protected from contact? What is the temperature limit for the heater's thermostat?

Topic: Length Limitation
Reference: IRC M1501.3

Category: Exhaust Systems
Subject: Clothes Dryers Exhaust

Code Text: *The maximum length of a clothes dryer exhaust duct shall not exceed 25 feet (7620 mm) from the dryer location to the wall or roof termination. The maximum length of the duct shall be reduced 2.5 feet (762 mm) for each 45-degree (0.79 rad) bend and 5 feet (1524 mm) for each 90-degree (1.6 rad) bend. The maximum length of the exhaust duct does not include the transition duct.*

Discussion and Commentary: Clothes dryer exhausts present a different problem than other exhaust systems because the exhaust air is laden with moisture and lint. The air should be vented to the outside and not discharged into an attic or crawl space because wood structural members could be adversely affected. Exhaust outlets must be equipped with a backdraft damper to prevent cold air, rain, snow, rodents and vermin from entering the vent.

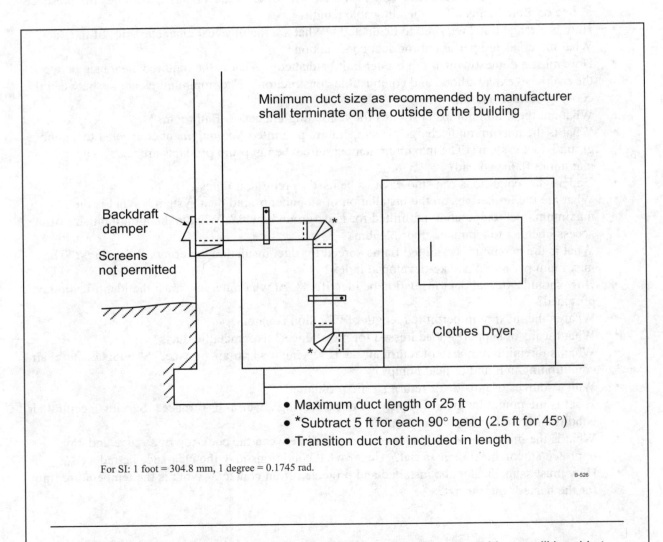

Minimum duct size as recommended by manufacturer shall terminate on the outside of the building

Backdraft damper

Screens not permitted

Clothes Dryer

- Maximum duct length of 25 ft
- *Subtract 5 ft for each 90° bend (2.5 ft for 45°)
- Transition duct not included in length

For SI: 1 foot = 304.8 mm, 1 degree = 0.1745 rad.

B-526

The length restrictions placed on the vent ensure that the dryer exhaust blower will be able to drive sufficient air volume to carry off the moisture. Only when the make and model of the dryer is known, or where a listed booster fan is installed, can the length be increased in conformance with the manufacturer's instructions.

2003 IRC Study Companion

Topic: Domestic Open-top Broiler Units
Reference: IRC M1504.1

Category: Exhaust Systems
Subject: Overhead Exhaust Hoods

Code Text: *Domestic open-top broiler units shall be provided with a metal exhaust hood, not less than 28 gage, with a clearance of not less than 0.25 inch (6.4 mm) between the hood and the underside of combustible materials or cabinets. A clearance of at least 24 inches (610 mm) shall be maintained between the cooking surface and the combustible material or cabinet.*

Discussion and Commentary: A minimal clearance is needed between the hood and any cabinets or similar items to ensure that the surface material of the hood does not come in direct contact with the combustible material. The vertical clearance protects overhead cabinets from ignition should a fire initiate on the cooktop. It is important that the length of the hood be at least that of the boiler unit to allow for complete coverage.

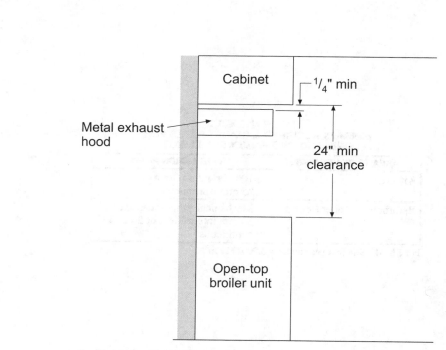

For SI: 1 inch = 25.4 mm.

B-527

A boiler unit exhaust hood must discharge to the exterior and be provided with a backdraft damper. A range hood must also exhaust directly to the outdoors unless a listed and labeled ductless range hood is installed in accordance with the manufacturer's instructions.

Topic: Kitchens and Bathrooms

Category: Exhaust Systems

Reference: IRC M1506

Subject: Mechanical Ventilation

Code Text: *Where toilet rooms and bathrooms are mechanically ventilated, the ventilation equipment shall be installed in accordance with* Section M1506. *Exhaust air from bathrooms and toilet rooms shall not be recirculated within a residence or to another dwelling unit. Ventilation systems shall be designed to have the capacity to exhaust the minimum air flow rate determined in accordance with Table 1506.3.*

Discussion and Commentary: The mechanical ventilation rates for bathrooms and water closet compartments are also set forth in Section R303.3 as an exception to the general requirement for natural ventilation. The fundamental provisions require such spaces to be provided with a minimum of $1^1/_2$ square feet of exterior openings.

TABLE M1506.3
MINIMUM REQUIRED EXHAUST RATES FOR
ONE- AND TWO-FAMILY DWELLINGS

AREA TO BE VENTILATED	VENTILATION RATES
Kitchens	100 cfm intermittent or 25 cfm continuous
Bathrooms—Toilet Rooms	Mechanical exhaust capacity of 50 cfm intermittent or 20 cfm continuous

For SI: 1 cubic foot per minute = 0.0004719 m³/s.

Kitchen ventilation can be addressed by two different methods, as is the case for bathroom ventilation. Where continuous kitchen ventilation is provided, a minimum rate of 25 cfm is required. If an intermittent exhaust system is utilized, it must be rated for at least 100 cfm.

Topic: Above-ground Duct Systems
Reference: IRC M1601.1.1

Category: Duct Systems
Subject: Duct Construction

Code Text: *Equipment connected to duct systems shall be designed to limit discharge air temperature to a maximum of 250°F (121°C). Factory-made air ducts shall be constructed of Class 0 or Class 1 materials as designated in Table M1601.1.1(1). Minimum thicknesses of metal duct material shall be as listed in Table M1601.1.1(2).*

Discussion and Commentary: The temperature limit for discharge air is based on the possible use of non-metallic ducts. Higher air temperatures could cause their rapid deterioration. The flame spread is strictly limited to 25 for Class 1 ducts and 0 for Class 0 ducts. Where metal ducts are used, they shall be of substantial construction to ensure that their structural integrity is maintained. Ducts fabricated from lighter gage materials will experience an unacceptable amount of distortion in use.

TABLE M1601.1.1(1)
CLASSIFICATION OF FACTORY-MADE AIR DUCTS

DUCT CLASS	MAXIMUM FLAME-SPREAD RATING
0	0
1	25

TABLE M1601.1.1(2)
GAGES OF METAL DUCTS AND PLENUMS USED FOR HEATING OR COOLING

TYPE OF DUCT	SIZE (inches)	MINIMUM THICKNESS (inch)	EQUIVALENT GALVANIZED SHEET GAGE	APPROXIMATE ALUMINUM B & S GAGE
Round ducts and enclosed rectangular ducts	14 or less over 14	0.013 0.016	30 28	26 24
Exposed rectangular ducts	14 or less over 14	0.016 0.019	28 26	24 22

For SI: 1 inch = 25.4 mm.

Gypsum products may be used in the fabrication of return air ducts or plenums, provided the air temperature is limited and condensation on the duct surfaces is not a concern. Excessive temperatures and high humidity are both potential causes of damage to the gypsum materials.

Topic: Above-ground Duct Systems **Category:** Duct Systems
Reference: IRC M1601.1.1, #7 **Subject:** Duct Construction

Code Text: *Stud wall cavities and the spaces between solid floor joists to be utilized as air plenums shall comply with the following conditions: 7.1) such cavities or spaces shall not be utilized as a plenum for supply air; 7.2) such cavities or spaces shall not be part of a required fire-resistance-rated assembly; 7.3) stud wall cavities shall not convey air from more than one floor level; and 7.4) stud wall cavities and joist space plenums shall be isolated from adjacent concealed spaces by tight-fitting fire blocking in accordance with Section R602.8.*

Discussion and Commentary: The use of stud spaces and panned joists space as return air ducts is acceptable since the air temperature is such not to cause a hazard or deterioration of the construction materials. Isolation of the return air plenum from other unused spaces is necessary so that debris will not be drawn into the furnace.

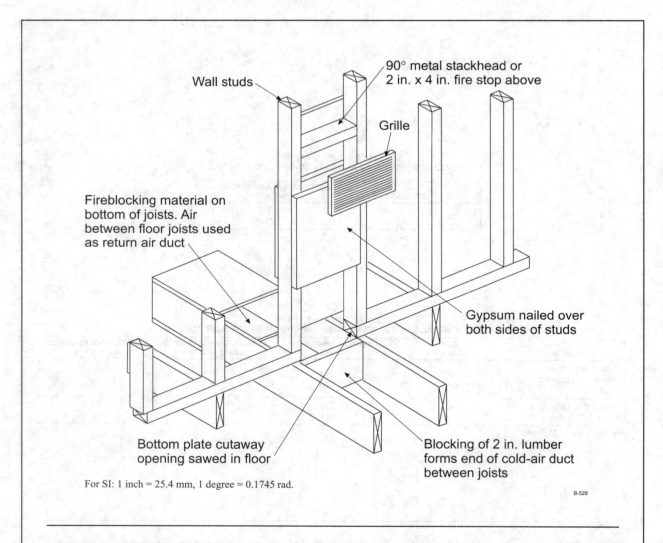

Wall studs

90° metal stackhead or
2 in. x 4 in. fire stop above

Grille

Fireblocking material on
bottom of joists. Air
between floor joists used
as return air duct

Gypsum nailed over
both sides of studs

Bottom plate cutaway
opening sawed in floor

Blocking of 2 in. lumber
forms end of cold-air duct
between joists

For SI: 1 inch = 25.4 mm, 1 degree = 0.1745 rad.

B-528

Often, walls or floor-ceiling assemblies such as dwelling unit separations or exterior walls adjacent to property lines must be of fire-resistive-rated construction. Such assemblies must not be used as air plenums.

Topic: Underground Duct Systems
Reference: IRC M1601.1.2

Category: Duct Systems
Subject: Duct Construction

Code Text: *Underground duct systems shall be constructed of approved concrete, clay, metal or plastic. The maximum duct temperature for plastic ducts shall not be greater than 150°F (66°C). Metal ducts shall be protected from corrosion in an approved manner or shall be completely encased in concrete not less than 2 inches (51 mm) in thickness. Nonmetallic ducts shall be installed in accordance with the manufacturer's installation instructions.*

Discussion and Commentary: Ducts fabricated from the materials listed above should not collapse when installed underground or within concrete. Ducts installed in or beneath concrete should be properly sealed and secured before the concrete pour. Otherwise, leakage of concrete into the ducts could block the flow of air when the system is placed in operation. For the same reason, underground ducts should be secured to avoid damage during backfilling.

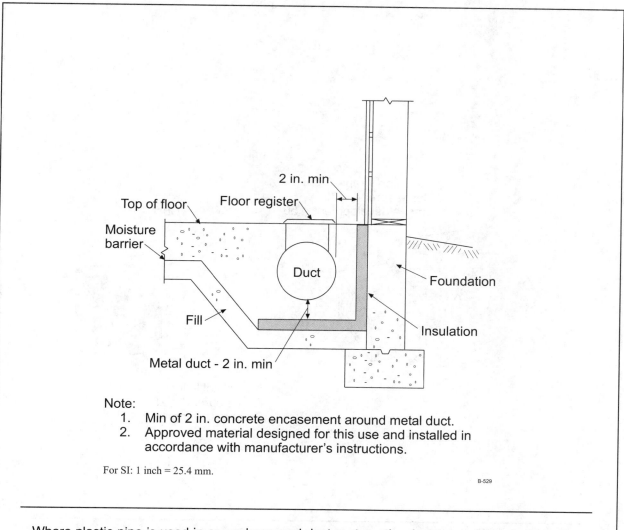

Note:
1. Min of 2 in. concrete encasement around metal duct.
2. Approved material designed for this use and installed in accordance with manufacturer's instructions.

For SI: 1 inch = 25.4 mm.

B-529

Where plastic pipe is used in an underground duct system, the duct temperature is limited to avoid the possibility of deterioration or distortion of the plastic materials. To ensure the adequacy of plastic ducts, specific material and structural criteria are referenced in several ASTM standards.

Topic: Duct Insulation Materials   **Category:** Duct Systems
Reference: IRC M1601.2.1   **Subject:** Duct Construction

Code Text: *Duct coverings and linings shall have a flame-spread index not greater than 25, and a smoke-developed index not greater than 50 when tested in accordance with ASTM E 84. External duct insulation and factory-insulated flexible ducts shall be legibly printed or identified at intervals not greater than 36 inches (914 mm) with the name of the manufacturer; the thermal resistance R-value at the specified installed thickness; and the flame spread and smoke-developed indexes of the composite materials.*

Discussion and Commentary: Ducts that pass through nonconditioned areas should be insulated to prevent excessive heat loss or heat gain in cooling installations. Insulation also prevents the cool surfaces of air-conditioning ducts from condensing moisture from the surrounding unconditioned air.

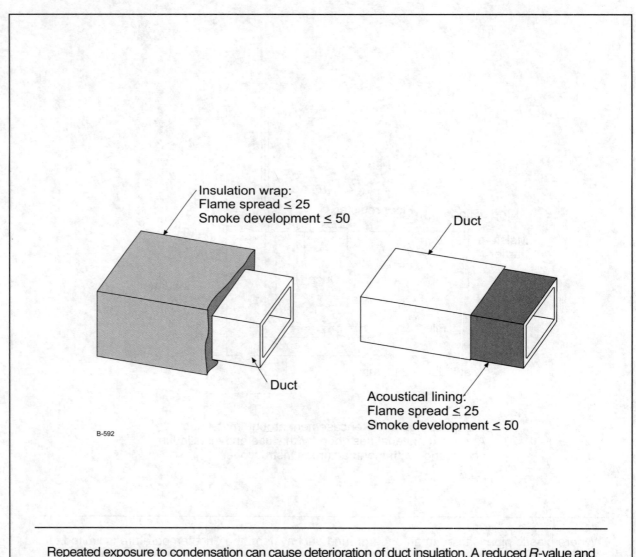

Repeated exposure to condensation can cause deterioration of duct insulation. A reduced *R*-value and duct failure are both possible. Therefore, vapor retarders are required over insulation on cooling supply ducts that pass through nonconditioned areas where condensation may occur.

Topic Installation
Reference: IRC M1601.3

Category: Duct Systems
Subject: Duct Construction

Code Text: *Joints of duct systems shall be made substantially airtight by means of tapes, mastics, gasketing or other approved closure systems. Duct connections to flanges of air distribution system equipment or sheet metal fittings shall be mechanically fastened. Crimp joints for round ducts shall have a contact lap of at least 1.5 inches (38 mm) and shall be mechanically fastened by means of at least three sheet metal screws or rivets equally spaced around the joint. Metal ducts shall be supported by 0.5-inch (12.7 mm) wide 18-gage metal straps or 12-gage galvanized wire at intervals not exceeding 10 feet (3048 mm) or other approved means.*

Discussion and Commentary: Duct leakage should be minimized to prevent energy loss. Adequate duct support is necessary to maintain alignment and limit deflection.

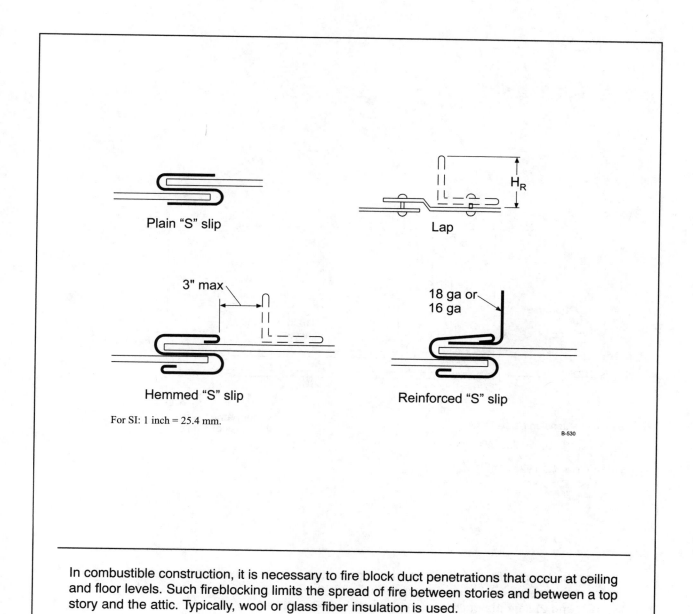

Plain "S" slip

Lap

H_R

3" max

Hemmed "S" slip

18 ga or 16 ga

Reinforced "S" slip

For SI: 1 inch = 25.4 mm.

B-530

In combustible construction, it is necessary to fire block duct penetrations that occur at ceiling and floor levels. Such fireblocking limits the spread of fire between stories and between a top story and the attic. Typically, wool or glass fiber insulation is used.

Topic: Under-floor Plenums
Reference: IRC M1601.4

Category: Duct Systems
Subject: Duct Construction

Code Text: *Fuel gas lines and plumbing waste cleanouts shall not be located within an under-floor space used as a supply plenum. The space shall be cleaned of loose combustible materials and scrap, and shall be tightly enclosed. The ground surface of the space shall be covered with a moisture barrier having a minimum thickness of 4 mils (0.102 mm). The under-floor space, including the sidewall insulation, shall be formed by materials having flame-spread ratings not greater than 200 when tested in accordance with ASTM E 84.*

Discussion and Commentary: A counterflow furnace is typically used to discharge conditioned air to the underfloor plenum. Floor registers are installed in openings cut through the floor for delivery of the conditioned air to the space above. Access to the plenum area must be provided by a minimum 18-inch by 24-inch (457 mm by 610 mm) opening.

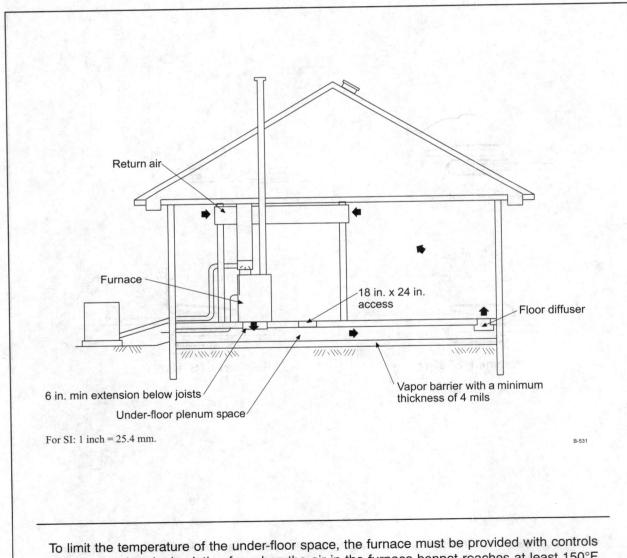

Return air

Furnace

18 in. x 24 in. access

Floor diffuser

6 in. min extension below joists

Under-floor plenum space

Vapor barrier with a minimum thickness of 4 mils

For SI: 1 inch = 25.4 mm.

B-531

To limit the temperature of the under-floor space, the furnace must be provided with controls that 1) start the air-circulating fan when the air in the furnace bonnet reaches at least 150°F (66°C), and 2) limit the outlet air temperature to 200°F (93°C).

Topic: Required Area
Reference: IRC M1602

Category: Duct Systems
Subject: Return Air

Code Text: *Return air shall be taken from inside the dwelling. Dilution of return air with outdoor air shall not be prohibited.*

Discussion and Commentary: Recirculation of air contaminated with objectionable odors, fumes or flammable vapors is prohibited because of potential health and safety hazards. Therefore, return air ducts must not serve kitchens, bathrooms, closets, attached garages or other dwelling units. Also, outside air intakes must be located carefully to preclude objectionable or contaminated air from being drawn into the air distribution system. Air intakes above the roof must be located a minimum of 10 feet laterally from a plumbing vent or fuel-burning appliance vent, unless the source of noxious fumes is more than 3 feet higher than the outside air intake. In that case, a lesser lateral separation is acceptable.

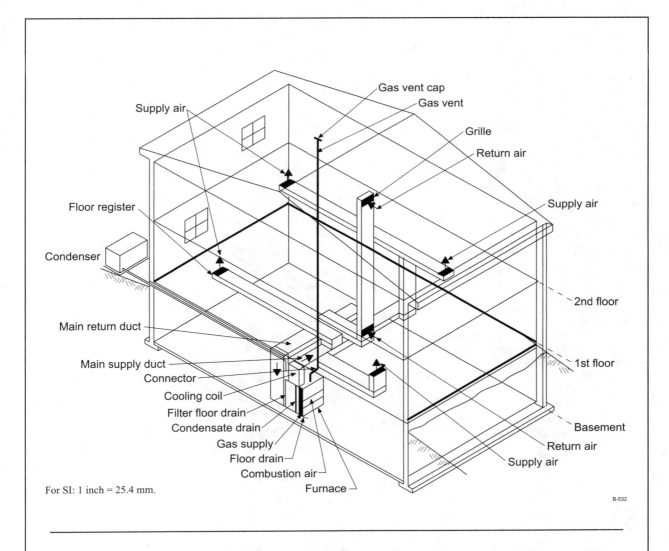

For SI: 1 inch = 25.4 mm.

B-532

To prevent the intrusion of animals and insects, it is necessary to provide a mesh screen having openings no greater than 1/2 inch. However, in order to maintain the necessary openings free and clear from obstruction, a minimum dimension of 1/4 inch is mandated.

Topic: Clearances

Category: Special Fuel-burning Equipment

Reference: IRC M1901.1

Subject: Ranges and Ovens

Code Text: *Freestanding or built-in ranges shall have a vertical clearance above the cooking top of not less than 30 inches (762 mm) to unprotected combustible material. Reduced clearances are permitted in accordance with the listing and labeling of the range hoods or appliances.*

Discussion and Commentary: Vertical clearance to combustibles is important because cooking range tops produce considerable heat. Additionally, overheated utensils, particulary those containing grease, may be the source of flame. An adequate clear space above the range also ensures that hoods, shelves and other potential obstructions do not encroach into the working space needed for cooking operations. It is also important that the installation of any household cooking appliance not interfere with combustion air.

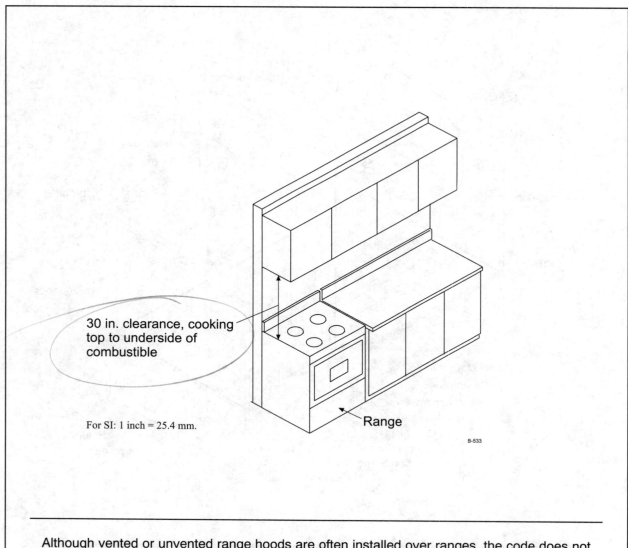

30 in. clearance, cooking top to underside of combustible

For SI: 1 inch = 25.4 mm.

Range

B-533

Although vented or unvented range hoods are often installed over ranges, the code does not require hoods to be installed. Where installed however, the hoods and exhaust methods must be in compliance with the code and the manufacturer's installation instructions.

Topic: Locations and protection; Controls
Reference: IRC M1902.1, M1902.4

Category: Special Fuel-burning Equipment
Subject: Sauna Heaters

Code Text: *Sauna heaters shall be protected from accidental contact by persons with a guard of material having a low thermal conductivity, such as wood. The guard shall have no substantial effect on the transfer of heat from the heater to the room. Sauna heaters shall be equipped with a thermostat that will limit room temperature to not greater than 194°F (90°C). Where the thermostat is not an integral part of the heater, the heat-sensing element shall be located within 6 inches (152 mm) of the ceiling.*

Discussion and Commentary: In a small space such as a sauna room, it is quite possible that contact with the heating element will occur unless the element is adequately guarded. The guard must be of a type that allows the heated air to circulate throughout the room, as concentrated heat and a combustible guard could create a fire hazard.

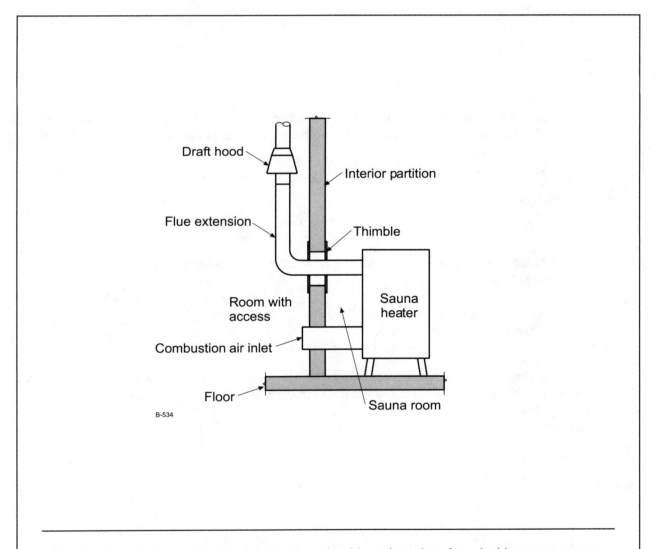

B-534

Combustion air for gas-fired sauna heaters should not be taken from inside sauna rooms, which are normally small, confined spaces that do not contain an adequate volume of combustion air. Combustion air and venting shall be in compliance with Chapters 17 and 18.

QUIZ

Study Session 11

1. Exhaust ducts for clothes dryers shall not _____.
   a. extend to the exterior of the dwelling
   b. be connected with sheet-metal screws
   c. be equipped with a backdraft damper
   d. be connected to flexible transition ducts

2. Unless modified by the manufacturer's installation instructions, what is the maximum permitted length of a clothes dryer exhaust duct when measured from the dryer location to the wall or roof termination?
   a. 18 feet
   b. 20 feet
   c. 25 feet
   d. 30 feet

3. Unless modified by the manufacturer's installation instructions or provided with a booster fan, what is the maximum permitted length of a clothes dryer exhaust duct that has one 90-degree bend and two 45-degree bends?
   a. 25 feet
   b. 20 feet
   c. $17^1/_2$ feet
   d. 15 feet

4. Which of the following requirements is not applicable to a duct serving a range hood?
   a. It shall have a smooth surface.
   b. It shall be air tight.
   c. It shall be equipped with a backdraft damper.
   d. It may terminate in a crawl space.

5. When serving a domestic kitchen cooking appliance equipped with a down draft exhaust system, a schedule 40 PVC exhaust duct shall extend a maximum of _____ above grade outside the building.
   a. 1 inch
   b. 6 inches
   c. 12 inches
   d. 36 inches

6. A domestic open-top broiler unit shall be provided with a minimum clearance of _____ between the hood and the underside of combustible material or cabinets.
   a. 0 inches; no clearance is required
   b. 1/4 inch
   c. 1/2 inch
   d. 1 inch

7. For a domestic open-top boiler unit, a minimum of _____ shall be maintained between the cooking surface and combustible material or the cabinet.
   a. 24 inches
   b. 28 inches
   c. 30 inches
   d. 32 inches

8. Heating equipment connected to an above-ground duct system shall be designed to limit discharge air temperature to a maximum of _____.
   a. 125°F
   b. 200°F
   c. 225°F
   d. 250°F

9. When serving heating, cooling or ventilation equipment, above-ground factory-made duct systems shall be constructed of materials classified as _____.
   a. Class 0 only
   b. Class 1 only
   c. Class 0 or Class 1
   d. Class 1 or Class 2

10. A 16-inch rectangular duct in an exposed location shall have a minimum equivalent galvanized sheet gage of _____.
    a. No. 22
    b. No. 26
    c. No. 28
    d. No. 30

11. Gypsum board may be used in the construction of a return air plenum, provided the maximum air temperature is _____ and the exposed surfaces are not subject to condensation.
    a. 125°F
    b. 150°F
    c. 180°F
    d. 210°F

12. What is the maximum flame spread index permitted for materials used to construct above-ground duct systems?
   a. 25
   b. 75
   c. 200
   d. 450

13. Where used in an underground duct system, the maximum permitted duct temperature for plastic ducts shall be _____.
   a. 125°F
   b. 150°F
   c. 180°F
   d. 210°F

14. When protected by concrete, metal ducts installed in an underground duct system shall be encased with a minimum thickness of _____.
   a. 1 inch
   b. 2 inches
   c. 3 inches
   d. 4 inches

15. Duct coverings and linings shall have a maximum flame spread index of _____ and a maximum smoke-developed index of _____.
   a. 25, 50
   b. 25, 200
   c. 50, 200
   d. 50, 450

16. Factory-insulated flexible ducts shall be identified at maximum intervals of _____ with the name of the manufacturer, the thermal resistance R-value, the flame spread index and the smoke-developed index.
   a. 3 feet
   b. 5 feet
   c. 6 feet
   d. 10 feet

17. Where vibration isolators are installed between mechanical equipment and metal ducts, they shall have a maximum length of _____.
   a. 6 inches
   b. 10 inches
   c. 12 inches
   d. 24 inches

18. Where metal ducts are supported by 0.5-inch metal straps, the straps shall be minimum _____ and located a maximum of _____ on center.
    a. 12-gage, 6 feet
    b. 18-gage, 6 feet
    c. 18-gage, 10 feet
    d. 22-gage, 10 feet

19. The under-floor space used as a supply plenum shall be formed by materials having a maximum flame-spread rating of _____.
    a. 25
    b. 50
    c. 75
    d. 200

20. Where the under-floor space is used as a supply plenum, a duct shall extend from the furnace supply outlet to a minimum of _____ inch(es) below the combustible framing.
    a. 1
    b. 3
    c. 6   M1601.4.3
    d. 8

21. Unless the discharge outlet is a minimum of _____ above the outside air inlet, outside air for a forced-air heating system shall be taken a minimum of _____ from an appliance vent outlet, a vent opening from a plumbing drainage system, or the discharge outlet of an exhaust fan.
    a. 2 feet, 6 feet
    b. 2 feet, 10 feet
    c. 3 feet, 6 feet
    d. 3 feet, 10 feet   1602.2 #1

22. Which one of the following spaces is acceptable as a source for return air for a forced-air heating system?
    a. sleeping room
    b. bathroom
    c. kitchen
    d. closet

23. Outdoor air inlets for a return air system shall be covered with screens having a minimum opening size of _____ and a maximum opening size of _____.
    a. 1/4 inch, 1/2 inch   M1602.3
    b. 1/4 inch, 3/4 inch
    c. 3/8 inch, 1/2 inch
    d. 3/8 inch, 3/4 inch

24. Built-in or freestanding ranges shall have a minimum vertical clearance of _____ above the cooking top to an unprotected combustible construction.
    a. 24 inches
    b. 30 inches *M1901.1*
    c. 32 inches
    d. 36 inches

25. Sauna heaters shall be equipped with a thermostat that will limit room temperature to a maximum of _____.
    a. 120°F
    b. 140°F
    c. 172°F
    d. 194°F

26. A mechanically-ventilated kitchen shall be provided with a minimum ventilation rate of _____ cfm where continuous ventilation is utilized.
    a. 20
    b. 25 *Table M 1506.3*
    c. 50
    d. 100

27. A bathroom ventilated by an intermittent mechanical means shall have a minimum mechanical exhaust capacity of _____ cfm.
    a. 20
    b. 25
    c. 50 *M1506.3*
    d. 100

28. The access through the floor to an under-floor plenum shall provide a minimum opening size of _____.
    a. 18 inches by 24 inches *M1601.4.4*
    b. 20 inches by 24 inches
    c. 22 inches by 30 inches
    d. 24 inches by 30 inches

29. Where the under-floor space is used as a supply plenum, the furnace shall be equipped with an approved automatic control that limits the outlet air temperature to a maximum of _____ F.
    a. 150 degrees
    b. 200 degrees *M1601.4.5*
    c. 225 degrees
    d. 250 degrees

30. Where the thermostat used to limit sauna room temperature is not an integral part of the sauna heater, the heat-sensing element shall be located a maximum of _____ inches below the ceiling.
    a. 3
    b. 4
    c. 6
    d. 8

INTERNATIONAL RESIDENTIAL CODE
Study Session 12
Chapters 17, 18, 20, 21, 22 and 23—Combustion Air, Chimneys and Vents, Boilers/Water Heaters, Hydronic Piping, Special Piping and Storage Systems, and Solar Systems

OBJECTIVE: To provide an understanding of the general provisions for combustion air, chimneys, vents, boilers, water heaters, hydronic piping, special piping and storage systems, and solar systems.

REFERENCE: Chapters 17, 18, 20, 21, 22 and 23, 2003 *International Residential Code*

KEY POINTS:

- How shall combustion air be obtained in buildings of unusually tight construction? Under what condition may combustion air be taken from inside a building through infiltration?
- Where are volume dampers prohibited?
- Which rooms in a dwelling unit are prohibited as sources for combustion air?
- What is considered the net free area of a metal louver when not specified by the manufacturer? A wood louver?
- What is considered a confined space? How many combustion air openings are required in such a space? Where are they to be located? What is the minimum area of each opening when communicating with the outdoors by vertical ducts? Horizontal ducts?
- What is the minimum allowable cross-sectional dimension of a rectangular air duct?
- Under what conditions may combustion air be taken from the attic? From an under-floor area?
- What are the minimum and maximum opening sizes permitted for screens used to cover outside combustion air openings?
- What venting method is to be used where two or more listed appliances are connected to a common natural draft venting system?
- What is the minimum thickness for a single-wall metal pipe connector?
- How far above the highest connected appliance outlet must a natural draft appliance vent terminate? How far above the bottom of a wall vent must a natural draft gas vent terminate?
- Where must a chimney connector enter a masonry chimney? How is the size of a chimney flue that is connected to more than one appliance calculated?
- Where are valves required for a boiler? What type of gauges are mandated? What is a boiler low-water cutoff?
- What requirements apply to expansion tanks used with hot water boilers?
- Where is the installation of fuel-fired water heaters prohibited? In what manner can access be provided to a water heater in an under-floor or attic area?
- What are the general provisions related to hydronic piping systems? How must floor heating systems be tested?
- How are special piping and storage systems to be installed?
- What special conditions apply to the installation of a solar energy system?

Topic: Prohibited Sources   **Category:** Combustion Air
Reference: IRC M1701.4   **Subject:** General Provisions

Code Text: *Combustion air ducts and openings shall not connect appliance enclosures with space in which the operation of a fan may adversely affect the flow of combustion air. Combustion air shall not be obtained from an area in which flammable vapors present a hazard. Fuel-fired appliances shall not obtain combustion air from any of the following rooms or spaces: 1) sleeping rooms, 2) bathrooms, or 3) toilet rooms.* See two exceptions.

Discussion and Commentary: Exhaust fans, as just one example, can produce negative pressures in a space, thereby reducing the availability of combustion air. Air that could have served as combustion air could be consumed instead as makeup air for an exhaust system. Openings to the outdoors and infiltration might not be able to provide the required combustion air if also providing makeup air for exhaust fans or appliances such as clothes dryers.

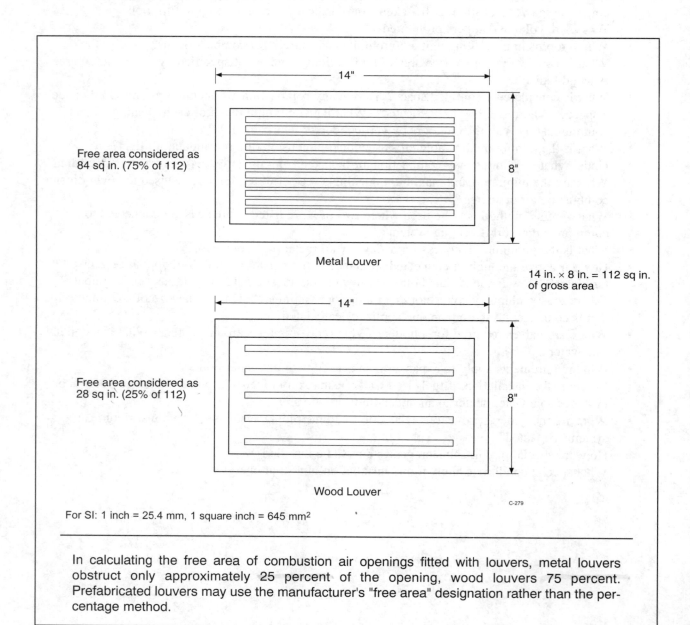

For SI: 1 inch = 25.4 mm, 1 square inch = 645 mm²

In calculating the free area of combustion air openings fitted with louvers, metal louvers obstruct only approximately 25 percent of the opening, wood louvers 75 percent. Prefabricated louvers may use the manufacturer's "free area" designation rather than the percentage method.

Topic: Confined Space **Category:** Combustion Air
Reference: IRC M1702.2 **Subject:** All Air From Inside the Building

Code Text: *Where the space in which the appliance is located does not meet the criteria specified in Section M1702.1 (unconfined space), two permanent openings to adjacent spaces shall be provided so that the combined volume of all spaces meets the criterion. One opening shall be within 12 inches (305 mm) of the top and one within 12 inches (305 mm) of the bottom of the space. Each opening shall have a free area equal to a minimum of 1 square inch per 1,000 Btu/h (2.20 mm²/W) input rating of all appliances installed within the space, but not less than 100 square inches (0.064 m²).*

Discussion and Commentary: If the volume of the space in which the fuel-burning equipment is installed is greater than 50 cubic feet per 1,000 Btu/h (4.83L/W) of the aggregate input rating of the equipment, then infiltration of air into a residence normally provides sufficient air for combustion, ventilation and draft hood dilution.

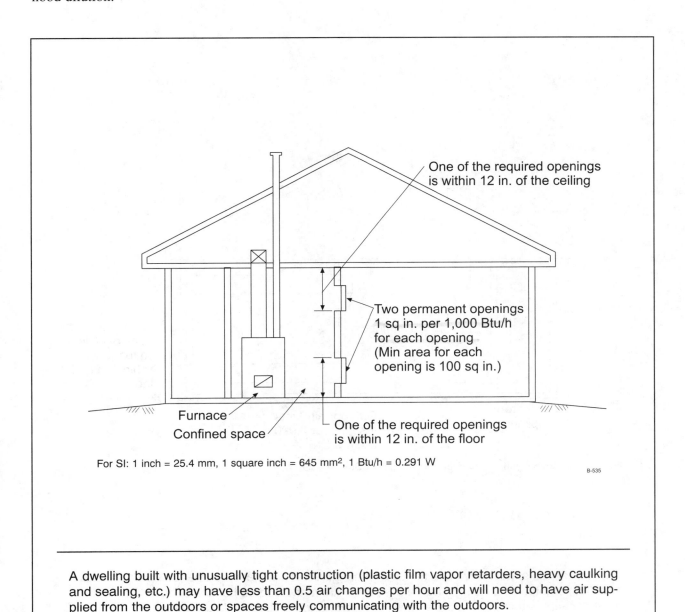

One of the required openings is within 12 in. of the ceiling

Two permanent openings 1 sq in. per 1,000 Btu/h for each opening (Min area for each opening is 100 sq in.)

Furnace
Confined space

One of the required openings is within 12 in. of the floor

For SI: 1 inch = 25.4 mm, 1 square inch = 645 mm², 1 Btu/h = 0.291 W

B-535

A dwelling built with unusually tight construction (plastic film vapor retarders, heavy caulking and sealing, etc.) may have less than 0.5 air changes per hour and will need to have air supplied from the outdoors or spaces freely communicating with the outdoors.

Topic: Two Openings or Ducts **Category:** Combustion Air
Reference: IRC M1703.1, M1703.2 **Subject:** All Air From Outdoors

Code Text: *Where the space in which fuel-burning appliances are located does not meet the criterion for indoor air specified in Section M1702, outside combustion air shall be provided through openings or ducts. One opening shall be within 12 inches (305 mm) of the top of the enclosure, and one within 12 inches (305 mm) of the bottom of the enclosure. Where communicating with the outdoors by means of vertical ducts, each opening shall have a free area of at least 1 square inch per 4,000 Btu/h (0.55 mm²/W) input rating of all appliances in the space.*

Discussion and Commentary: With two openings located as required, the warm air rises and is vented to the outdoors through the upper duct. The lower duct draws in colder, heavier air from the outdoors, which is heated by the appliances and rises toward the ceiling. Thus, the two openings create a natural draft that provides a constant source of combustion air.

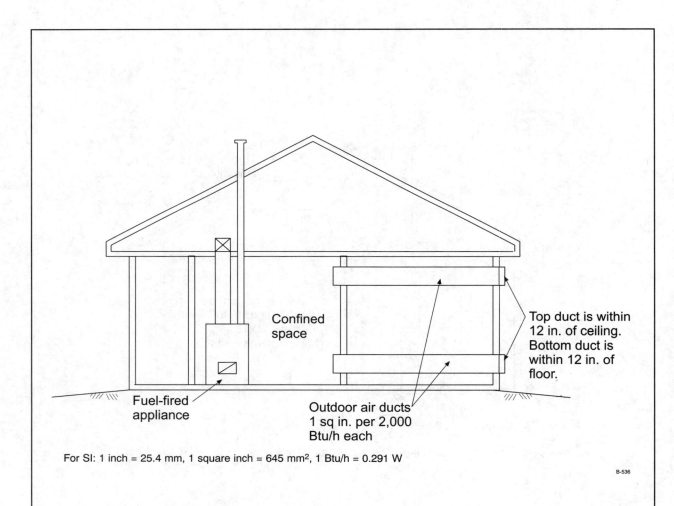

Confined space

Top duct is within 12 in. of ceiling. Bottom duct is within 12 in. of floor.

Fuel-fired appliance

Outdoor air ducts 1 sq in. per 2,000 Btu/h each

For SI: 1 inch = 25.4 mm, 1 square inch = 645 mm², 1 Btu/h = 0.291 W

B-536

Horizontal ducts must have a minimum opening area of 1 square inch per 2,000 Btu/h (0.275 mm²/W) of total input ratings. Larger openings are mandated on horizontal ducts than on vertical ducts because the draft is somewhat less and the resistance is greater.

Topic: Attic Combustion Air
Reference: IRC M1703.3

Category: Combustion Air
Subject: All Air From Outdoors

Code Text: For *combustion air taken from an attic area, 1) the attic ventilation shall be sufficient to provide the required volume of combustion air; 2) the combustion air opening shall be provided with a metal sleeve extending from the appliance enclosure to at least 6 inches (152 mm) above the top of the ceiling joists and ceiling insulation; 3) an inlet air duct within an outlet air duct shall be an acceptable means of supplying attic combustion air to an appliance room provided that the inlet duct extends at least 12 inches (305 mm) above the top of the outlet duct in the attic space; and 4) the end of ducts that terminate in an attic shall not be screened.*

Discussion and Commentary: A primary concern when taking combustion air from an attic space is that insulation is usually present. The minimum 6-inch extension of the metal sleeve should ensure against insulation from entering the sleeve, while the prohibition against screening eliminates the potential for the screen to be blocked by loose insulation.

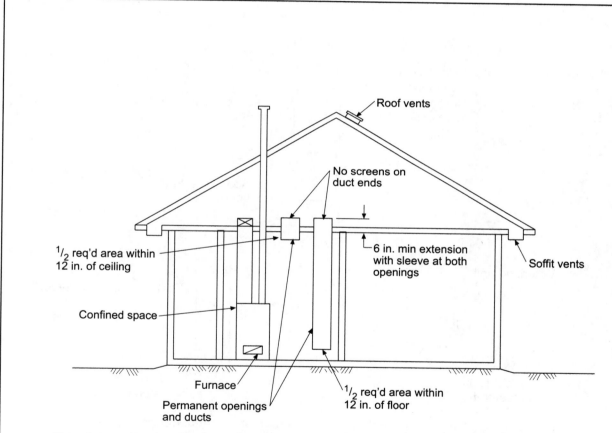

Note: Attic should be sufficiently vented for combustion air to be taken from attic.

For SI: 1 inch = 25.4 mm.

B-538

Where a ventilated crawl space is to be used to obtain combustion air, the free area of the openings between the crawl space and the outdoors must be at least twice the area required for the combustion air openings.

Topic: Vent Dampers **Category:** Chimneys and Vents
Reference: IRC M1802.2 **Subject:** Vent Components

Code Text: *Manually operated dampers shall not be installed except in connectors or chimneys serving solid-fuel-burning appliances. Automatically operated dampers shall conform to UL 17* (Vent or Chimney Connector Dampers for Oil-fired Appliances) *and be installed in accordance with the terms of their listing and label. The installation shall prevent firing of the burner when the damper is not opened to a safe position.*

Discussion and Commentary: A manual damper installed in a solid-fuel-burning appliance is usually opened manually during the process of starting a fire. If the damper is not opened, smoke will spill back into the room, alerting the user. For gas or liquid-fuel-fired appliances, however, the user may not be aware of a partially or completely closed damper, and a hazardous condition could develop.

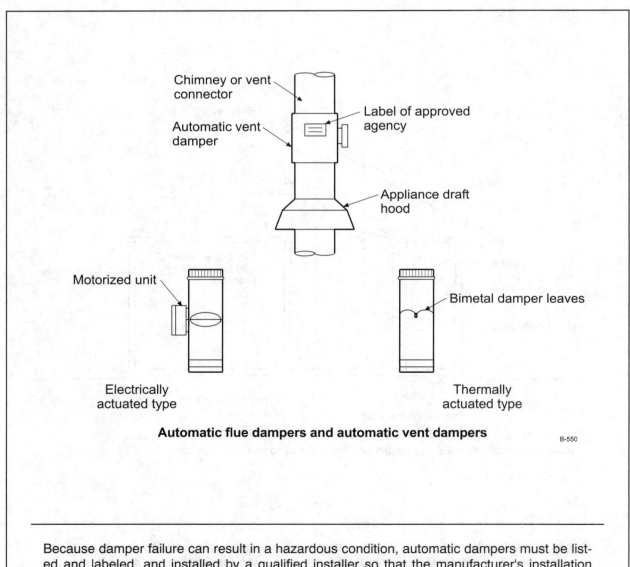

Automatic flue dampers and automatic vent dampers

B-550

Because damper failure can result in a hazardous condition, automatic dampers must be listed and labeled, and installed by a qualified installer so that the manufacturer's installation instructions and the terms of the listing are followed.

Topic: Termination
Reference: IRC M1804.2

Category: Chimneys and Vents
Subject: Vents

Code Text: *Vents passing through a roof shall extend through flashing and terminate in accordance with the manufacturer's installation requirements. Decorative shrouds shall not be installed at the termination of vents except where such shrouds are listed and labeled for use with the specific venting system and are installed in accordance with the manufacturer's installation instructions.*

Discussion and Commentary: The vent system must be physically protected to prevent damage to the vent and to prevent combustibles from coming into contact with or being placed too close to the vents. This protection is usually provided by enclosing the vent in chases, shafts or cavities during the building construction. Physical protection is not required in the room or space where the vent originates (at the appliance connection) and is not required in such locations as attics that are not occupied or used for storage.

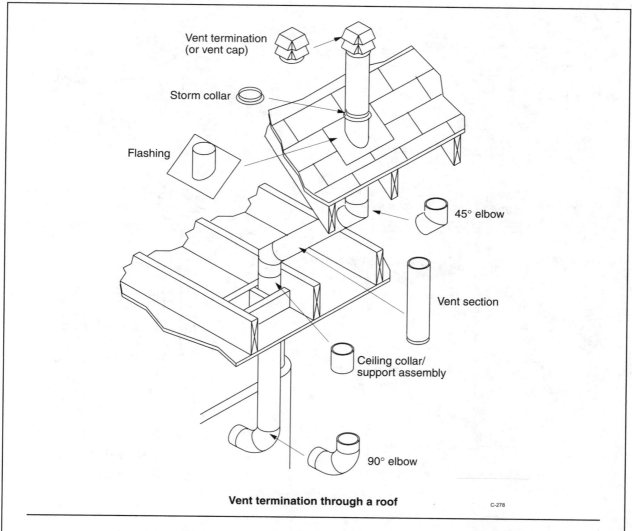

Vent termination through a roof

C-278

To maintain adequate venting action, termination of gravity vents must be at least 5 feet above the vent collar of the appliance and at least 12 feet above the bottom of a wall furnace. The bottom of the vent terminal for a direct-vent appliance must be at least 12 inches above grade.

Topic: Termination
Reference: IRC M1804.2

Code Text: *Vents for natural draft appliances shall terminate at least 5 feet (1524 mm) above the highest connected appliance outlet, and natural draft gas vents serving wall furnaces shall terminate at an elevation at least 12 feet (3658 mm) above the bottom of the furnace. Type L vents shall terminate with a listed and labeled cap in accordance with the vent manufacturer's installation instructions not less than 2 feet (610 mm) above the roof and not less than 2 feet (610 mm) above any portion of the building within 10 feet (3048 mm).*

Discussion and Commentary: Natural draft appliances must terminate at a high enough point to ensure that the venting system will perform properly under all conditions. Type L vents are used for oil-burning appliances and are designed for higher temperature flue gases than those encountered in gas-fired appliances.

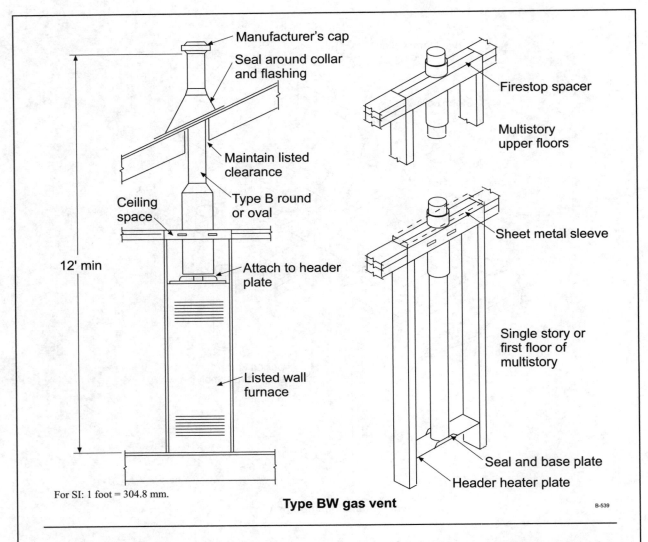

For SI: 1 foot = 304.8 mm.

Type BW gas vent

B-539

Unless the vent is an integral part of a listed and labeled appliance, an individual vent for a single appliance must have a cross-sectional area equivalent to or greater than that of the connector to the appliance, with a minimum area of 7 square inches.

Topic: Installation; Pressure-relief
Reference: IRC M2001.1, M2002.4

Category: Boilers/Water Heaters
Subject: Boilers

Code Text: *In addition to the requirements of the IRC, the installation of boilers shall conform to the manufacturer's instructions. The manufacturer's rating data, the nameplate and operating instructions of a permanent type shall be attached to the boiler. Boilers shall have all controls set, adjusted and tested by the installer. Boilers shall be equipped with pressure-relief valves with minimum rated capacities for the equipment served. Pressure-relief valves shall be set at the maximum rating of the boiler. Discharge shall be piped to drains by gravity to within 18 inches (457 mm) of the floor or to an open receptor.*

Discussion and Commentary: To ensure a safe and proper installation, boilers must be constructed to the appropriate standards and installed in accordance with the manufacturer's instructions.

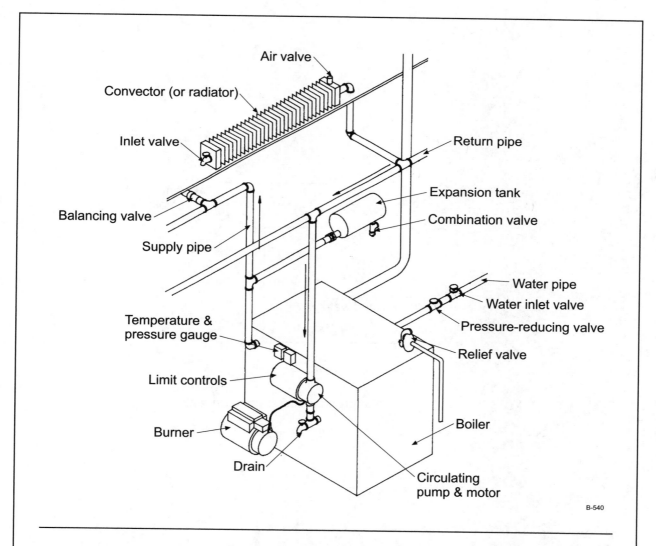

The installer must provide a complete set of boiler operating instructions for use by the owner. It is wise to attach such instructions directly to the boiler for immediate access. In addition, a complete control diagram must be furnished and located with the operating instructions.

Topic: Expansion Tanks
Reference: IRC M2003

Category: Boilers/Water Heaters
Subject: Boilers

Code Text: *Hot water boilers shall be provided with expansion tanks. Nonpressurized expansion tanks shall be securely fastened to the structure or boiler and supported to carry twice the weight of the tank filled with water. Pressurized expansion tanks shall be consistent with the volume and capacity of the system. Tanks shall be capable of withstanding a hydrostatic test pressure of two and one-half times the allowable working pressure of the system.*

Discussion and Commentary: Inactive hydronic heating systems are filled with water at room temperature. When the boiler is fired, the temperature of the water increases, causing the volume of the water to increase. Without an expansion tank, hydrostatic pressure would rapidly build within the boiler and piping system, activating the pressure-relief valve. An expansion tank provides space for the water to expand as it is heated and keeps the water pressure in the normal operating range while the boiler is operating.

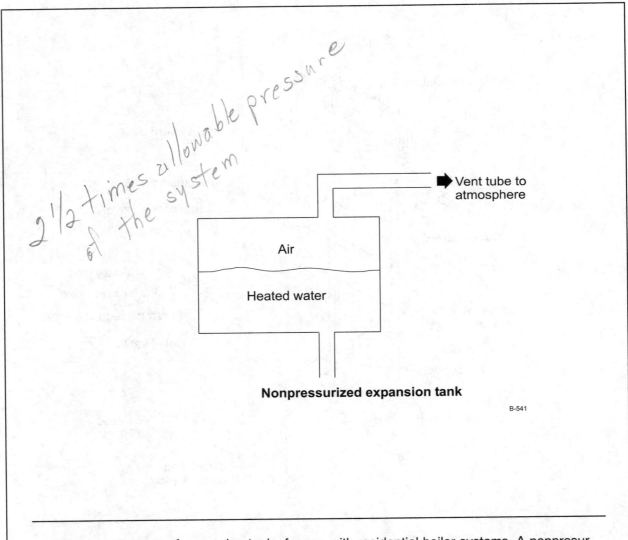

2½ times allowable pressure of the system

Vent tube to atmosphere

Air

Heated water

Nonpressurized expansion tank

B-541

There are two types of expansion tanks for use with residential boiler systems. A nonpressurized tank is simply a cylinder filled with air at atmosphere pressure, i.e., vented to the atmosphere. A pressurized tank is configured as a sealed cylinder divided by a flexible diaphragm.

Topic: Installation
Reference: IRC M2101

Category: Hydronic Piping
Subject: Piping Systems

Code Text: *Hydronic piping shall conform to Table M2101.1. Approved piping, valves, fittings and connections shall be installed in accordance with the manufacturer's installation instructions. The potable water system shall be protected from backflow in accordance with the provisions listed in Section P2902. Piping shall be installed so that piping, connections and equipment shall not be subjected to excessive strains or stresses. Provisions shall be made to compensate for expansion, contraction, shrinkage and structural settlement.*

Discussion and Commentary: A variety of piping materials is acceptable for use in hydronic piping systems. Because of its high resistance to corrosion and strength, crosslinked polyethylene (PEX) tubing is the only type of material that should be used underground. Plastic piping is only allowed in hydronic systems where the maximum water temperature will not exceed 180°F (82°C). Higher temperatures can weaken the pipe and may lead to bursting.

TABLE M2101.1
HYDRONIC PIPING MATERIALS

MATERIAL	USE CODE[a]	STANDARD[b]	JOINTS	NOTES
Brass pipe	1	ASTM B 43	Brazed, welded, threaded, mechanical and flanged fittings	
Brass tubing	1	ASTM B 135	Brazed, soldered and mechanical fittings	
Chlorinated poly (vinyl chloride) (CPVC) pipe and tubing	1, 2, 3	ASTM D 2846	Solvent cement joints, compression joints and threaded adapters	
Copper pipe	1	ASTM B 42, B 302	Brazed, soldered and mechanical fittings threaded, welded and flanged	
Copper tubing (type K, L or M)	1, 2	ASTM B 75, B 88, B 251, B 306	Brazed, soldered and flared mechanical fittings	Joints embedded in concrete
Crosslinked polyethylene (PEX) tubing	2, 3	ASTM F 876, F 877	Mechanical compression	Install in accordance with manufacturer's instructions.
Crosslinked polyethylene/ aluminum/crosslinked polyethylene-(PEX-AL-PEX) pressure pipe	1, 2	ASTM F 1281 or CAN/ CSA B137.10	Mechanical, crimp/insert	Install in accordance with manufacturer's instructions.
Plastic fittings PEX		ASTM F 1807		
Polybutylene (PB) pipe and tubing	1, 2, 3	ASTM D 3309	Heat-fusion, crimp/insert and compression	Joints in concrete shall be heat-fused.
Polyethylene (PE) pipe, tubing and fittings (for ground source heat pump loop systems)	1, 2, 4	ASTM D 2513; ASTM D 3350; ASTM D 2513; ASTM D 3035; ASTM D 2447; ASTM D 2683; ASTM F 1055; ASTM D 2837; ASTM D 3350; ASTM D 1693	Heat-fusion	
Soldering fluxes	1	ASTM B 813	Copper tube joints	
Steel pipe	1, 2	ASTM A 53; A 106	Brazed, welded, threaded, flanged and mechanical fittings	Joints in concrete shall be welded. Galvanized pipe shall not be welded or brazed.
Steel tubing	1	ASTM A 254	Mechanical fittings, welded	

For SI: °C = [(°F)-32]/1.8.

a. Use code:
  1. Above ground.
  2. Embedded in radiant system.
  3. Temperatures below 180°F only.
  4. Low temperature (below 130°F) applications only.
b. Standards as listed in Chapter 43.

Pipe openings through concrete or masonry must be sleeved to compensate for expansion of the piping due to hot water. As the flow of water causes the pipe to move and expand when the boiler cycles, sufficient clearance is necessary to avoid stress damage to the piping.

Topic: Installation

Category: Special Piping and Storage Systems

Reference: IRC M2201.1, M2201.2

Subject: Oil Tanks

Code Text: *Supply tanks shall be listed and labeled and shall conform to UL 58 for underground tanks and UL 80 for inside tanks. The maximum amount of fuel oil stored above ground or inside of a building shall be 660 gallons (2498 L). Supply tanks (located within a building) larger than 10 gallons (38 L) shall be placed not less than 5 feet (1524 mm) from any fire or flame either within or external to any fuel-burning appliance. Tanks installed outside above ground shall be a minimum of 5 feet (1524 mm) from an adjoining property line. Such tanks shall be suitably protected from the weather and from physical damage.*

Discussion and Commentary: The cross connection of two supply tanks on the same level is permitted if the total capacity of the system does not exceed 660 gallons. This procedure, which uses gravity flow from one tank to the other, provides an option for installation purposes.

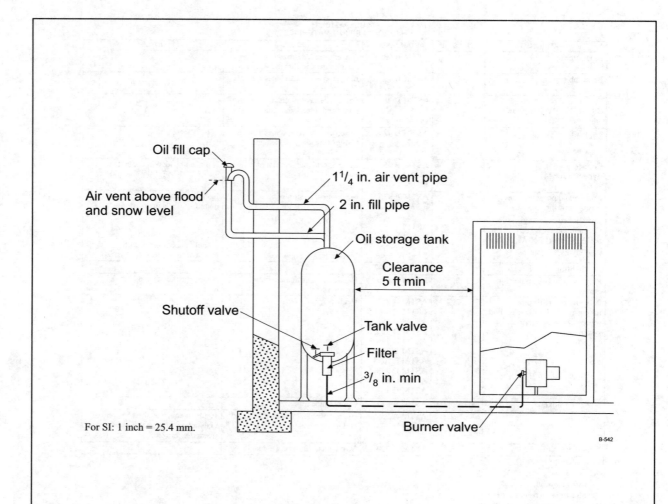

For SI: 1 inch = 25.4 mm.

B-542

Underground tanks must be at least 1 foot (305 mm) from the wall of a basement, pit, or property line to ensure that settlement of surrounding materials does not place a stress on the tank or piping. Tanks must also be covered with at least 12 inches (305 mm) of earth for physical protection.

Topic: Installation
Reference: IRC M2301.2

Category: Solar Systems
Subject: Solar Energy Systems

Code Text: *Solar energy collectors, controls, dampers, fans, blowers and pumps shall be accessible for inspection, maintenance, repair and replacement. Where mounted on or above the roof coverings, the collectors and supporting structure shall be constructed of noncombustible materials or fire-retardant-treated wood equivalent to that required for the roof construction. Roof and wall penetrations shall be flashed and sealed to prevent entry of water, rodents and insects in accordance with Chapter 9.*

Discussion and Commentary: Developed due to an emphasis placed on conservation of our natural resources, solar energy systems allow for an alternative method in providing space heating or cooling, domestic water heating, and pool heating. The effectiveness of these systems depends on the size of the solar array, the energy requirements of the dwelling and the availability of sunlight.

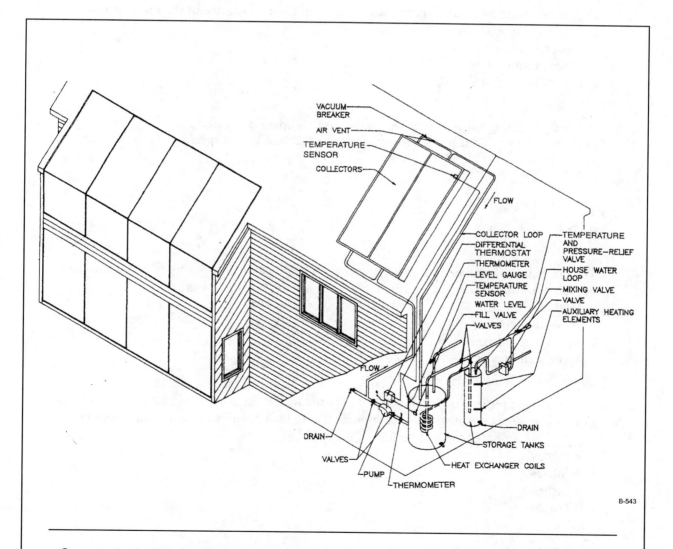

B-543

Some water heaters and boilers are designed to operate with "solar assistance." These units heat water by ordinary methods and have additional connections for solar preheating. Preheated water reduces the amount of energy that is consumed with ordinary heating methods.

QUIZ

Study Session 12

1. In a building of ordinary tightness, the combustion air for fuel-burning appliances may be obtained from infiltration where the volume of the room or space is a minimum _____ cubic feet per 1,000 Btu/h.
    a. 20
    b. 25
    c. 40
    d. 50

2. In general, combustion air is permitted to be taken from which one of the following spaces?
    a. sleeping room
    b. toilet room
    c. storage room
    d. bathroom

3. When determining the free area for combustion air openings, a free area of _____ of the gross area is to be used for metal louvers unless the free area is specified by the louver manufacturer.
    a. 25%
    b. 40%
    c. 60%
    d. 75%

4. The gross area of a wood louver used to cover a combustion air opening is 240 square inches. For the purpose of determining combustion air, what is the net free area?
    a. 60 square inches
    b. 96 square inches
    c. 144 square inches
    d. 180 square inches

5. Where normal infiltration is not adequate to provide the necessary combustion air, a minimum of two permanent openings shall be provided, with the openings located a maximum of _____ from the top and bottom of the space.
    a. 4 inches
    b. 6 inches
    c. 8 inches
    d. 12 inches

6. The two combustion air openings required to serve a confined space shall each have a minimum free area of _____ square inches.
   a. 60
   b. 80
   c. 100
   d. 120

7. Where all air is taken from inside the building, the net free area for each of two required combustion air openings shall be a minimum of 1 square inch per _____ Btu/h input rating of all appliances in a confined space.
   a. 1,000
   b. 2,000
   c. 3,000
   d. 4,000

8. Where combustion air is taken from the outdoors by means of vertical ducts, each opening shall have a free area of at least 1 square inch per _____ Btu/h of total input rating of all appliances in the space.
   a. 1,000
   b. 2,000
   c. 3,000
   d. 4,000

9. Where combustion air is taken from the outdoors by means of horizontal ducts, each opening shall have a free area of at least 1 square inch per _____ Btu/h of total input rating of all appliances in the space.
   a. 1,000
   b. 2,000
   c. 3,000
   d. 4,000

10. What is the minimum cross-sectional dimension permitted for a rectangular combustion air duct?
    a. 3 inches
    b. 4 inches
    c. 6 inches
    d. 8 inches

11. Outside combustion air openings shall be covered with a corrosion-resistant screen or equivalent protection having a minimum opening size of _____ inch and a maximum opening size of _____ inch.
    a. $1/8$, $1/4$
    b. $1/4$, $1/2$
    c. $1/2$, $3/4$
    d. $1/2$, 1

12. Where passing through a wall or partition, a single-wall metal pipe used as a vent connector shall be guarded by a ventilated metal thimble a minimum of _____ inches larger in diameter than the vent connector.
   a. 2
   b. 3
   c. 4
   d. 6

13. For other than direct vent appliances, the vent termination for a mechanical draft system shall be mounted a minimum of _____ feet horizontally from an oil tank or gas meter.
   a. 3
   b. 4
   c. 5
   d. 10

14. Where combustion air is taken from the attic, the combustion air opening shall be provided with a metal sleeve that extends from the appliance enclosure to a minimum height of _____ above the top of the ceiling joists and insulation.
   a. 3 inches
   b. 4 inches
   c. 6 inches
   d. 12 inches

15. Where combustion air is taken from under-floor areas, the required free opening area to the outside shall be a minimum of _____ the required combustion air opening.
   a. one-half
   b. 60%
   c. twice
   d. three times

16. Where an 8-inch-diameter single wall metal pipe is used as a vent connector, it shall have a minimum galvanized sheet metal gage of _____.
   a. No. 22
   b. No. 24
   c. No. 26
   d. No. 28

17. Vent and chimney connectors shall have a minimum slope of _____ rise per foot of run.
   a. $^1/_8$ inch
   b. $^1/_4$ inch
   c. $^3/_8$ inch
   d. $^1/_2$ inch

18. The horizontal run of a listed vent connector to a natural draft chimney shall be a maximum of _____ of the height of the vertical portion of the chimney above the connector.
   a. 50%
   b. 75%
   c. 100%
   d. 125%

19. What is the minimum required clearance between a Type L vent piping connector serving a solid-fuel appliance and any combustible materials?
   a. 3 inches
   b. 6 inches
   c. 9 inches
   d. 18 inches

20. A natural gas draft vent serving a wall furnace shall terminate a minimum height of _____ above the bottom of the furnace.
   a. 5 feet
   b. 8 feet
   c. 10 feet
   d. 12 feet

21. Chimney flues connected to more than one appliance shall be sized based on the area of the largest connector plus _____ of the areas of additional chimney connectors.
   a. 50%
   b. 75%
   c. 100%
   d. 125%

22. The pressurized expansion tank for a boiler shall be capable of withstanding a minimum hydrostatic test pressure _____ the allowable working pressure of the system.
   a. equal to
   b. twice
   c. two and one-half times
   d. three times

23. Hydronic piping to be embedded in concrete shall be tested by applying a minimum hydrostatic pressure of _____ for a minimum time period of _____.
   a. 50 psi, 15 minutes
   b. 50 psi, 30 minutes
   c. 100 psi, 15 minutes
   d. 100 psi, 30 minutes

24. Fill piping for oil tanks shall terminate outside of a building a minimum distance of _____ from any building opening at the same or lower level.
   a. 12 inches
   b. 2 feet
   c. 5 feet
   d. 10 feet

25. Solar energy systems shall be equipped with means to limit the maximum water temperature of the system fluid entering or exchanging heat with any pressurized vessel inside the dwelling to a maximum of _____.
   a. 120°F
   b. 150°F
   c. 180°F
   d. 210°F

26. Fuel-fired water heaters shall not be installed in a _____.
   a. bathroom
   b. bedroom
   c. kitchen
   d. storage closet

27. PEX tubing used for hydronic piping shall be supported at maximum intervals of _____ when installed in a horizontal position.
   a. 2 feet, 8 inches
   b. 4 feet
   c. 6 feet, 6 inches
   d. 8 feet

28. Hydronic piping shall be tested hydrostatically at a minimum pressure of _____ psi for a minimum of _____ minutes.
   a. 5, 15
   b. 10, 30
   c. 60, 30
   d. 100, 15

29. A fuel oil storage tank installed inside of a building shall have a maximum capacity of _____ gallons.
   a. 330
   b. 660
   c. 1,000
   d. 2,000

30.   Pressure at the fuel oil supply inlet to an appliance shall be a maximum of _____ psi.
        a. 3
        b. 5
        c. 10
        d. 15

INTERNATIONAL RESIDENTIAL CODE
Study Session 13
Chapter 24—Fuel Gas

OBJECTIVE: To provide an understanding of the general provisions for fuel gas piping systems, fuel gas utilization equipment and related accessories, venting systems and combustion air configurations.

REFERENCE: Chapter 24, 2003 *International Residential Code*

KEY POINTS:

- In what locations are the installation of fuel-fired appliances prohibited? What concerns are applicable to outdoor installations?
- How shall combustion, ventilation and dilution air be provided in buildings of unusually tight construction?
- How shall openings be provided for a confined space where all combustion air is taken from inside the building? From outside the building? From both inside and outside the building? In what manner are combustion air ducts regulated?
- Where must equipment and appliances having an ignition source be located in a private garage?
- How may the required clearances between combustible materials and chimneys; vents; fuel gas appliances, devices, and equipment; and kitchen exhaust equipment be reduced?
- What identifying mark is to be used on exposed gas piping? What type of pipe material is exempt from the labeling requirement?
- How is the maximum gas demand determined? How is the gas piping to be sized?
- What materials are acceptable for use as gas piping? When are materials allowed to be used again?
- When is a protective coating mandated for gas piping? How are field threading operations regulated? What methods are acceptable for pipe joints?
- Where is gas piping prohibited? What are the limitations for gas piping in concealed locations? How must such piping be protected where installed through a foundation or basement wall?
- What methods of protection against physical damage are identified for gas piping? How should pipe be protected against corrosion? What is the minimum required burial depth?
- How shall changes in direction be made for metallic pipe? For plastic pipe?
- What are the general conditions for the inspection and testing of fuel gas piping?
- Where must gas shutoff valves be located? Where must appliance shutoff valves be installed? Equipment shutoff valves for a fireplace?
- What is the maximum permitted length for an appliance fuel connector serving a range or domestic clothes dryer? Serving other appliances?
- At what maximum intervals must fuel gas piping be supported?
- What methods are available for terminating a gas vent? Where are decorative shrouds permitted?
- Where must chimneys serving gas-fired appliances terminate? At what minimum height must gas vents terminate above a roof? Where must a Type B or Type L gas vent terminate? A Type B-W vent? A single-wall metal pipe?
- How are vented wall furnaces to be located? Floor furnaces? Unit heaters?
- What are prohibited sources for outside or return air for a forced-air heating system?

Topic: Protection Against Damage
Reference: IRC G2415.5

Category: Fuel Gas
Subject: Piping System Installation

Code Text: *In concealed locations, where piping other than black or galvanized steel is installed through holes or notches in wood studs, joists, rafters or similar members less than 1 inch (25 mm) from the nearest edge of the member, the pipe shall be protected by shield plates. Shield plates shall be a minimum of 0.0625-inch-thick (.6 mm) steel, shall cover the area of the pipe where the member is notched or bored, and shall extend a minimum of 4 inches above sole plates, below top plates and to each side of a stud, joist or rafter.*

Discussion and Commentary: The provision is intended to minimize the possibility that nails or screws will be driven into the gas pipe or tube. Because nails and screws sometimes miss the stud, rafter, joist, or sole or top plates, the shield plates must extend beyond the stud itself.

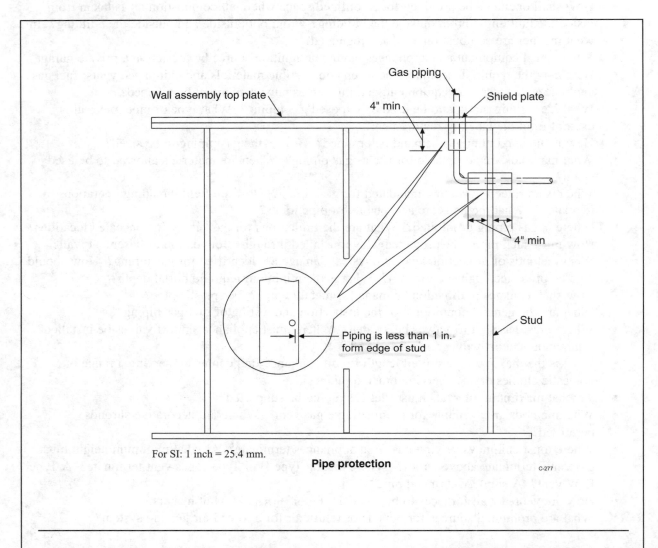

For SI: 1 inch = 25.4 mm.

Pipe protection

C-277

Piping must not be installed in any solid concrete or masonry floor construction. The potential for pipe damage due to slab settlement, cracking or the corrosive action of the floor material makes it imperative that the gas piping be installed in channels or casings, or similarly protected.

Topic: Design and Installation
Reference: IRC G2418.2, G2424

Category: Fuel Gas
Subject: Piping Support

Code Text: *Piping shall be supported with pipe hooks, metal pipe straps, bands, brackets or hangers suitable for the size of piping, of adequate strength and quality, and located at intervals so as to prevent or damp out excessive vibration. Piping shall be anchored to prevent undue strains on connected equipment and shall not be supported by other piping. Supports, hangers, and anchors shall be installed so as not to interfere with the free expansion and contraction of the piping between anchors.*

Discussion and Commentary: Suspended gas piping must be supported properly by hangers and straps to prevent straining or bending, which could cause leaks and create a hazardous condition. The supporting material should be compatible with the piping material to avoid electrolytic action.

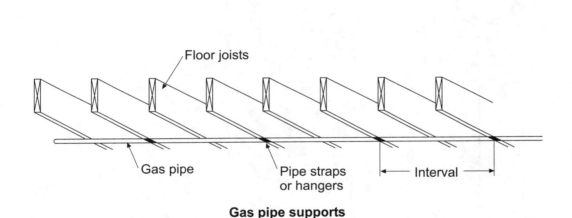

Gas pipe supports

B-544

TABLE G2424.1
SUPPORT OF PIPING

STEEL PIPE, NOMINAL SIZE OF PIPE (inches)	SPACING OF SUPPORTS (feet)	NOMINAL SIZE OF TUBING SMOOTH-WALL (inch O.D.)	SPACING OF SUPPORTS (feet)
$1/2$	6	$1/2$	4
$3/4$ or 1	8	$5/8$ or $3/4$	6
$1^1/4$ or larger (horizontal)	10	$7/8$ or 1 (horizontal)	8
$1^1/4$ or larger (vertical)	Every floor level	1 or Larger (vertical)	Every floor level

For SI: 1 inch = 25.4 mm, 1 foot = 304.8 mm.

Buried piping should be laid on a solid bed. This could be 4 to 6 inches of sand, gravel or tamped earth. The backfill should not contain any large rocks or frozen chunks of earth that might impose loads that could damage the pipe.

Topic: Sediment Trap

Reference: IRC G2419.4

Category: Fuel Gas

Subject: Drips and Sloped Piping

Code Text: *Where a sediment trap is not incorporated as a part of the gas utilization equipment, a sediment trap shall be installed downstream of the equipment shut-off valve as close to the inlet of the equipment as practical. The sediment trap shall be either a tee fitting with a capped nipple in the bottom opening of the run of the tee or other device approved as an effective sediment trap. Illuminating appliances, ranges, clothes dryers and outdoor grills need not be so equipped.*

Discussion and Commentary: Sediment traps cause the gas flow to change direction 90 degrees at the sediment collection point, thus causing solid or liquid contaminants to drop out of the gas flow. Sediment traps prevent debris from entering the gas controls where such contaminants could cause hazardous appliance malfunction.

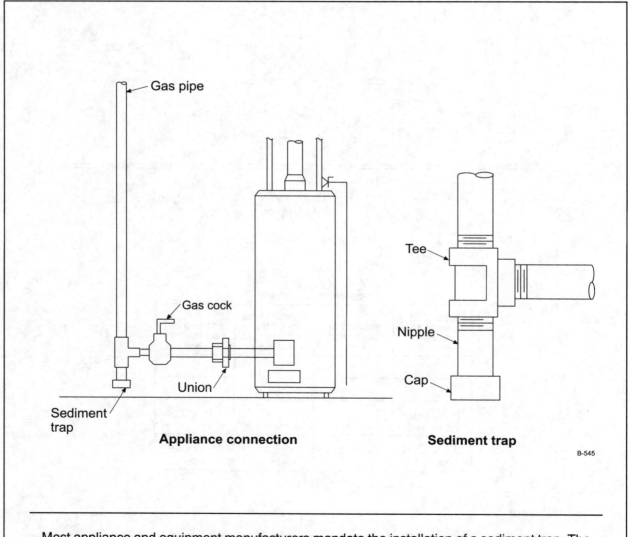

Gas pipe

Gas cock

Union

Sediment trap

Appliance connection

Tee

Nipple

Cap

Sediment trap

B-545

Most appliance and equipment manufacturers mandate the installation of a sediment trap. The code requires compliance with manufacturers' instructions and listings, so it is consistent that the installation of a trap is a basic code provision.

Topic: Equipment Shutoff Valve
Reference: IRC G2420.5

Category: Fuel Gas
Subject: Gas Shutoff Valves

Code Text: *Each appliance shall be provided with a shutoff valve separate from the appliance. The shutoff valve shall be located in the same room as the appliance, not further than 6 feet (1829 mm) from the appliance, and shall be installed upstream from the union, connector or quick disconnect device it serves. Such shutoff valves shall be provided with access.* See exception for vented decorative appliances.

Discussion and Commentary: The fuel gas connection to every gas appliance must be provided with an individual gas shutoff valve to permit maintenance, repair, replacement or temporary disconnection. The shutoff valve must be adjacent to the appliance, conspicuously located, and within reach to permit it to be easily located and operated in the event of an emergency. For example, a gas cock serving an appliance on the first floor could not be located in the basement.

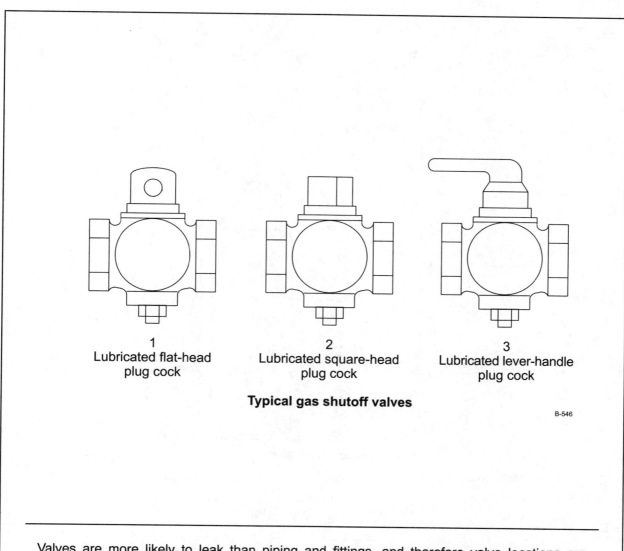

1
Lubricated flat-head
plug cock

2
Lubricated square-head
plug cock

3
Lubricated lever-handle
plug cock

Typical gas shutoff valves

B-546

Valves are more likely to leak than piping and fittings, and therefore valve locations are restricted. Locating fuel gas valves in concealed spaces unduly restricts access to the valve in the event of an emergency or servicing the equipment to which the valve is attached.

Topic: Fuel Connectors

Reference: IRC G2422.1.2

Category: Fuel Gas

Subject: Appliance Connections

Code Text: *Connectors shall have an overall length not to exceed 3 feet (914 mm), except for range and domestic clothes dryer connectors, which shall not exceed 6 feet (1829 mm) in length. Connectors shall not be concealed within, or extended through, walls, floors, partitions, ceilings or appliance housings. A shutoff valve not less than the nominal size of the connector shall be installed ahead of the connector.*

Discussion and Commentary: Flexible (semirigid) appliance fuel connectors are primarily for use with cooking ranges and clothes dryers where the gas connection is located behind the appliance and some degree of flexibility is necessary to facilitate the hook-up. Longer lengths for range and dryer installations allow the connector to be coiled or otherwise arranged to allow limited movement of the appliance without stressing the connector. Rigid connectors are not practical for such movable appliances.

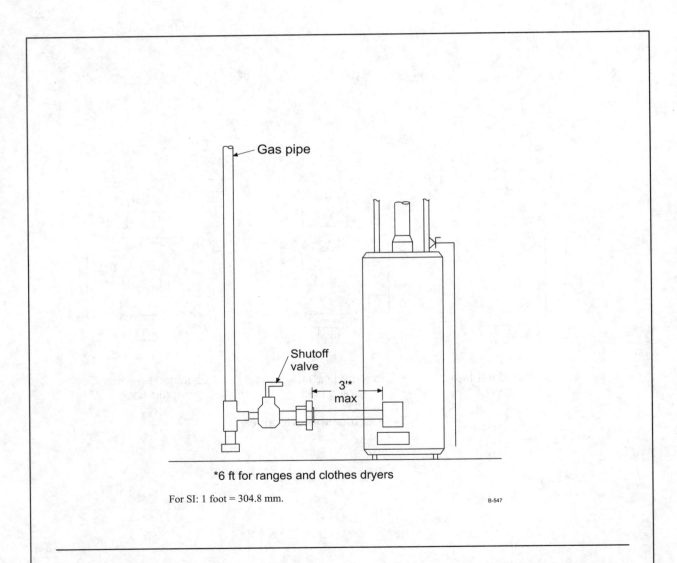

Gas pipe

Shutoff valve

3'* max

*6 ft for ranges and clothes dryers

For SI: 1 foot = 304.8 mm.

B-547

Flexible connectors, usually constructed of brass or stainless steel, are labeled with tags of metal rings placed over the tubing. The manufacturer's installation instructions must be followed to protect the connector from damage and to prevent leakage.

Topic: Factory-built Chimneys
Reference: IRC G2427.5.1

Category: Fuel Gas
Subject: Venting of Equipment

Code Text: *Factory-built chimneys shall be installed in accordance with their listing and the manufacturers' instructions. Factory-built chimneys used to vent appliances that operate at positive vent pressure shall be listed for such application.*

Discussion and Commentary: All prefabricated chimney systems must bear the label of an approved agency. The label states information such as the type of appliance the chimney was tested for use with, a reference to the manufacturer's installation instructions and the minimum required clearances to combustibles. The manufacturer's instructions contain sizing criteria and the requirements for every aspect of a factory-built chimney installation, which include component assembly, clearances to combustibles, support, terminations, connections, protection from damage and fireblocking.

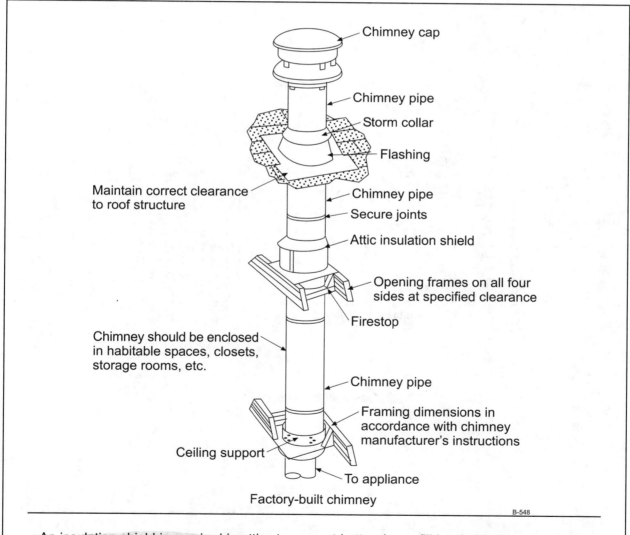

An insulation shield is required in attics to prevent batt or loose-fill insulation from contacting the chimney. The attic shield maintains the airspace clearance between the chimney and the insulation material to reduce the chimney temperatures transmitted from the chimney surfaces.

Topic: Gas Vent Termination
Reference: IRC G2427.6.5

Category: Fuel Gas
Subject: Venting of Equipment

Code Text: *A gas vent shall terminate 1) above the roof surface with a listed cap or listed roof assembly. Gas vents 12 inches (305 mm) in size or smaller with listed caps shall be permitted to be terminated in accordance with Figure G2426.6.5, provided that such vents are at least 8 feet (2438 mm) from a vertical wall or similar obstruction. All other gas vents shall terminate not less than 2 feet (610 mm) above the highest point where they pass through the roof and at least 2 feet (610 mm) higher than any portion of a building within 10 feet (3048 mm).* See three other termination methods.

Discussion and Commentary: Figure G2426.6.6 requires greater vent height above the roof as the roof approaches being a vertical surface. The greater the roof pitch, the greater effect of the wind striking the roof surface. However, the height should be limited as much as possible because of several concerns. Vent piping exposed to the outdoors encourages condensation in cold weather, could require guy wires or braces depending on height and is considered aesthetically unattractive.

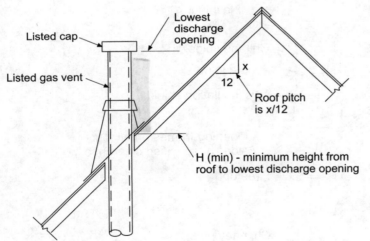

Roof pitch	H (min) ft	m
Flat to 6/12	1.0	0.30
6/12 to 7/12	1.25	0.38
Over 7/12 to 8/12	1.5	0.46
Over 8/12 to 9/12	2.0	0.61
Over 9/12 to 10/12	2.5	0.76
Over 10/12 to 11/12	3.25	0.99
Over 11/12 to 12/12	4.0	1.22
Over 12/12 to 14/12	5.0	1.52
Over 14/12 to 16/12	6.0	1.83
Over 16/12 to 18/12	7.0	2.13
Over 18/12 to 20/12	7.5	2.27
Over 20/12 to 21/12	8.0	2.44

Gas vent termination

B-549

It is a common misapplication for code users to apply chimney termination height requirements to vents, thereby causing vents to extend above roofs much higher than required in many cases. For roof pitches up to a $^6/_{12}$ pitch, only a 1-foot extension is necessary.

Topic: Common Connector/Manifold
Reference: IRC G2427.10.3.4

Category: Fuel Gas
Subject: Venting of Equipment

Code Text: *Where two or more gas appliances are vented through a common vent connector or vent manifold, the common vent connector or vent manifold shall be located at the highest level consistent with available headroom and the required clearance to combustible materials and shall be sized in accordance with Section G2428 or other approved engineering methods.*

Discussion and Commentary: A manifold is defined as a vent or connector that is a lateral (horizontal) extension of the lower end of a common vent. Manifolds must be installed as high as possible to provide for the most appliance connector rise, which is always beneficial for venting performance. Connector rise takes advantage of the energy of hot gases discharging directly from the appliance and develops flow velocity and draft in addition to that produced by the common vent.

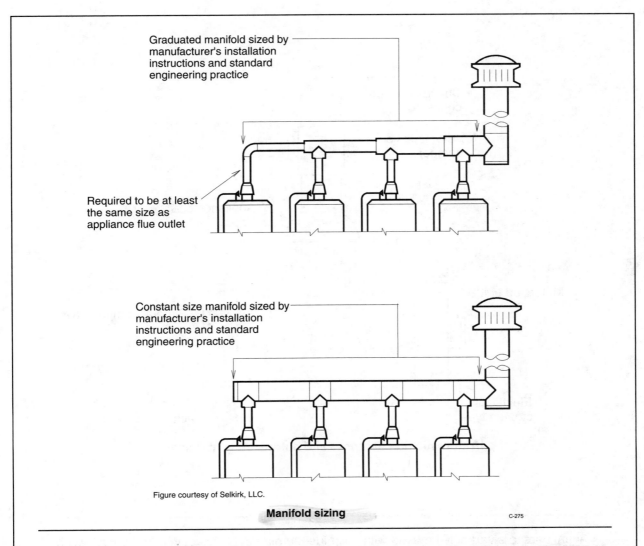

Graduated manifold sized by manufacturer's installation instructions and standard engineering practice

Required to be at least the same size as appliance flue outlet

Constant size manifold sized by manufacturer's installation instructions and standard engineering practice

Figure courtesy of Selkirk, LLC.

Manifold sizing

C-275

An option is available where there are only two draft-hood-equipped appliances. This alternative method is intended for use only with the alternative sizing method of Section G2427.6.8.1 and is not compatible with Section G2428.

Topic: Automatic Dampers   **Category:** Fuel Gas
Reference: IRC G2427.14   **Subject:** Venting of Equipment

Code Text: *An automatically operated vent damper shall be of a listed type.*

Discussion and Commentary: The installation of an automatic damper must be in strict accordance with the manufacturer's installation instructions. A malfunctioning or improperly installed flue damper could cause the appliance to malfunction and discharge the products of combustion directly into the building interior. Because automatic damper failure could result in a hazardous condition, automatic dampers are required to be listed and labeled. Automatic flue dampers are energy-saving devices designed to close off or restrict an appliance flue passageway when the appliance is not operating and is in its "off" cycle. Such devices save energy by trapping residual heat in a heat exchanger after the burners shut off and by preventing the escape of conditioned room air up the vent. Thus, the appliance efficiency can be boosted, and building air infiltration can be reduced.

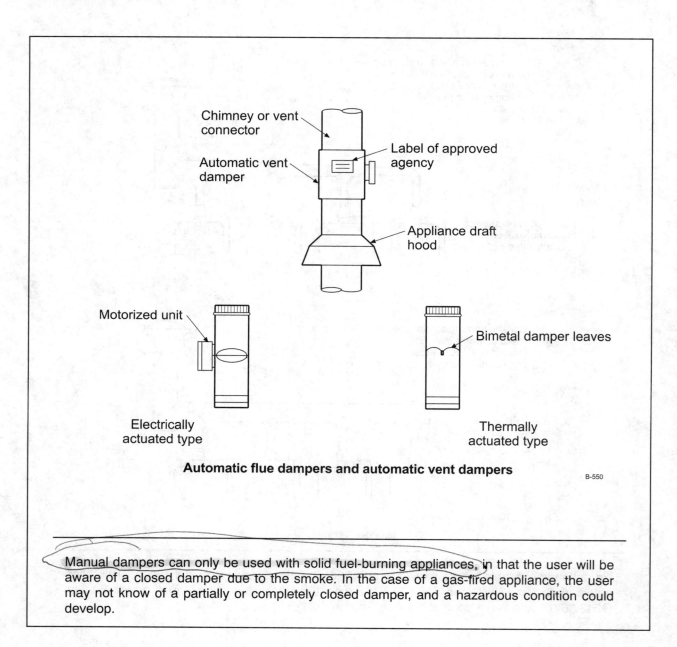

Automatic flue dampers and automatic vent dampers

B-550

Manual dampers can only be used with solid fuel-burning appliances, in that the user will be aware of a closed damper due to the smoke. In the case of a gas-fired appliance, the user may not know of a partially or completely closed damper, and a hazardous condition could develop.

Topic: Flame Safeguard Device
Reference: IRC G2432.2

Category: Fuel Gas
Subject: Decorative Appliances for Fireplaces

Code Text: *Decorative appliances for installation in approved solid fuel-burning fireplaces, with the exception of those tested in accordance with ANSI Z21.84, shall utilize a direct ignition device, an ignitor or a pilot flame to ignite the fuel at the main burner, and shall be equipped with a flame safeguard device. The flame safeguard device shall automatically shut off the fuel supply to a main burner or group of burners when the means of ignition of such burners becomes inoperative.*

Discussion and Commentary: The typical flame safeguard device used with gas log set appliances is a combination manual control valve, pressure regulator, pilot feed and magnetic pilot safety mechanism with a thermocouple generator. If the pilot flame is extinguished, the drop in thermocouple output voltage will cause the control valve to "lock out" in the closed position, thereby preventing the flow of gas to the main burner and the pilot burner.

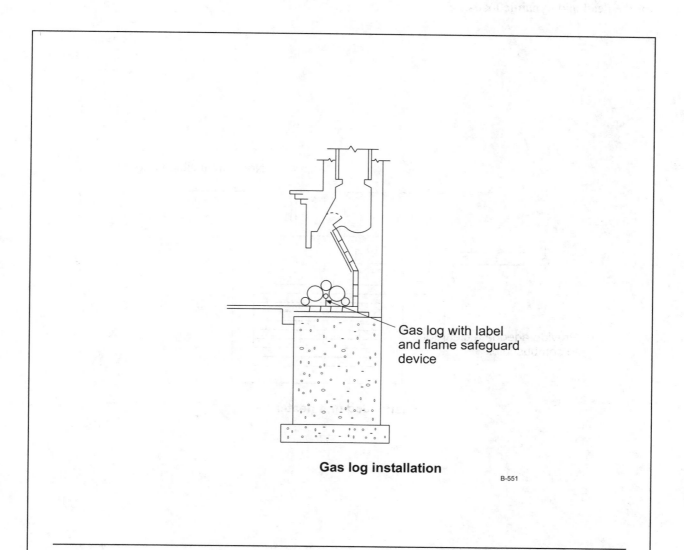

Gas log with label and flame safeguard device

Gas log installation

B-551

Gas-fired decorative log sets and log lighters are examples of accessory appliances designed for installation in solid fuel-burning fireplaces. Gas log sets provide some radiant heat; however, their primary function is to create an aesthetically pleasant simulation of a wood log fire.

Topic: Support and Clearance
Reference: IRC G2444.2, G2444.4

Category: Fuel Gas
Subject: Unit Heaters

Code Text: *Suspended-type unit heaters shall be supported by elements that are designed and constructed to accommodate the weight and dynamic loads. Hangers and brackets shall be of noncombustible material. Suspended-type unit heaters shall be installed with clearances to combustible materials of not less than 18 inches (457 mm) at the sides, 12 inches (305 mm) at the bottom and 6 inches (152 mm) above the top where the unit heater has an internal draft hood or 1 inch (25.4 mm) above the top of the sloping side of the vertical draft hood.*

Discussion and Commentary: As with all suspended fuel-fired appliances, a support failure could result in a fire, explosion or injury to building occupants. The supports themselves must be properly designed. Equally important are the structural members to which the supports are attached, such as rafters, beams, joists and purlins. All brackets, pipes, rods, angle iron, structural members and fasteners must be designed for the dead and dynamic loads.

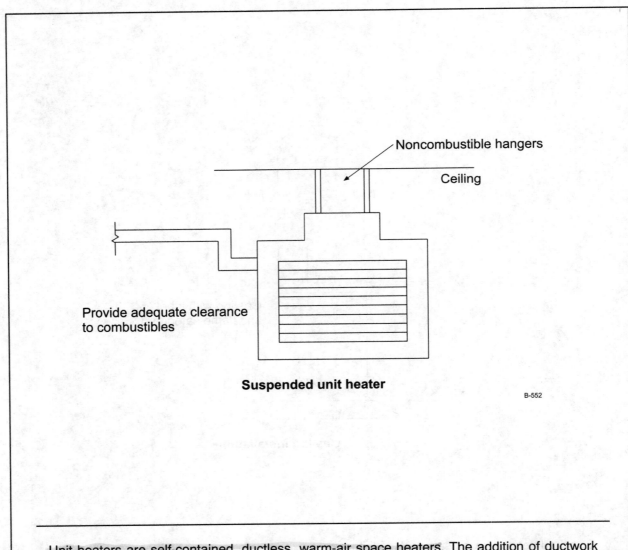

Noncombustible hangers

Ceiling

Provide adequate clearance
to combustibles

Suspended unit heater

B-552

Unit heaters are self-contained, ductless, warm-air space heaters. The addition of ductwork could create a hazard by restricting air flow through the heater. In residential applications, garages and workshops are the most common locations for unit heaters.

Topic: Oxygen-depletion Safety System
Reference: IRC G2445.6

Category: Fuel Gas
Subject: Unvented Room Heaters

Code Text: *Unvented room heaters shall be equipped with an oxygen-depletion-sensitive safety shutoff system. The system shall shut off the gas supply to the main and pilot burners when the oxygen in the surrounding atmosphere is depleted to the percent concentration specified by the manufacturer, but not lower than 18 percent. The system shall not incorporate field adjustment means capable of changing the set point at which the system acts to shut off the gas supply to the room heater.*

Discussion and Commentary: The oxygen depletion sensor is basically a pilot burner that is extremely sensitive to the oxygen content in the combustion air. If the oxygen content (approximately 80 percent) drops to a predetermined level, the pilot flame will destabilize and extinguish or become incapable of sufficiently heating the thermocouple or thermopile generator, resulting in main gas control valve shutdown and lockout.

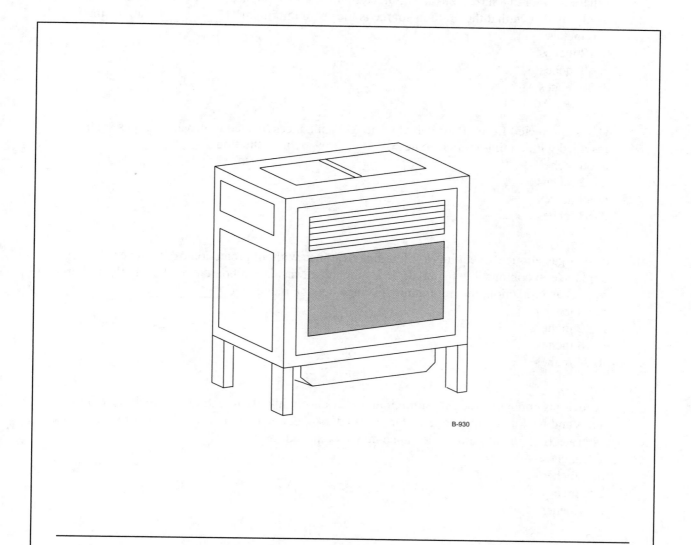

B-930

Unvented room heaters are limited to a maximum input rating of 40,000 Btu/h. When determining the maximum unvented room heater rating allowed, the aggregate input rating of all unvented appliances in a room is limited to 20 Btu/h per cubic foot of volume of the room.

QUIZ

Study Session 13

1. In general, fuel-fired appliances are permitted to be located in which one of the following spaces?
   a. sleeping rooms
   b. storage closets
   c. bathrooms
   d. garages

2. Where located in a hazardous location, equipment and appliances having an ignition source shall be elevated such that the ignition source is located a minimum of _____ above the floor.
   a. 6 inches
   b. 12 inches
   c. 18 inches
   d. 30 inches

3. Unless protected from motor vehicle damage, appliances located in private garages shall be installed with a minimum clearance of _____ above the floor.
   a. 60 inches
   b. 72 inches
   c. 80 inches
   d. 84 inches

4. A fuel-gas appliance requires a minimum clearance without protection of 36 inches above the appliance to combustible material. If $1/2$-inch thick insulation board over 1-inch glass fiber batts is used as protection, the minimum clearance may be reduced to _____.
   a. 9 inches
   b. 12 inches
   c. 18 inches
   d. 24 inches

5. A fuel-gas appliance requires a minimum clearance without protection of 9 inches from the sides and back to combustible material. If 0.024 sheet metal with a ventilated air space is used as protection, the minimum clearance may be reduced to _____.
   a. 2 inches
   b. 3 inches
   c. 4 inches
   d. 6 inches

6. For other than black steel pipe, exposed gas piping shall be identified by a yellow label spaced at maximum intervals of _____.
   a. 5 feet
   b. 6 feet
   c. 8 feet
   d. 10 feet

7. Unless special conditions are met, the maximum design operating pressure for gas piping systems located inside buildings shall be _____.
   a. 5 psig
   b. 10 psig
   c. 15 psig
   d. 20 psig

8. The operating pressure for undiluted LP-Gas systems shall be a maximum of _____.
   a. 5 psig
   b. 10 psig
   c. 15 psig
   d. 20 psig

9. Copper tubing to be used with gases shall comply with standard _____ of ASTM B 88 or ASTM B 280.
   a. Type K only
   b. Type L only
   c. Type K or Type L
   d. Type L or Type M

10. For the field threading of 1-inch metallic pipe, a length of _____ (approximate) shall be provided for the threaded portion.
    a. $^3/_4$ inch
    b. $^7/_8$ inch
    c. 1 inch
    d. $1^1/_8$ inches

11. In which of the following locations is the installation of gas piping not specifically prohibited?
    a. in a circulating air duct
    b. through a clothes chute
    c. in an elevator shaft
    d. adjacent to an exterior stairway

12. For gas piping other than black or galvanized steel installed in a concealed location, a shield plate shall be provided for protection against physical damage where a bored hole is located a maximum of _____ from the edge of the wood member.
   a. $^5/_8$ inch
   b. 1 inch
   c. $1^1/_4$ inch
   d. $1^3/_8$ inch

13. In general, underground gas piping shall be installed a minimum depth of _____ below grade.
   a. 6 inches
   b. 12 inches
   c. 18 inches
   d. 24 inches

14. The unthreaded portion of gas piping outlets shall extend a minimum of _____ through finished ceilings and walls.
   a. 1 inch
   b. 2 inches
   c. 3 inches
   d. 4 inches

15. The tracer wire required adjacent to underground nonmetallic gas piping shall be a minimum of _____ wire size.
   a. 14 AWG
   b. 16 AWG
   c. 18 AWG
   d. 20 AWG

16. The inside radius of a bend in a metallic gas pipe shall be a minimum of _____ the outside diameter of the pipe.
   a. six times
   b. eight times
   c. twelve times
   d. twenty-five times

17. Gas piping shall be tested at a minimum pressure _____ the proposed maximum working pressure, but not less than 3 psig.
   a. equal to
   b. of one and one-half times
   c. of twice
   d. of two and one-half times

18. The minimum duration of a pressure test for gage piping shall be _____.
    a. 10 minutes
    b. 15 minutes
    c. 20 minutes
    d. 30 minutes

19. Piping for other than dry gas conditions shall be sloped a minimum of _____ to prevent traps.
    a. $\frac{1}{8}$ inch in 10 feet
    b. $\frac{1}{8}$ inch in 15 feet
    c. $\frac{1}{4}$ inch in 10 feet
    d. $\frac{1}{4}$ inch in 15 feet

20. A gas shutoff valve for an appliance shall be located in the same room as the appliance, with a maximum distance of _____ between the valve and the appliance.
    a. 3 feet
    b. 4 feet
    c. 6 feet
    d. 10 feet

21. Except for range and domestic clothes dryer connectors, appliance connectors shall have a maximum overall length of _____.
    a. 3 feet
    b. 5 feet
    c. 6 feet
    d. 10 feet

22. Supports for $\frac{3}{4}$-inch steel pipe shall be provided at maximum intervals of _____.
    a. 4 feet
    b. 6 feet
    c. 8 feet
    d. 12 feet

23. Appliance connectors where connected to a masonry chimney shall connect at a point a minimum of _____ above the lowest portion of the interior of the chimney flue.
    a. 12 inches
    b. 18 inches
    c. 30 inches
    d. 36 inches

24. A sauna room shall be provided with a minimum _____ ventilation opening located near the top of the door into the sauna room.
   a. 4-inch by 8-inch
   b. 4-inch by 12-inch
   c. 8-inch by 8-inch
   d. 8-inch by 12-inch

25. Unvented room heaters shall be limited to a maximum input rating of _____.
   a. 35,000 Btu/h
   b. 40,000 Btu/h
   c. 50,000 Btu/h
   d. 65,000 Btu/h

26. Gas-fired equipment and appliances, where suspended above grade level, shall be provided with a minimum clearance of _____ inches from adjoining grade.
   a. 2
   b. 4
   c. 6
   d. 8

27. Fuel gas piping installed across a roof surface shall be elevated above the roof a minimum of _____ inches.
   a. $1^1/_2$
   b. 2
   c. $3^1/_2$
   d. 6

28. Which of the following test mediums shall not be used in the testing of a fuel gas piping system?
   a. air
   b. carbon dioxide
   c. nitrogen
   d. oxygen

29. One-inch O.D. smooth-wall tubing used as fuel gas piping shall be supported at maximum intervals of _____ feet.
   a. 4
   b. 6
   c. 8
   d. 10

30. Where a post-mounted illuminating appliance is installed on a steel pipe post having a height of 30 inches, what is the minimum size Schedule 40 steel pipe required?

    a. $^3/_4$-inch

    b. 1-inch

    c. $1^1/_2$- inch

    d. $2^1/_2$-inch

INTERNATIONAL RESIDENTIAL CODE
Study Session 14

Chapters 25, 26 and 27—General Plumbing
Requirements and Plumbing Fixtures

OBJECTIVE: To provide an understanding of the general requirements for the installation of plumbing systems, including inspections and tests; piping installation, protection and support; and plumbing fixtures.

REFERENCE: Chapters 25, 26 and 27, 2003 *International Residential Code*

KEY POINTS:

- By what method must the building sewer system be tested and inspected? The DWV system? The water-supply system?
- Where installed through framing members in concealed locations, how shall piping be protected? How should it be protected from breakage where passing through or under walls?
- How shall a soil pipe, waste pipe or building drain be installed where passing under a footing or through a foundation wall?
- What are the installation limitations for piping installed in areas subject to freezing temperatures? To what minimum depth must water service pipe be installed? How shall trenches be located in relationship to footings or bearing walls?
- How must piping be installed when located in a trench? How is backfilling to be accomplished? How is trenching installed parallel to footings to be located?
- What are the general provisions for the support of piping? What is the maximum horizontal spacing between supports for various types of pipe material? Between vertical supports?
- How are exterior openings for pipes to be waterproofed?
- How must piping be listed and identified? When is third-party testing required? Third-party certification?
- What is the minimum required diameter for fixture tail pieces serving sinks, dishwashers, laundry tubs, bathtubs and similar fixtures? For bidets, lavatories and similar fixtures?
- What type of fixture connection must be provided with an access panel or utility space? What size of access opening must be provided?
- What are the general requirements for the installation of plumbing fixtures regarding support? Watertightness?
- What horizontal side clearances are mandated for water closets and bidets? What minimum clearance is required at the front of water closets, bidets and lavatories?
- What is the minimum required extension for a standpipe? The maximum allowable extension? Under what conditions may a laundry tray waste connect into a standpipe for an automatic clothes washer drain?
- What is the minimum permitted size for a shower compartment? Which direction must hinged shower doors swing? How are shower receptors to be constructed and installed?
- How are water closets regulated in regard to flushing devices, water supplies and flush valves?
- What is the minimum waste outlet size for a lavatory? A bathtub? A sink? A laundry tub? A food-waste grinder? A floor drain?
- How may a sink and dishwasher discharge through a single trap? A sink, dishwasher and food grinder?

Topic: Required Testing
Reference: IRC P2503.1, P2503.5

Category: Plumbing Administration
Subject: Inspection and Tests

Code Text: *New plumbing work and parts of existing systems affected by new work or alterations shall be inspected by the building official to ensure compliance with the requirements of the IRC. DWV systems shall be tested on completion of the rough piping installation by water or air with no evidence of leakage. After the plumbing fixtures have been set and their traps filled with water, their connections shall be tested and proved gas tight and/or water tight.*

Discussion and Commentary: All plumbing and drainage work must be tested in an appropriate manner to verify that it is leak free. The building sewer must be tested with a minimum 10-foot head of water. Both rough plumbing and finished plumbing are tested with a variety of methods available. Water supply piping and any backflow prevention assemblies must also be inspected and tested.

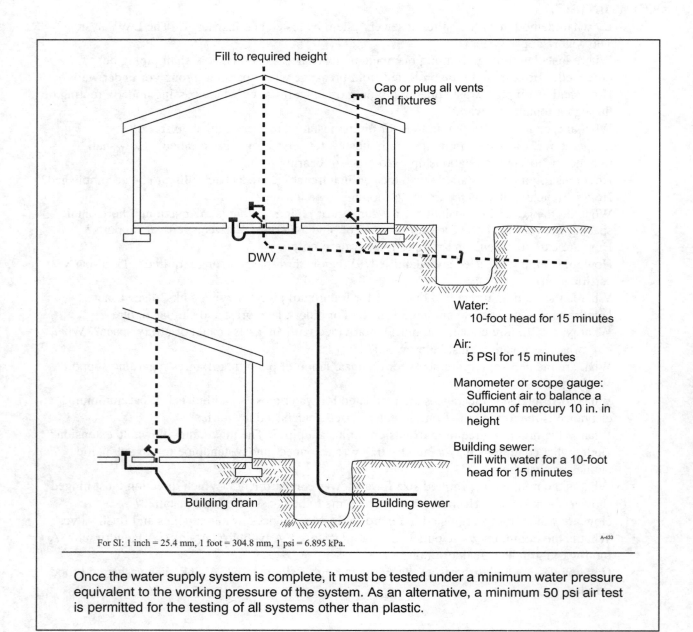

Fill to required height

Cap or plug all vents and fixtures

DWV

Water:
 10-foot head for 15 minutes

Air:
 5 PSI for 15 minutes

Manometer or scope gauge:
 Sufficient air to balance a column of mercury 10 in. in height

Building sewer:
 Fill with water for a 10-foot head for 15 minutes

Building drain

Building sewer

For SI: 1 inch = 25.4 mm, 1 foot = 304.8 mm, 1 psi = 6.895 kPa.

A-433

Once the water supply system is complete, it must be tested under a minimum water pressure equivalent to the working pressure of the system. As an alternative, a minimum 50 psi air test is permitted for the testing of all systems other than plastic.

Topic: Breakage and Corrosion
Reference: IRC P2603.3

Category: General Plumbing Requirements
Subject: Structural and Piping Protection

Code Text: *Pipes passing through or under walls shall be protected from breakage. Pipes passing through concrete or cinder walls and floors, cold-formed steel framing or other corrosive material shall be protected against external corrosion by a protective sheathing or wrapping or other means that will withstand any reaction from lime and acid of concrete, cinder or other corrosive material. Sheathing or wrapping shall allow for expansion and contraction of piping to prevent any rubbing action. Minimum wall thickness of material shall be 0.025 inch (0.64 mm).*

Discussion and Commentary: Pipes made from brass, copper, cast iron and steel are subject to corrosion when exposed to the lime and acid of concrete and cinder or other corrosive material such as soil; therefore, a coating or wrapping is necessary. Typical protective coatings include coal tar wrapper with paper, epoxy or plastic coatings.

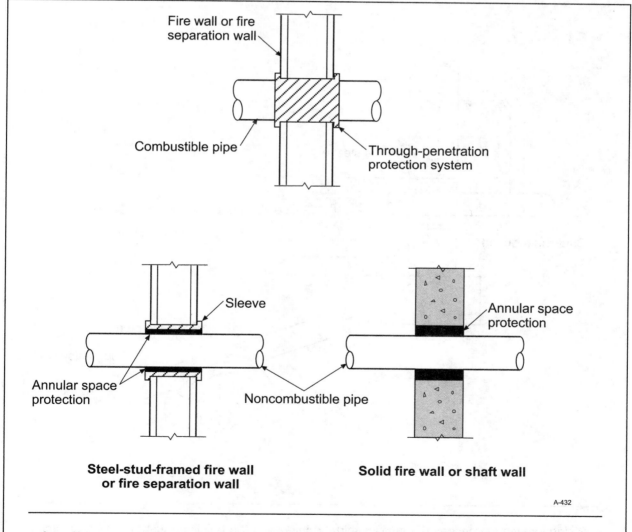

Annular spaces between sleeves and pipes must be filled or tightly caulked. Where the penetration of a pipe occurs in a fire-resistive assembly, protection must comply with Section R317.3 for through-penetrations. This may require a listed firestop system.

Topic: Pipes Through Footings or Foundation Walls
Reference: IRC P2603.5

Category: General Plumbing Requirements
Subject: Structural and Piping Protection

Code Text: *Any pipe that passes under a footing or through a foundation wall shall be provided with a relieving arch; or there shall be built into the masonry wall a pipe sleeve two pipe sizes greater than the pipe passing through.*

Discussion and Commentary: Piping installed within or under a footing or foundation wall must be structurally protected from any transferred loading from the foundation wall or footing. This protection may be provided through a relieving arch or a pipe sleeve. When a sleeve is used, it should be sized such that it is two pipe sizes larger than the penetration pipe. This space will allow for any differential movement of the pipe. Providing structural protection to the piping system ensures that the piping will not be subjected to the undue stresses that may cause the piping to rupture and leak.

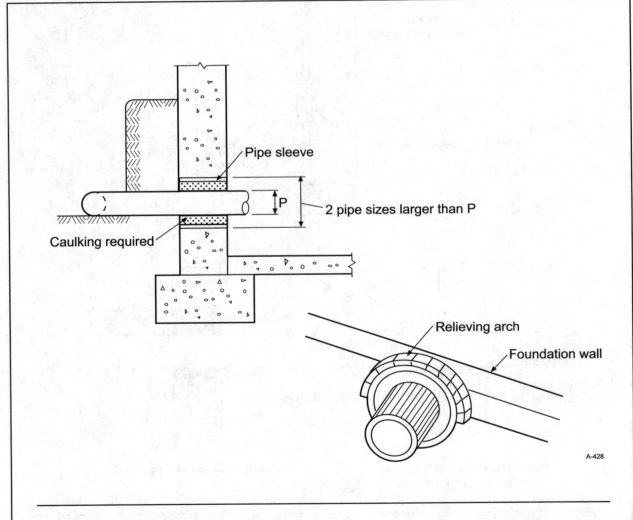

Piping located in concealed spaces within the interior of the building must also be protected from damage if installed through holes or notches in studs, joists and similar framing members. Where the hole or notch is less than $1^1/_2$ inches from the nearest member edge, steel protective shield plates are required.

Topic: Freezing
Reference: IRC P2603.6

Category: General Plumbing Requirements
Subject: Structural and Piping Protection

Code Text: *In localities having a winter design temperature of 32°F (0°C) or lower, . . . a water, soil or waste pipe shall not be installed outside of a building, in exterior walls, in attics or crawl spaces, or in any other place subjected to freezing temperature unless adequate provision is made to protect it from freezing by insulation or heat or both.*

Discussion and Commentary: Occupied spaces utilize the building's comfort heating system to assist in protecting pipes from freezing. Where the temperature of the air surrounding the insulation remains low for a significant period of time, such as in a vacant building, insulation alone will not provide adequate protection without the addition of heat. In such conditions, the water in the pipe will freeze regardless of the amount of insulation used, and heat must be provided—often in the form of electric resistance heat tape or cable inside the insulation next to the piping.

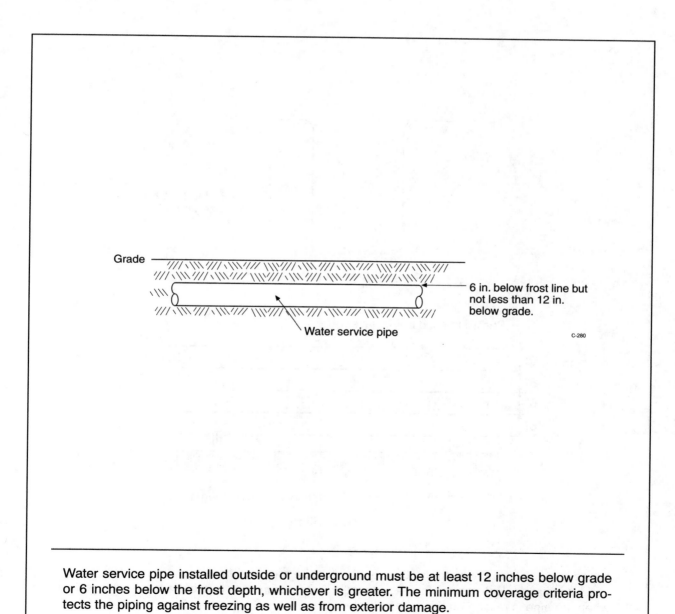

Water service pipe installed outside or underground must be at least 12 inches below grade or 6 inches below the frost depth, whichever is greater. The minimum coverage criteria protects the piping against freezing as well as from exterior damage.

Topic: Trenching and Bedding

Reference: IRC P2604.1

Category: General Plumbing Requirements

Subject: Trenching and Backfilling

Code Text: *Piping shall be installed in trenches so that the piping rests on solid and continuous bearing. When over excavated, the trench shall be backfilled to the proper grade with compacted earth, sand, fine gravel or similar granular material. Piping shall not be supported on rocks or blocks at any point. Rocky or unstable soil shall be over excavated by two or more pipe diameters and brought to the proper grade with suitable compacted granular material.*

Discussion and Commentary: Pipe is protected from damage by proper backfilling techniques. Backfilling is accomplished in 6-inch layers, with each layer being tamped into place. The material used for backfilling must be evenly distributed to avoid any pipe movement. Large rocks or frozen chunks of dirt must not be located within 6 inches of the pipe.

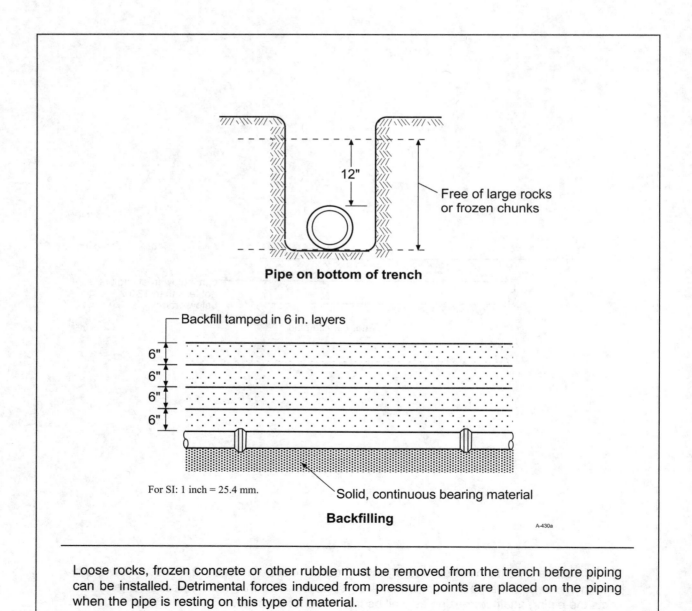

12"

Free of large rocks or frozen chunks

Pipe on bottom of trench

Backfill tamped in 6 in. layers

6"
6"
6"
6"

For SI: 1 inch = 25.4 mm.

Solid, continuous bearing material

Backfilling

A-430a

Loose rocks, frozen concrete or other rubble must be removed from the trench before piping can be installed. Detrimental forces induced from pressure points are placed on the piping when the pipe is resting on this type of material.

Topic: Protection of Footings

Reference: IRC P2604.4

Category: General Plumbing Requirements

Subject: Trenching and Backfilling

Code Text: *Trenching installed parallel to footings shall not extend below the 45-degree (0.79 mm) bearing plane of the bottom edge of a wall or footing.*

Discussion and Commentary: A footing requires a minimum load-bearing area to distribute the weight of the building. That plane extends downward at approximately a 45-degree angle from the base of the footing. The trench containing the piping must not be located below that load-bearing plane, to avoid undermining the building footing. The key to the provision is the term "parallel." Where extending a water or sewer pipe beneath a footing, such installation is usually perpendicular to the foundation and footing. To avoid undermining the load-bearing capacity of the soil, extend the trench beyond the load-bearing plane.

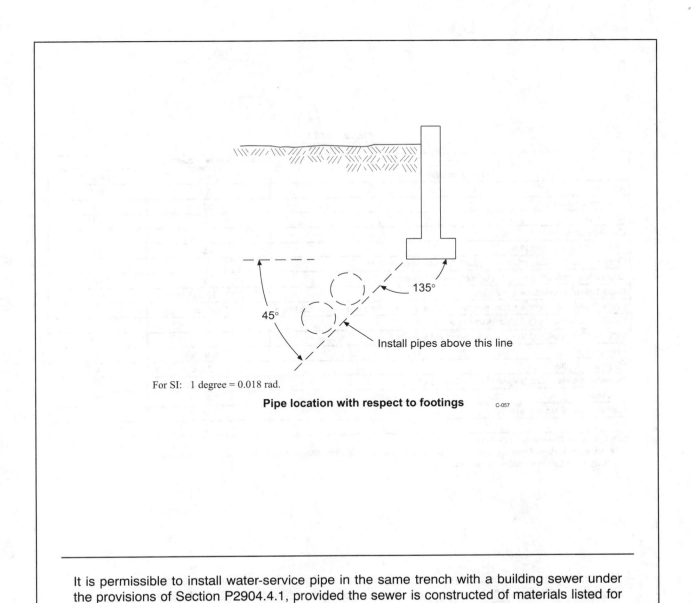

For SI: 1 degree = 0.018 rad.

Pipe location with respect to footings   C-057

It is permissible to install water-service pipe in the same trench with a building sewer under the provisions of Section P2904.4.1, provided the sewer is constructed of materials listed for underground use within a building. Otherwise, various methods of separation are described.

Topic: General Provisions **Category:** General Plumbing Requirements
Reference: IRC P2605.1 **Subject:** Support

Code Text: *Piping shall be supported so as to ensure alignment and prevent sagging, and allow movement associated with the expansion and contraction of the piping system. Piping in the ground shall be laid on a firm bed for its entire length, except where support is otherwise provided. Hangers and strapping shall be of approved material that will not promote galvanic action. Rigid support sway bracing shall be provided at changes in direction greater than 45 degrees (0.39 rad) for pipe sizes 4 inches (102 mm) and larger. Piping shall be supported at distances not to exceed those indicated in Table P2605.1.*

Discussion and Commentary: For larger pipes, hangers alone may not be sufficient to resist the forces created by water movement within the piping. Therefore, rigid bracing is required to restrict or eliminate lateral movement of both horizontal and vertical piping.

**TABLE P2605.1
PIPING SUPPORT**

PIPING MATERIAL	MAXIMUM HORIZONTAL SPACING (feet)	MAXIMUM VERTICAL SPACING
ABS pipe	4	10[b]
Aluminum tubing	10	15
Brass pipe	10	10
Cast-iron pipe	5[a]	15
Copper or copper alloy pipe	12	10
Copper or copper alloy tubing ($1^1/_4$ inch diameter and smaller)	6	10
Copper or copper alloy tubing ($1^1/_2$ inch diameter and larger)	10	10
Cross-linked polyethylene (PEX) pipe	2.67 (32 inches)	10[b]
Cross-linked polyethylene/aluminum/cross-linked polyethylene (PEX-AL-PEX) pipe	2.67 (32 inches)	4[b]
CPVC pipe or tubing (1 inch in diameter and smaller)	3	10[b]
CPVC pipe or tubing ($1^1/_4$ inch in diameter and larger)	4	10[b]
Lead pipe	Continuous	4
PB pipe or tubing	2.67 (32 inches)	4
Polyethylene/aluminum/polyethylene (PE-AL-PE) pipe	2.67 (32 inches)	4[b]
PVC pipe	4	10[b]
Stainless steel drainage systems	10	10[b]
Steel pipe	12	15

For SI: 1 inch = 25.4 mm, 1 foot = 304.8 mm.

a. The maximum horizontal spacing of cast-iron pipe hangers shall be increased to 10 feet where 10-foot lengths of pipe are installed.

b. Midstory guide for sizes 2 inches and smaller.

Depending on the type of pipe material, the maximum vertical and horizontal spacing between supports can vary. Whereas supports must occur at frequent intervals for horizontal piping due to the potential for sagging, vertical piping typically requires support only at each story height.

Topic: Minimum Size
Reference: IRC P2703.1

Category: Plumbing Fixtures
Subject: Tail Pieces

Code Text: *Fixture tail pieces shall be not less than 1.5 inches (38 mm) in diameter for sinks, dishwashers, laundry tubs, bathtubs and similar fixtures, and not less than 1.25 inches (32 mm) in diameter for bidets, lavatories and similar fixtures.*

Discussion and Commentary: Tail pieces are short lengths of pipe attached directly to a fixture by means of a flange for connection to other piping or traps. Tail pieces, as well as traps, continuous wastes, waste fittings and overflow fittings, must be of No. 20 gage minimum thickness when constructed of seamless drawn brass. Plastic tubular fittings are permitted when in conformance with ASTM F 409. The minimum size of $1^1/_2$ inches for sinks, dishwashers, laundry tubs and bathtubs is consistent with the minimum sizes for fixture drains, traps and trap arms.

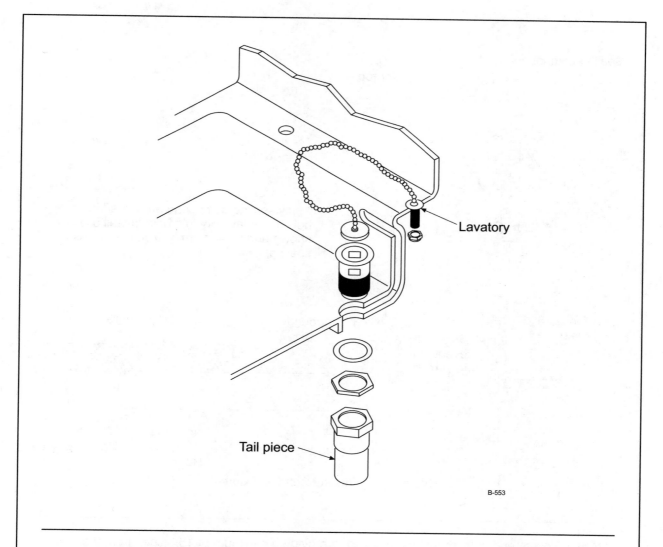

Lavatory

Tail piece

B-553

Fixture fittings, faucets and plumbing fixtures require smooth impervious surfaces with no surface defects that could reduce the efficiency, effectiveness or sanitary requirements of the code. Table P2701.1 lists the appropriate material standards for various plumbing items.

Topic: General Provisions
Reference: IRC P2705.1

Category: Plumbing Fixtures
Subject: Installation

Code Text: *Floor-outlet or floor-mounted fixtures shall be secured to the drainage connection and to the floor, when so designed, by screws, bolts, washers, nuts and similar fasteners of copper, brass or other corrosion-resistant material. Wall-hung fixtures shall be rigidly supported so that strain is not transmitted to the plumbing system. Where fixtures come in contact with walls, the contact area shall be water tight. Plumbing fixtures shall be functionally accessible. The location of piping, fixtures or equipment shall not interfere with the operation of windows or doors.*

Discussion and Commentary: The installation of fixtures shall be accomplished based on structural, sanitation and operational considerations.

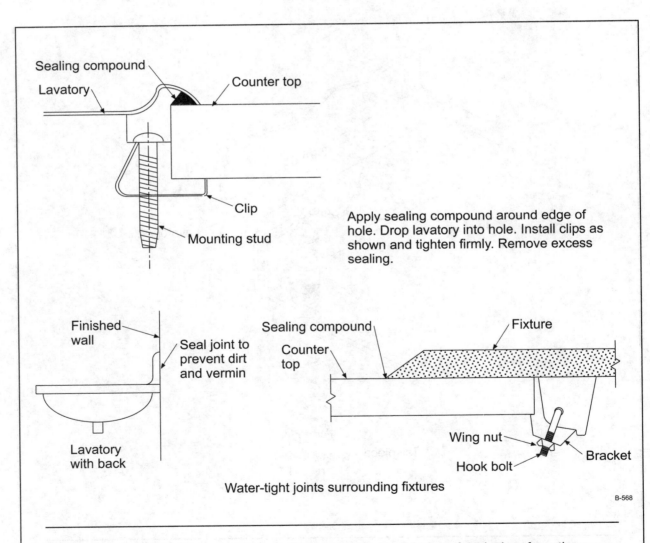

Apply sealing compound around edge of hole. Drop lavatory into hole. Install clips as shown and tighten firmly. Remove excess sealing.

Water-tight joints surrounding fixtures

B-568

Water closets and bidets must be installed to provide a minimum of 15 inches from the centerline to adjacent walls, creating a minimum compartment size of 30 inches. In front of the water closet, a minimum of 21 inches is mandated. This clearance also applies to lavatories and bidets.

Topic: Standpipes
Reference: IRC P2706.2

Category: Plumbing Fixtures
Subject: Waste Receptors

Code Text: *Standpipes shall extend a minimum of 18 inches (457 mm) and a maximum of 42 inches (1067 mm) above the trap weir. Access shall be provided to all standpipe traps and drains for rodding. A laundry tray waste line is permitted to connect into a standpipe for the automatic clothes washer drain. The standpipes shall not be less than 30 inches (762 mm) as measured from the crown weir. The outlet of the laundry tray shall be a maximum horizontal distance of 30 inches (762 mm) from the standpipe trap.*

Discussion and Commentary: A standpipe can serve as an indirect waste receptor. The established minimum receptor height aims to provide a minimal retention capacity and head pressure to prevent overflow. This height limitation is especially necessary when an indirect waste pipe receives a high rate of discharge, such as the pumped discharge from a clothes washer.

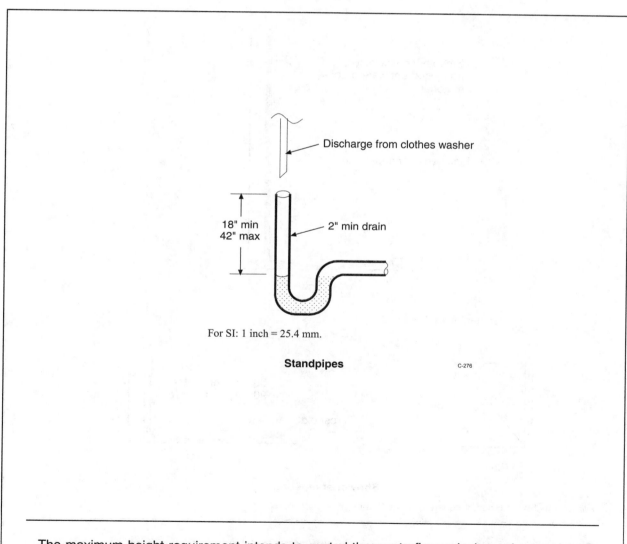

For SI: 1 inch = 25.4 mm.

Standpipes C-276

The maximum height requirement intends to control the waste flow velocity at the trap inlet. Excessive velocity promotes trap siphonage. Although the standpipe itself can be concealed within construction, the point of indirect connection must be accessible to drain cleaning equipment.

Topic: General Provisions
Reference: IRC P2708.1

Category: Plumbing Fixtures
Subject: Showers

Code Text: *Shower compartments shall have at least 900 square inches (0.581 m2) of interior cross-sectional area. Shower compartments shall not be less than 30 inches (752 mm) in minimum dimension measured from the finished interior dimension of the shower compartment, exclusive of fixture valves, shower heads, soap dishes, and safety grab bars or rails. The minimum required area and dimension shall be measured from the finished interior dimension at a height equal to the top of the threshold at a point tangent to its centerline and shall be continued to a height not less than 70 inches (1778 mm) above the shower drain outlet. See exception for showers with fold-down seats.*

Discussion and Commentary: The dimensional criteria facilitate safety and comfort based on typical body movement while showering. Fold-down seats can encroach into the required dimensions, provided the 900 square inches are available when the seats are folded.

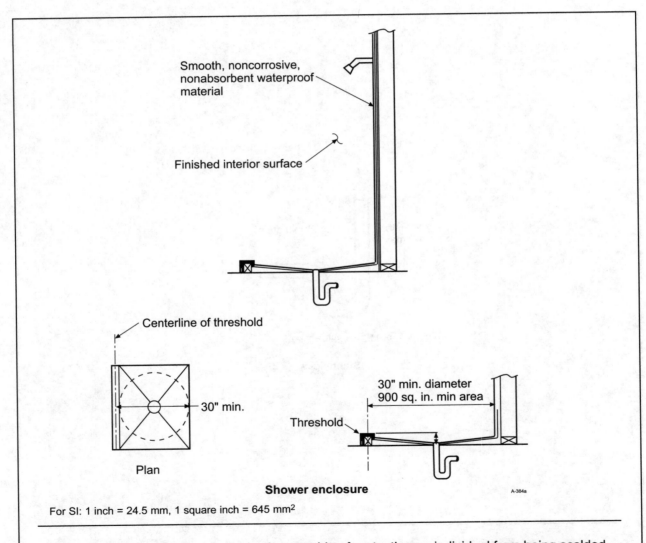

Smooth, noncorrosive, nonabsorbent waterproof material

Finished interior surface

Centerline of threshold

30" min.

Plan

30" min. diameter
900 sq. in. min area

Threshold

Shower enclosure

A-384a

For SI: 1 inch = 24.5 mm, 1 square inch = 645 mm²

Every shower must have a control valve capable of protecting an individual from being scalded. These devices are also required to protect against rapid temperature fluctuation by automatically maintaining the discharge temperature to plus or minus 3°F of the selected temperature.

2003 IRC Study Companion

Topic: Construction
Reference: IRC P2709.1

Category: Plumbing Fixtures
Subject: Shower Receptors

Code Text: *Shower receptors shall have a finished curb threshold not less than 1 inch (25.4 mm) below the sides and back of the receptor. The curb shall be not less than 2 inches (51 mm) and not more than 9 inches (229 mm) in depth when measured from the top of the curb to the top of the drain. The finished floor shall slope uniformly toward the drain not less than one-fourth unit vertical in 12 units horizontal (2-percent slope) nor more than 0.5 inch (12.7 mm), and floor drains shall be flanged to provide a watertight joint in the floor.*

Discussion and Commentary: When a shower is built in place, a shower pan is required under the shower's finished floor. Various materials are used for shower pans, including sheet lead, sheet copper, polyethylene sheet, chlorinated polyethylene sheet and preformed ABS. The shower pan is designed to collect any water that leaks through the shower floor.

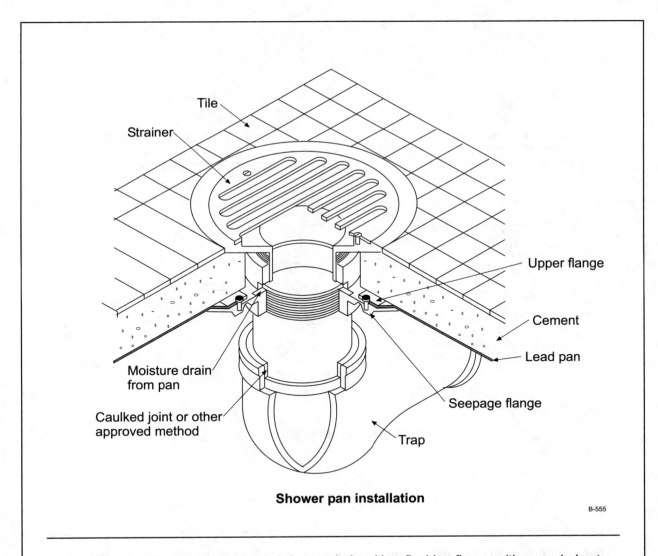

Shower pan installation

B-555

The shower pan must connect to the shower drain with a flashing flange with weep holes to allow the leaking water to enter the drainage system. A clamping ring or similar device is used to clamp down the waterproof membrane and make a watertight joint.

Topic: Sink and Dishwasher

Reference: IRC P2717.2, P2717.3

Category: Plumbing Fixtures

Subject: Dishwashing Machines

Code Text: *A sink and dishwasher are permitted to discharge through a single 1.5-inch (38 mm) trap. The discharge pipe from the dishwasher shall be increased to a minimum of 0.75 inch (19.1 mm) in diameter and shall be connected with a wye fitting to the sink tailpiece. The dishwasher waste line shall rise and be securely fastened to the underside of the counter before connecting to the sink tailpiece. The combined discharge from a sink, dishwasher and waste grinder is permitted to discharge through a single 1.5 inch (38 mm) trap.*

Discussion and Commentary: Dishwashing machines may connect directly to the drainage system. An indirect connection by air gap or air break is not required. The direct connection is acceptable because the potable water supply is adequately protected against backflow.

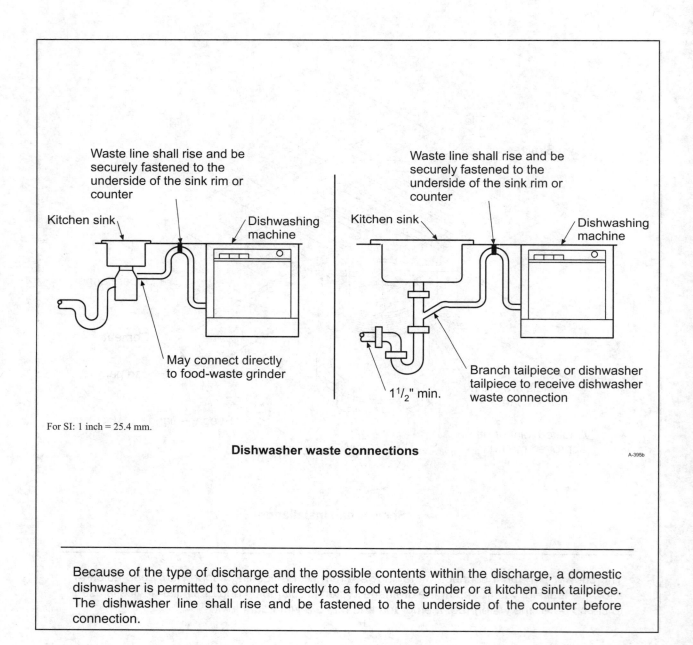

For SI: 1 inch = 25.4 mm.

Dishwasher waste connections

A-395b

Because of the type of discharge and the possible contents within the discharge, a domestic dishwasher is permitted to connect directly to a food waste grinder or a kitchen sink tailpiece. The dishwasher line shall rise and be fastened to the underside of the counter before connection.

QUIZ

Study Session 14

1. Testing of the building sewer shall be accomplished with a minimum _____ head of water with the ability to maintain such pressure for a minimum of _____.
    a. 8-foot, 10 minutes
    b. 8-foot, 15 minutes
    c. 10-foot, 10 minutes
    d. 10-foot, 15 minutes

2. When tested by air, the DWV system shall be tested at _____ with the ability to maintain such pressure for a minimum of _____.
    a. 5 psi, 10 minutes
    b. 5 psi, 15 minutes
    c. 10 psi, 10 minutes
    d. 10 psi, 15 minutes

3. The water-supply system shall be proved tight under a water pressure not less than the working pressure of the system, or by a minimum _____ air test for other than plastic piping.
    a. 5 psi
    b. 10 psi
    c. 20 psi
    d. 50 psi

4. Where copper piping is to pass through holes bored in wood studs, the pipe shall be protected by shield plates except for those holes that are a minimum of _____ from the nearest edge.
    a. $5/8$ inch
    b. $3/4$ inch
    c. $1 3/8$ inches
    d. $1 1/2$ inches

5. Shield plates required to protect piping installed through notches or bored holes in framing members shall extend a minimum of _____ above sole plates and below top plates.
    a. $1/2$ inch
    b. 1 inch
    c. 2 inches
    d. 3 inches

6. Unless provided with a relieving arch, a building drain passing through a foundation wall shall pass through a pipe sleeve sized _____ greater than the pipe passing through.
    a. 1 inch
    b. 2 inches
    c. one pipe size
    d. two pipe sizes

7.   Under all circumstances, water service pipe shall be installed a minimum of _____ deep.
      a. 6 inches
      b. 12 inches
      c. 24 inches
      d. 30 inches

8.   Where the frost line is at 30 inches, water service pipe shall be installed a minimum of _____ below grade.
      a. 30 inches
      b. 32 inches
      c. 36 inches
      d. 42 inches

9.   Where unstable soil conditions exist in an area to be trenched for a piping installation, the trench shall be overexcavated by a minimum of _____.
      a. 4 inches
      b. 6 inches
      c. one pipe diameter
      d. two pipe diameters

10.   The backfilling of a trench where piping is installed shall be done in maximum _____ layers of loose earth.
      a. 4-inch
      b. 6-inch
      c. 8-inch
      d. 12-inch

11.   One-half-inch copper water pipe installed horizontally shall be supported at maximum intervals of _____.
      a. 3 feet
      b. 4 feet
      c. 6 feet
      d. 12 feet

12.   PB pipe shall be supported at maximum intervals of _____ when installed horizontally.
      a. 32 inches
      b. 36 inches
      c. 60 inches
      d. 72 inches

13.   A fixture tail piece for a laundry tub shall be a minimum of _____ in diameter.
      a. 1 inch
      b. $1^1/_4$ inches
      c. $1^1/_2$ inches
      d. 2 inches

14. A fixture with a concealed slip-joint connection shall be provided with an access panel or utility space a minimum of _____ in its smallest dimension.
   a. 12 inches
   b. 18 inches
   c. 30 inches
   d. 36 inches

15. Standpipes shall extend a minimum of _____ and a maximum of _____ above the trap weir.
   a. 18 inches, 30 inches
   b. 18 inches, 42 inches
   c. 24 inches, 30 inches
   d. 24 inches, 42 inches

16. The outlet of a laundry tray shall be a maximum horizontal distance of _____ from the standpipe trap.
   a. 18 inches
   b. 30 inches
   c. 42 inches
   d. 60 inches

17. Shower compartments shall have a minimum interior cross-sectional area of _____ square inches.
   a. 900
   b. 1,080
   c. 1,296
   d. 1,440

18. The curb for a shower receptor shall have a minimum depth of _____ and a maximum depth of _____ when measured from the top of the curb to the top of the drain.
   a. 1 inch, 6 inches
   b. 1 inch, 9 inches
   c. 2 inches, 6 inches
   d. 2 inches, 9 inches

19. A lavatory shall have a waste outlet a minimum of _____ in diameter.
   a. 1 inch
   b. $1^1/_4$ inches
   c. $1^1/_2$ inches
   d. 2 inches

20. Unless an approved alternative is provided, flush valve seats in tanks for flushing water closets shall be a minimum of _____ above the flood-level rim of the bowl to which it is connected.
   a. $^1/_2$ inch
   b. 1 inch
   c. $1^1/_2$ inches
   d. 2 inches

21. Bathtubs shall have outlets and overflows a minimum of _____ in diameter.
   a. 1 inch
   b. $1^1/_4$ inches
   c. $1^1/_2$ inches
   d. 2 inches

22. Sinks shall have a waste outlet a minimum of _____ in diameter.
   a. $1^1/_4$ inches
   b. $1^1/_2$ inches
   c. 2 inches
   d. $3^1/_2$ inches

23. Food waste grinders shall be connected to a drain having a minimum diameter of _____.
   a. $1^1/_4$ inches
   b. $1^1/_2$ inches
   c. 2 inches
   d. $3^1/_2$ inches

24. Floor drains shall have waste outlets a minimum of _____ in diameter.
   a. $1^1/_4$ inches
   b. $1^1/_2$ inches
   c. 2 inches
   d. $3^1/_2$ inches

25. What is the minimum size drain required for a macerating toilet system?
   a. $^3/_4$ inch
   b. $1^1/_4$ inches
   c. $1^1/_2$ inches
   d. 2 inches

26. Where a smoke test is utilized to verify gas tightness of the DWV system, a pressure equivalent to a 1-inch water column shall be applied and maintained for a minimum time period of _____ minutes.
   a. 10
   b. 15
   c. 30
   d. 60

27. Where an air test of the rough plumbing installation requires a gauge pressure of 5 psi, the testing gauge shall have maximum increments of _____ psi.
   a. 0.10
   b. 0.50
   c. 1.00
   d. 2.00

28. Trenches used for the installation of piping that are excavated parallel to footings shall extend a maximum of _____ below the bearing plane of the bottom edge of the footing.
   a. 30 degrees
   b. 45 degrees
   c. 12 inches
   d. 30 inches

29. Which of the following materials and products shall be certified by an approved third-party certification agency?
   a. plumbing fixtures
   b. special waste system components
   c. subsoil drainage system components
   d. water distribution system safety devices

30. There shall be a minimum of _____ inches clearance in front of a water closet to any wall, fixture or door.
   a. 18
   b. 21
   c. 24
   d. 27

INTERNATIONAL RESIDENTIAL CODE
Study Session 15

Chapters 28, 29 and 30—Water Heaters, Water Supply and Distribution, and Sanitary Drainage

OBJECTIVE: To provide an understanding of the general provisions for the installation of water heaters, the supply and distribution of water, and the materials, design, construction and installation of sanitary drainage systems.

REFERENCE: Chapters 28, 29 and 30, 2003 *International Residential Code*

KEY POINTS:

- When is a water heater required to be installed in a pan? What type of pan is mandated? How big must it be? How shall it be drained?
- What special condition applies where a water heater having an ignition source is installed in a garage?
- What special conditions are applicable where a combination water heater-space heating system is utilized?
- What type of relief valve or valves is required for a water heater? How is a pressure-relief valve regulated? A temperature-relief valve?
- How is the water heater discharge pipe to be installed? What is the minimum required size of the discharge pipe? Where is the termination point to be located?
- How shall backflow preventers be applied?
- What special conditions exist where both potable and nonpotable water distribution systems are installed? How shall such systems be identified?
- How is the potable water supply system to be protected? What methods of backflow protection are acceptable?
- What is the minimum permitted air gap for various fixtures? Where shall the critical level of an atmospheric-type vacuum breaker be set?
- How must the potable water supply to boilers be protected? To heat exchangers? To lawn irrigation systems? To automatic fire sprinkler systems? To solar systems?
- What are the maximum water consumption flow rates and quantities for various plumbing fixtures and fittings?
- What is the minimum permitted static pressure at the building entrance for water service? The maximum static pressure? How should thermal expansion be addressed?
- What is the minimum size for a water service pipe? How should the water distribution system be sized?
- How should water service pipe be installed? What special requirements apply where the building sewer piping is installed in the same trench? What special concerns are addressed in the installation of water distribution pipe?
- What types of joints and connections are prohibited in the DWV system? How are approved joints to be made?
- How is the load on the DWV-system piping determined? What type of fittings are to be used?
- Where shall cleanouts be located? What should be done at changes in direction? What clearances are needed in front of cleanouts? What is the minimum size of a cleanout?
- What are the conditions of installation for sewage ejectors or sewage pumps?

Topic: Required Pan

Reference: IRC P2801.5

Category: Water Heaters

Subject: General Provisions

Code Text: *Where water heaters or hot water storage tanks are installed in locations where leakage of the tanks or connections will cause damage, the tank or water heater shall be installed in a galvanized steel pan having a minimum thickness of 24 gage (0.016 inch)(0.4 mm) or other pans for such use. The pan shall be not less than 1.5 inches (38 mm) deep and shall be of sufficient size and shape to receive all dripping and condensate from the tank or water heater. The pan shall be drained by an indirect waste pipe having a minimum diameter of $^3/_4$ inch (19 mm) or the outlet diameter of the relief valve, whichever is larger.*

Discussion and Commentary: The pan catches water from a leaking tank, a leaking connection to the tank or condensate from the tank, and it may also be used as the indirect waste receptor for the relief valve discharge.

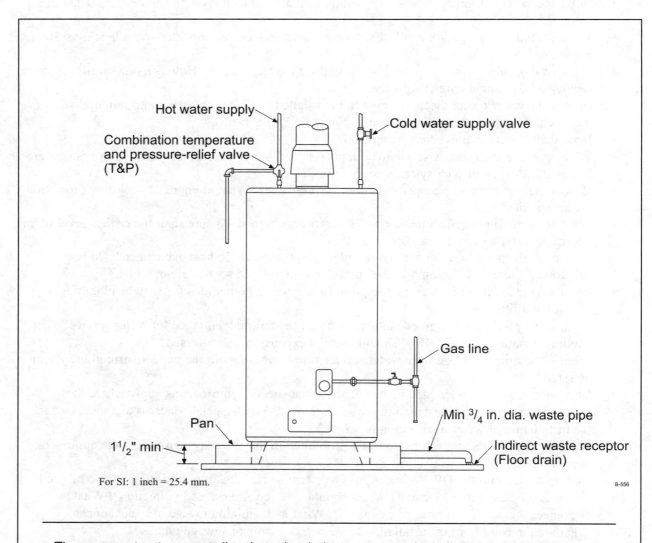

For SI: 1 inch = 25.4 mm.

B-556

The pan must not connect directly to the drainage system. An indirect waste pipe is necessary to prevent backflow from the drainage system into the pan. Because the relief valve discharge piping typically terminates in the pan, the waste pipe must be as large as the discharge pipe.

2003 IRC Study Companion

Topic: Garage Installations
Reference: IRC P2801.6

Category: Water Heaters
Subject: General Provisions

Code Text: *Water heaters having an ignition source shall be elevated such that the source of ignition is not less than 18 inches (457 mm) above the garage floor.*

Discussion and Commentary: Unless properly elevated, a fuel-burning water heater installed in a garage poses a serious fire hazard. A garage is commonly used as a storage area for gasoline and common household items such as paint, paint thinner and solvents. Fumes from these flammable liquids tend to collect near the floor and produce vapors denser than air. The 18-inch requirement is intended to prevent an energy or heat source from becoming a source of ignition. The ignition source could be a pilot flame, burner, burner igniter or electrical component capable of producing a spark or an arc. The 18-inch elevation must be increased where required by the manufacturer's installation instructions.

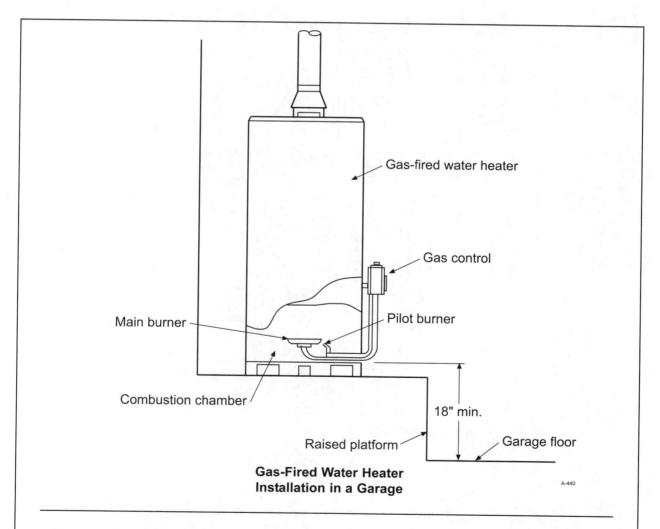

**Gas-Fired Water Heater
Installation in a Garage**

A-440

The provision addresses all types of water heaters— not just those that are gas- or oil-fired. It applies equally to gas-fired water heaters of the direct vent type. The requirement would not apply only in the unlikely event that the water heater were a true sealed combustion chamber type. Only if, in the unlikely event, the water heater were a true sealed combustion chamber type, the requirement would not apply.

Topic: Relief Valves   **Category:** Water Heaters
Reference: IRC P2803.1, P2803.3, P2803.4   **Subject:** Relief Valves

Code Text: *Appliances and equipment used for heating water or storing hot water shall be protected by 1) a separate pressure-relief valve and a separate temperature-relief valve; or 2) a combination pressure- and temperature-relief valve. Pressure-relief valves . . . shall be set to open at least 25 psi (172 kPa) above the system pressure but not over 150 psi (1034 kPa). The relief-valve setting shall not exceed the tank's rated working pressure. Temperature-relief valves . . . shall be installed such that the temperature-sensing element monitors the water within the top 6 inches (152 mm) of the tank. The valve shall be set to open at a maximum temperature of 210°F (99°C).*

Discussion and Commentary: A combination temperature- and pressure-relief valve or separate temperature-relief and pressure-relief valves protect the water heater against possible explosion.

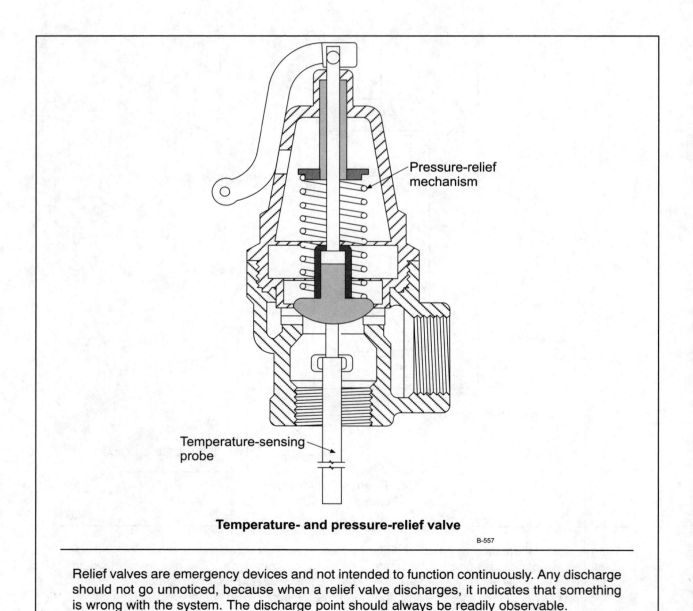

Pressure-relief
mechanism

Temperature-sensing
probe

Temperature- and pressure-relief valve

B-557

Relief valves are emergency devices and not intended to function continuously. Any discharge should not go unnoticed, because when a relief valve discharges, it indicates that something is wrong with the system. The discharge point should always be readily observable.

Topic: General Provisions
Reference: IRC P2902.1

Category: Water Supply and Distribution
Subject: Protection of Potable Water Supply

Code Text: *A potable water supply system shall be designed and installed in such a manner as to prevent contamination from nonpotable liquids, solids or gases being introduced into the potable water supply. Connections shall not be made to a potable water supply in a manner that could contaminate the water supply or provide a cross-connection between the supply and source of contamination unless an approved backflow-prevention device is provided. Cross-connections between an individual water supply and a potable public water supply shall be prohibited.*

Discussion and Commentary: Arguably, the most important aspect of the plumbing provisions of the IRC is the protection of potable water systems. History is filled with local and widespread occurrences of sickness and disease caused by not safeguarding the water supply. It is imperative that the potable water supply be maintained in a safe-for-drinking condition at all times and at all outlets.

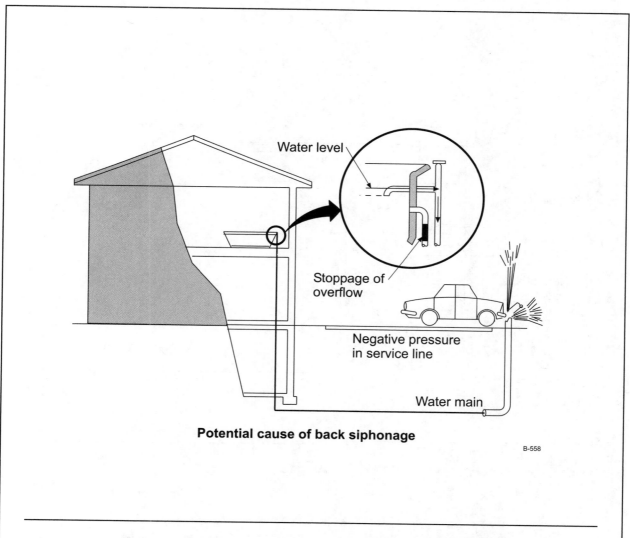

Water level

Stoppage of overflow

Negative pressure in service line

Water main

Potential cause of back siphonage

B-558

As a general condition, all dwelling units must be provided with potable water. If the building contains both a potable and nonpotable water-distribution system, each system needs to be clearly identified by an appropriate method, such as a color marking or metal tags.

Topic: Air Gaps
Reference: IRC P2902.2.1

Category: Water Supply and Distribution
Subject: Protection of Potable Water Supply

Code Text: *The minimum air gap shall be measured vertically from the lowest end of a water supply outlet to the flood level rim of the fixture or receptor into which such potable water outlets discharge. The minimum required air gap shall be twice the diameter of the effective opening of the outlet, but in no case less than the values specified in Table P2902.2.1. An air gap is required at the discharge point of a relief valve or piping.*

Discussion and Commentary: An air gap is not a device and has no moving parts. It is, therefore, the most effective and dependable method of preventing backflow. The potable water opening or outlet is terminated at an elevation above the level of the source of contamination. If the appropriate minimum air gap is provided, the potable water could only be contaminated if the entire area or room flooded to a depth that would submerge the potable water opening or outlet.

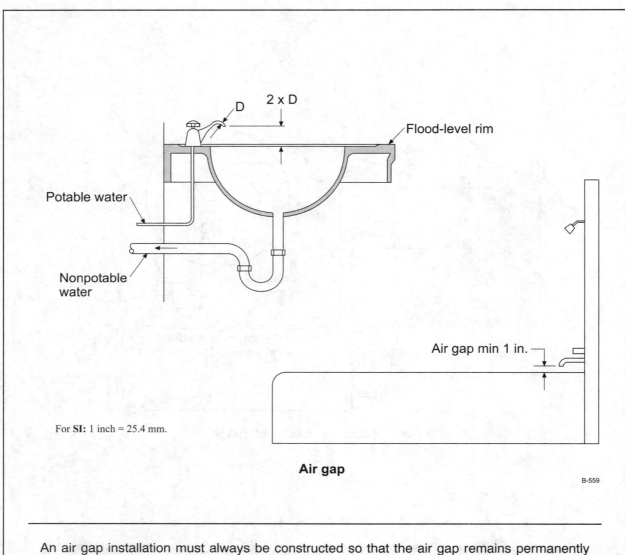

For **SI:** 1 inch = 25.4 mm.

Air gap

B-559

An air gap installation must always be constructed so that the air gap remains permanently fixed. Air gaps may be fabricated from commercially available plumbing components or obtained as separate units and integrated into plumbing and piping systems.

Topic: Design Criteria
Reference: IRC P2903.1, P2903.2

Category: Water Supply and Distribution
Subject: Water-supply System

Code Text: *The water service and water distribution systems shall be designed and pipe sizes shall be selected such that under conditions of peak demand, the capacities at the point of outlet discharge shall not be less than shown in Table P2903.1. The maximum water consumption flow rates and quantities for all plumbing fixtures and fixture fittings shall be in accordance with Table P2903.2.*

Discussion and Commentary: The water distribution system must be capable of supplying a sufficient volume of potable water at pressures adequate to enable all plumbing fixtures, devices and appurtenances to function properly and without undue noise. Each fixture has a given flow rate and minimum water supply pressure needed to properly operate the fixture.

TABLE P2903.1
REQUIRED CAPACITIES AT POINT OF OUTLET DISCHARGE

FIXTURE AT POINT OF OUTLET	FLOW RATE (gpm)	FLOW PRESSURE (psi)
Bathtub	4	8
Bidet	2	4
Dishwasher	2.75	8
Laundry tub	4	8
Lavatory	2	8
Shower	3	8
Shower, temperature controlled	3	20
Sillcock, hose bibb	5	8
Sink	2.5	8
Water closet, flushometer tank	1.6	15
Water closet, tank, close coupled	3	8
Water closet, tank, one-piece	6	20

For SI: 1 gallon per minute = 3.785 L/m, 1 pound per square inch = 6.895 kPa.

TABLE P2903.2
MAXIMUM FLOW RATES AND CONSUMPTION FOR PLUMBING FIXTURES AND FIXTURE FITTINGS[b]

PLUMBING FIXTURE OR FIXTURE FITTING	PLUMBING FIXTURE OR FIXTURE FITTING
Lavatory faucet	2.2 gpm at 60 psi
Shower head[a]	2.5 gpm at 80 psi
Sink faucet	2.2 gpm at 60 psi
Water closet	1.6 gallons per flushing cycle

For SI: 1 gallon per minute = 3.785 L/m, 1 pound per square inch = 6.895 kPa.

a. A handheld shower spray is also a shower head.

b. Consumption tolerances shall be determined from referenced standards.

The fixtures listed in Table P2903.2 function in a cyclical manner or consume water continuously for the duration of the fixture use. Water is conserved by limiting the flow rate for manually controlled fixtures and by limiting the volume per cycle usage for cyclically operating fixtures.

Topic: Determining Fixture Units
Reference: IRC P2903.6

Category: Water Supply and Distribution
Subject: Water Supply System

Code Text: *Supply loads in the building water-distribution system shall be determined by total load on the pipe being sized, in terms of water-supply fixture units (w.s.f.u.), as shown in Table P2903.6, and gallon per minute (gpm) flow rates. For fixtures not listed, choose a w.s.f.u. value of a fixture with similar flow characteristics.*

Discussion and Commentary: A prescriptive method for estimating the water demand for a building has evolved through the years and has been accepted as a standard method of sizing water systems. The method is based upon weighing fixtures in accordance with their water supply load-producing effects on the water distribution system. This approach couples the flow characteristics of the various fixtures with probability curves of simultaneous flushing of the fixtures.

TABLE P2903.6
WATER-SUPPLY FIXTURE-UNIT VALUES FOR VARIOUS PLUMBING FIXTURES AND FIXTURE GROUPS

TYPE OF FIXTURES OR GROUP OF FIXTURES	WATER-SUPPLY FIXTURE-UNIT VALUE (w.s.f.u.)		
	Hot	Cold	Combined
Bathtub (with/without overhead shower head)	1.0	1.0	1.4
Clothes washer	1.0	1.0	1.4
Dishwasher	1.4	—	1.4
Full-bath group with bathtub (with/without shower head) or shower stall	1.5	2.7	3.6
Half-bath group (water closet and lavatory)	0.5	2.5	2.6
Hose bibb (sillcock)[a]	—	2.5	2.5
Kitchen group (dishwasher and sink with/without garbage grinder)	1.9	1.0	2.5
Kitchen sink	1.0	1.0	1.4
Laundry group (clothes washer standpipe and laundry tub)	1.8	1.8	2.5
Laundry tub	1.0	1.0	1.4
Lavatory	0.5	0.5	0.7
Shower stall	1.0	1.0	1.4
Water closet (tank type)	—	2.2	2.2

For SI: 1 gallon per minute = 3.785 L/m.

a. The fixture unit value 2.5 assumes a flow demand of 2.5 gpm, such as for an individual lawn sprinkler device. If a hose bibb/sill cock will be required to furnish a greater flow rate, the equivalent fixture-unit value may be obtained from Table P2903.6 or Table P2903.7.

The water service pipe must be sized in accordance with Section P2903.7, but in no case may it be less in size than $3/4$ inch. Using a prescriptive approach, the size of the water distribution system including the service pipe, meter and main distribution pipe can be determined.

Topic: Valves
Reference: IRC P2903.9

Category: Water Supply and Distribution
Subject: Water-supply System

Code Text: *Each dwelling unit shall be provided with an accessible main shutoff valve near the entrance of the water service. The valve shall be of a full-open type having nominal restriction to flow, with provision for drainage such as a bleed orifice or installation of a separate drain valve. A readily accessible full-open valve shall be installed in the cold-water supply pipe to each water heater at or near the water heater. Valves serving individual fixtures, appliances, risers and branches shall be provided with access. An individual shutoff valve shall be required on the fixture supply pipe to each plumbing fixture other than bathtubs and showers.*

Discussion and Commentary: All gate valves and "rated" globe valves are fullway. Gate valves are easily identified by observing the stem when opening the valve. All gate valve stems have stationary stems, unlike globe valves, which have rising stems. Nonrated globe valves are not fullway and are referred to as compression stops.

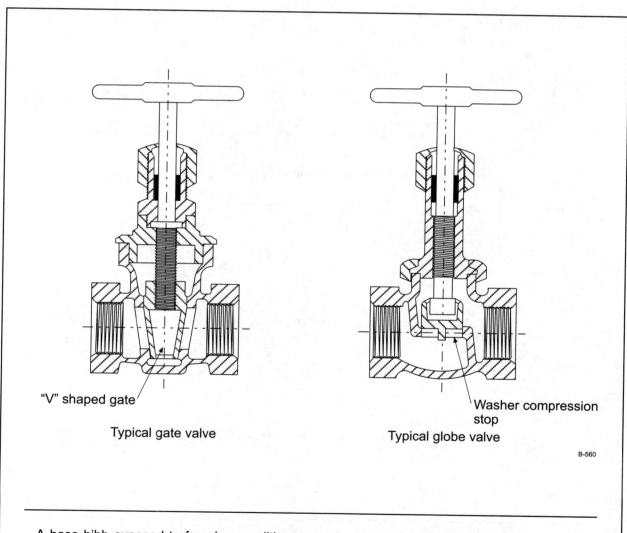

"V" shaped gate

Typical gate valve

Washer compression stop

Typical globe valve

B-560

A hose bibb exposed to freezing conditions must be equipped with a stop and waste valve inside the building so that it can be drained or controlled during freezing periods. A frost-proof bibb installed where the stem extends into conditioned space is not required to be valved separately.

Topic: Fittings
Reference: IRC P3002.3

Category: Sanitary Drainage
Subject: Materials

Code Text: *Fittings shall be approved and compatible with the type of piping being used and shall be of a sanitary or DWV design for drainage and venting. Waterpipe fittings shall be permitted for use in engineer designed systems where the design indicates compliance with Section P3101.2.1. Drainage fittings shall have a smooth interior waterway of the same diameter as the piping served. Threaded drainage pipe fittings shall be of the recessed drainage type, black or galvanized. Drainage fittings shall be designed to maintain one-fourth unit vertical in 12 units horizontal (2-percent slope) grade.*

Discussion and Commentary: Sanitary, DWV, or water-pipe fittings are acceptable in the dry section of the vent systems. Drainage fittings should have no ledges, shoulders or reductions that can retard or obstruct drainage flow in the piping.

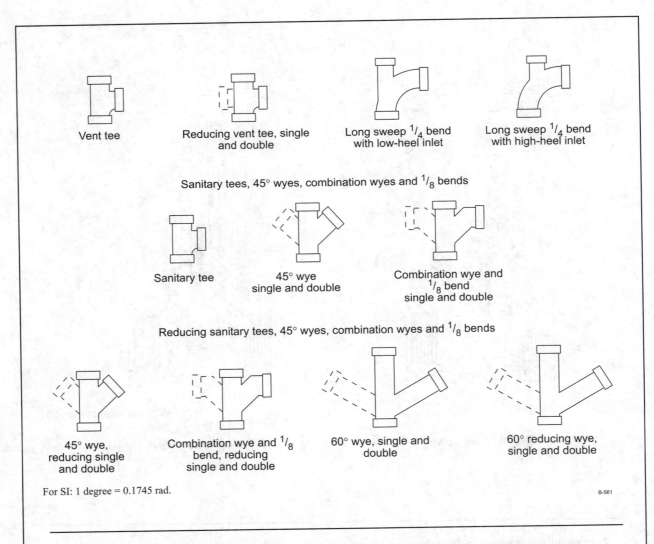

Vent tee

Reducing vent tee, single and double

Long sweep $^1/_4$ bend with low-heel inlet

Long sweep $^1/_4$ bend with high-heel inlet

Sanitary tees, 45° wyes, combination wyes and $^1/_8$ bends

Sanitary tee

45° wye single and double

Combination wye and $^1/_8$ bend single and double

Reducing sanitary tees, 45° wyes, combination wyes and $^1/_8$ bends

45° wye, reducing single and double

Combination wye and $^1/_8$ bend, reducing single and double

60° wye, single and double

60° reducing wye, single and double

For SI: 1 degree = 0.1745 rad.

B-561

Only materials approved for the specific type of installation should be used for DWV systems. These materials are available in several levels of quality, which vary with regard to characteristics and properties; thus, each may be used only in accordance with its referenced standard.

Topic: Cleanouts
Reference: IRC P3005.2

Category: Sanitary Drainage
Subject: Drainage System

Code Text: *Cleanouts shall be installed not more than 100 feet (30 480 mm) apart in horizontal drainage lines. Cleanouts shall be installed at each change of direction of the drainage system greater than 45 degrees, except not more than one cleanout shall be required in each 40 feet (12 192 mm) of run regardless of change in direction.*

Discussion and Commentary: Because a cleanout is designed to provide access into the drainage system, the cleanout itself must be accessible, regardless if the piping it serves is concealed or in a location not readily accessible. The cleanout is required to extend up to or above the finished grade either inside or outside the building. A cleanout must be accessible for use wherever it is located in the drainage system and must not be covered by any permanent, nonaccessible construction.

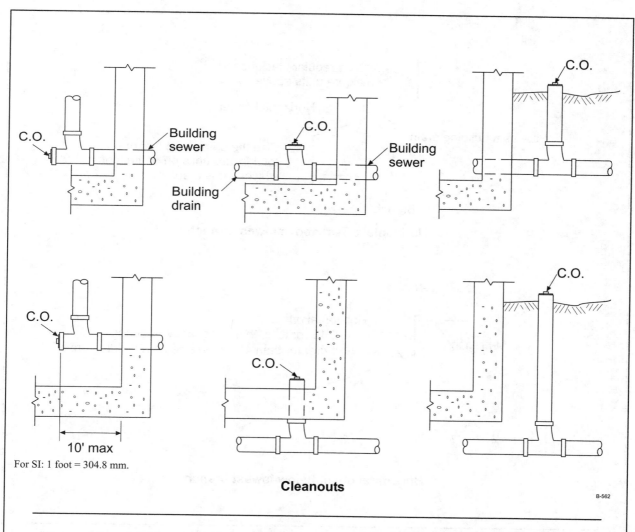

For SI: 1 foot = 304.8 mm.

Cleanouts

B-562

For a cleanout to be used, there must be adequate clear space provided to facilitate the cleaning operation. For pipes 3 inches or larger, at least 18 inches must be provided in front of the cleanout, measured perpendicular to the opening. Only 12 inches are needed on smaller pipes.

Topic: Vertical and Horizontal Offsets
Reference: IRC P3006

Category: Sanitary Drainage
Subject: Sizing of Drain Pipe Offsets

Code Text: *An offset in a vertical drain, with a change of direction of 45 degrees (0.79 rad) or less from the vertical, shall be sized as a straight vertical drain. A stack with an offset of more than 45 degrees (0.79 rad) from the vertical shall be sized in accordance with three listed conditions. In soil or waste stacks below the lowest horizontal branch, there shall be no change in diameter required if the offset is made at an angle not greater than 45 degrees (0.79 rad) from the vertical. If an offset greater than 45 degrees (0.79 rad) from the vertical is made, the offset and stack below it shall be sized as a building drain.*

Discussion and Commentary: Depending on the conditions involved, there are various methods for sizing offsets in drainage piping. The most complicated of the methods involves horizontal offsets above the lowest branch.

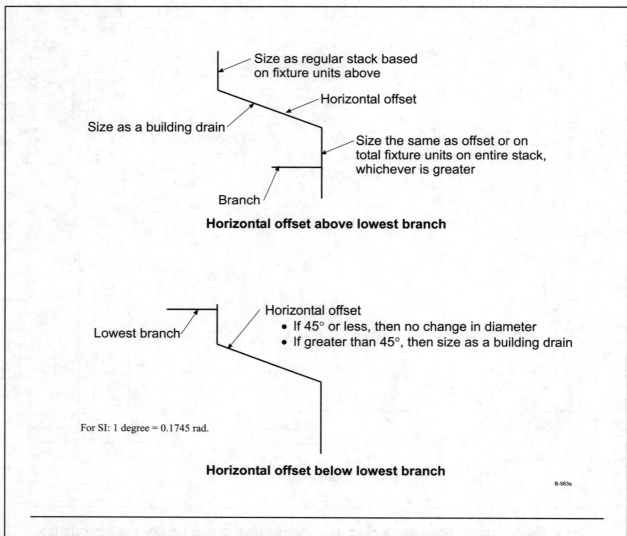

Size as regular stack based on fixture units above

Horizontal offset

Size as a building drain

Size the same as offset or on total fixture units on entire stack, whichever is greater

Branch

Horizontal offset above lowest branch

Horizontal offset
• If 45° or less, then no change in diameter
• If greater than 45°, then size as a building drain

Lowest branch

For SI: 1 degree = 0.1745 rad.

Horizontal offset below lowest branch

B-563a

Horizontal offsets below the lowest branch should not change size if the offset makes an angle less than 45 degrees from the vertical. Where the offset exceeds 45 degrees from the vertical, it should be sized as a building drain in accordance with Table P3005.4.2.

Topic: Sewage Ejectors or Sewage Pumps
Reference: IRC P3007.1, P3007.2

Category: Sanitary Drainage
Subject: Sumps and Ejectors

Code Text: *A sewage ejector, sewage pump, or grinder pump receiving discharge from a water closet shall have minimum discharge velocity of 1.9 feet per second (0.579 m/s) throughout the discharge piping to the point of connection with a gravity building drain, gravity sewer or pressure sewer system. Building drains which cannot be discharged to the sewer by gravity flow shall be discharged into a tightly covered and vented sump from which the contents shall be lifted and discharged into the building gravity drainage system by automatic pumping equipment.*

Discussion and Commentary: The ejector pump must be capable of passing $1\frac{1}{2}$-inch-diameter solids. A 2-inch backwater valve is required to prohibit sewage from backflowing into the basement. Where in compliance with ASME A112.3.4 or CSA B45.9, macerating toilet systems must be permitted as an alternate to the sewage pump or ejector system.

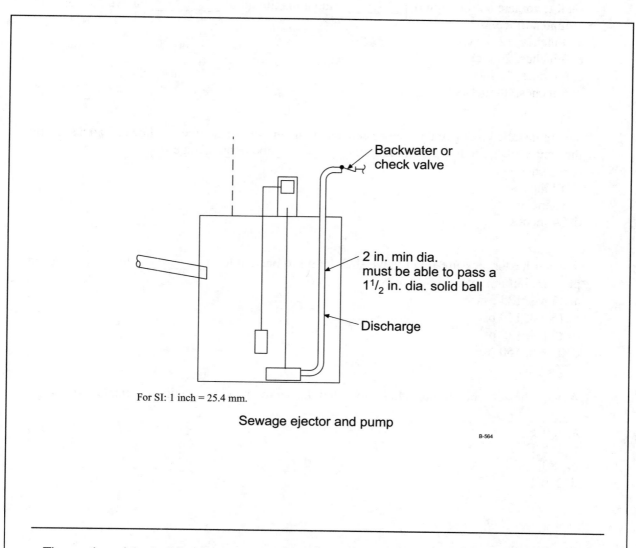

For SI: 1 inch = 25.4 mm.

Sewage ejector and pump

B-564

The portion of the building drain below the sewer grade should be installed and vented in the same manner as the gravity system. Vents should be independent of the vents serving the gravity system and must be carried up to open air.

QUIZ

Study Session 15

1. A required overflow pan for a water heater shall be a minimum of _____ in depth and of sufficient size and shape to collect all dripping and condensate.
   a. 1 inch
   b. $1^1/_2$ inches
   c. 2 inches
   d. 3 inches

2. Where the termination of a drain pan for a water heater extends to the exterior of the building, it shall terminate a minimum of _____ and a maximum of _____ above the adjacent ground surface.
   a. 3 inches, 12 inches
   b. 3 inches, 24 inches
   c. 6 inches, 12 inches
   d. 6 inches, 24 inches

3. Where installed in a garage, water heaters having an ignition source shall be elevated such that the source of ignition is a minimum of _____ above the garage floor.
   a. 6 inches
   b. 12 inches
   c. 18 inches
   d. 24 inches

4. A water heater pressure-relief valve shall be set to open at least _____ above the system pressure but not over _____.
   a. 15 psi, 125 psi
   b. 15 psi, 150 psi
   c. 25 psi, 125 psi
   d. 25 psi, 150 psi

5. A water heater temperature-relief valve shall be set to open at a maximum temperature of _____.
   a. 150°F
   b. 165°F
   c. 180°F
   d. 210°F

6.   Where located close to a wall, what is the minimum required air gap for a lavatory with an effective opening of $1/_2$ inch?
a. 1 inch
b. $1^1/_2$ inch
c. 2 inches
d. 3 inches

7.   Approved deck-mounted vacuum breakers shall be installed with the critical level a minimum of _____ above the flood level rim.
a. 1 inch
b. 2 inches
c. 4 inches
d. 6 inches

8.   In determining the peak demand for the water service and water distribution service, a flow rate of _____ and a flow pressure of _____ shall be used for a bathtub at the point of outlet discharge.
a. 3 gpm, 8 psi
b. 3 gpm, 20 psi
c. 4 gpm, 8 psi
d. 5 gpm, 8 psi

9.   A water closet shall have a maximum consumption of _____ gallons per flushing cycle.
a. 1.0
b. 1.6
c. 2.0
d. 2.3

10.   A handheld shower spray shall have a maximum flow rate of _____.
a. 2.2 gpm at 60 psi
b. 2.2 gpm at 80 psi
c. 2.5 gpm at 60 psi
d. 2.5 gpm at 80 psi

11.   What are the minimum and maximum static pressures required for a water service?
a. 40 psi, 80 psi
b. 40 psi, 100 psi
c. 50 psi, 80 psi
d. 50 psi, 100 psi

12.   When sizing a water distribution system, a tank-type water closet shall have a water-supply fixture unit value of _____.
a. 1.0
b. 1.4
c. 2.2
d. 2.7

13. Where the minimum available water pressure is 45 psi, distribution piping with a maximum developed length of 200 feet and serving 32 fixture units shall have a minimum diameter of
_____.
a. $1/2$ inch
b. $3/4$ inch
c. 1 inch
d. $1^1/_4$ inches

14. Pipe and fittings used in the water supply system shall have a maximum of _____ lead.
a. 1 percent
b. 2 percent
c. 5 percent
d. 8 percent

15. Bends of polyethlene pipe shall have a minimum installed radius of pipe curvature of _____ pipe diameters.
a. 15
b. 20
c. 25
d. 30

16. Hot-water-distribution piping within dwelling units shall have a minimum pressure rating of _____ at 180°F.
a. 80 psi
b. 100 psi
c. 110 psi
d. 125 psi

17. What is the minimum required radius for bends in copper tubing?
a. four tube diameters
b. six tube diameters
c. 4 inches
d. 6 inches

18. What is the calculated load on DWV-system piping for 2 bath groups, a kitchen group, a laundry group and a clothes washer standpipe?
a. 15 d.f.u.
b. 16 d.f.u.
c. 18 d.f.u.
d. 20 d.f.u.

19. Cleanouts shall be installed a maximum of _____ apart in horizontal drainage lines.
a. 40 feet
b. 60 feet
c. 80 feet
d. 100 feet

20. The minimum size of a cleanout serving a 3-inch pipe shall be _____.
    a. 3 inches
    b. 2$^1/_2$ inches
    c. 2 inches
    d. 1$^1/_2$ inches

21. Regardless of the number of changes in direction in a drainage system, only one cleanout is required for each _____ of run.
    a. 25 feet
    b. 40 feet
    c. 50 feet
    d. 75 feet

22. What is the minimum required clearance in front of a cleanout for a 2-inch drainage pipe?
    a. 12 inches
    b. 18 inches
    c. 24 inches
    d. 30 inches

23. Three-inch horizontal drainage piping shall be installed at a uniform slope of _____ per foot.
    a. $^1/_8$ inch
    b. $^3/_{16}$ inch
    c. $^1/_4$ inch
    d. $^5/_{16}$ inch

24. Below grade drain lines shall be a minimum of _____ in diameter.
    a. 1$^1/_2$ inches
    b. 2$^1/_2$ inches
    c. 3 inches
    d. 4 inches

25. What is the maximum number of fixture units allowed to be connected to a 3-inch-diameter vertical drain stack?
    a. 10
    b. 12
    c. 20
    d. 48

26. A water heater drainage pipe shall terminate atmospherically a maximum of _____ inch(es) above the floor.
    a. 1
    b. 2
    c. 4
    d. 6

27. A hose connection backflow preventer shall conform to which of the following applicable standards?
   a. ASSE 1012
   b. ASSE 1020
   c. ASSE 1052
   d. ASSE 1056

28. Water service pipe installed underground and outside of a structure shall have a minimum working pressure rating of _____ psi at 73°F.
   a. 100
   b. 140
   c. 160
   d. 175

29. Back-to-back water closet connections to double sanitary tee patterns are permitted provided the horizontal developed length between the outlet of the water closet and the connection to the double sanitary tee is a minimum of _____ inches.
   a. 18
   b. 24
   c. 30
   d. 36

30. The discharge pipe of a macerating toilet system shall be a minimum of _____ inch(es) in diameter.
   a. $3/4$
   b. $1^1/4$
   c. $1^1/2$
   d. 2

INTERNATIONAL RESIDENTIAL CODE
Study Session 16
Chapters 31 and 32—Vents and Traps

OBJECTIVE: To provide an understanding of the installation requirements for piping, tubing and fittings for vent systems; vent sizes and connections; and fixture traps.

REFERENCE: Chapters 31 and 32, 2003 *International Residential Code*

KEY POINTS:
- What is the purpose of a vent piping system?
- What is the difference between a "vent stack" and a "stack vent?" How are they regulated where used as the main vent? Where must a vent stack connect to a drainage stack? Where is the termination to be located?
- At what minimum height must an open vent pipe terminate above a roof? What if the roof is used for a purpose other than weather protection?
- What is the minimum required size for a vent extension through a roof or wall in areas subject to cold climates? Where must any increase in size occur?
- How must a vent terminal be located in relationship to a door, openable window or similar air intake? To an adjoining building?
- What is the maximum distance allowed between a fixture trap and its vent? What is the maximum fall due to pipe slope? Under what conditions is a vertical leg permitted within a fixture drain of a waste fixture?
- How many traps or trapped fixtures may be vented by a common vent? How is the connection regulated where the drains connect at the same level? At different levels?
- Under what conditions is a wet vent permitted? What are the limitations of a wet vent? How is a vertical wet vent regulated?
- How is a waste stack to be sized?
- Where is circuit venting acceptable? Where shall the connection be located? What is the maximum slope of the vent section of the horizontal branch drain?
- Where may a combination waste and vent system be used? How shall it be installed? What is the minimum size of the vent pipe?
- What fixtures are permitted to be vented by an island fixture venting system? Where is the connection to be located?
- What is the minimum required diameter of various vents and venting methods?
- How are air admittance valves regulated? Where are such valves permitted? Where should they be located?
- How shall traps be designed? How shall they be constructed? What types of traps are prohibited?
- What is the minimum required liquid seal for a trap? The maximum allowable trap seal? What special provisions apply to floor drains?
- Where shall a trap be located in relationship to the fixture outlet? What is the maximum vertical distance between the fixture outlet and the trap weir?
- What is the minimum trap size for various plumbing fixtures?

Topic: Vent Required and Vent Connection
Reference: IRC P3102.1, P3102.2

Category: Vents
Subject: Vent Stacks and Stack Vents

Code Text: *Every building shall have a vent stack or a stack vent. Such vent shall run undiminished in size and as directly as possible from the building drain through to the open air above the roof. Every vent stack shall connect to the base of the drainage stack. The vent stack shall connect at or below the lowest horizontal branch. Where the vent stack connects to the building drain, the connection shall be located within 10 pipe diameters downstream of the drainage stack.*

Discussion and Commentary: A vent stack is a vertical vent pipe installed to provide circulation of air to and from the drainage system that extends through one or more stories. A stack vent is the extension of a soil or waste stack above the highest horizontal drain connected.

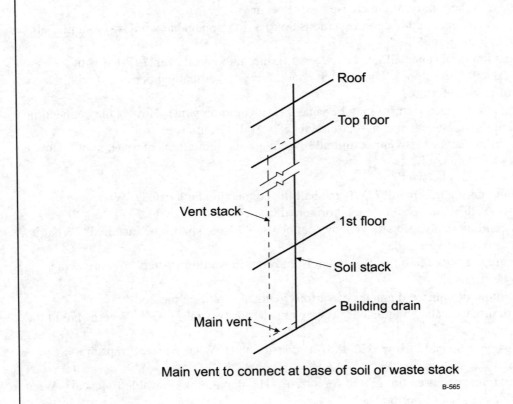

Main vent to connect at base of soil or waste stack

B-565

Vent pipes must be sized and arranged to provide a free flow of air to prevent either back pressure or siphoning action resulting in a loss of fixture trap seals. The pressure within a venting system must remain at or near atmospheric pressure.

Topic: Roof Extension
Reference: IRC P3103.1

Category: Vents
Subject: Vent Terminals

Code Text: *All open vent pipes which extend through a roof shall be terminated at least [number] inches above the roof or [number] inches above the anticipated snow accumulation, except that where a roof is to be used for any purpose other than weather protection, the vent extensions shall be run at least 7 feet (2134 mm) above the roof.*

Discussion and Commentary: The term *[number]* in the quotation above indicates that termination height is regulated by local conditions. It is anticipated that the roof will be unoccupied, so the only concern is that gases and odors will discharge well above the roof surface. However, if the roof is used as an occupiable space, such as an entertainment deck or observation platform, the termination point must be above the level of individuals occupying the roof.

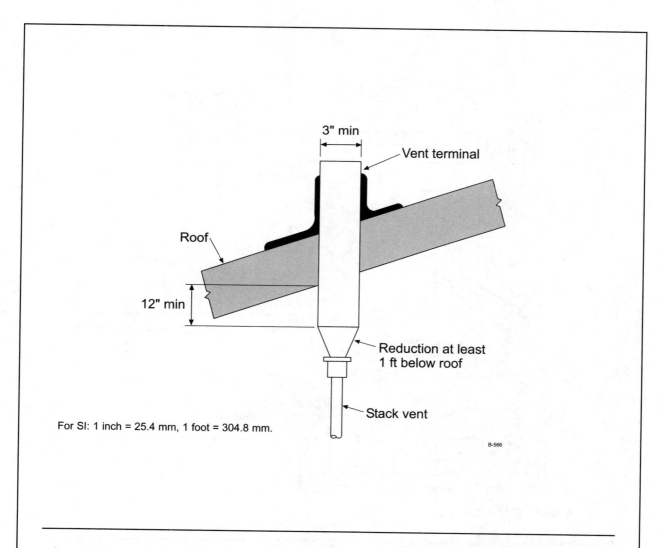

For SI: 1 inch = 25.4 mm, 1 foot = 304.8 mm.

In cold climates, the terminal opening may become closed with frost. This is caused when the flow of warm, moist air rising through the vent stack comes in contact with the frigid outside air, forming frost on the vent's interior. The chance of closure is reduced by increasing the vent size.

Topic: Location

Reference: IRC P3103.5

Category: Vents

Subject: Vent Terminals

Code Text: *An open vent terminal from a drainage system shall not be located less than 4 feet (1219 mm) directly beneath any door, openable window, or other air intake opening of the building or of an adjacent building, nor shall any such vent terminal be within 10 feet (3048 mm) horizontally of such an opening unless it is at least 2 feet (610 mm) above the top of such opening.*

Discussion and Commentary: The vent terminal opening should not be located directly beneath any opening where sewer gas and odors can enter the building. Sewer gases are undesirable, unhealthy and potentially explosive. For the same reasons, the proximity of any vent terminal to openings within a 10-foot horizontal distance, whether in the same building or an adjacent building, are also regulated.

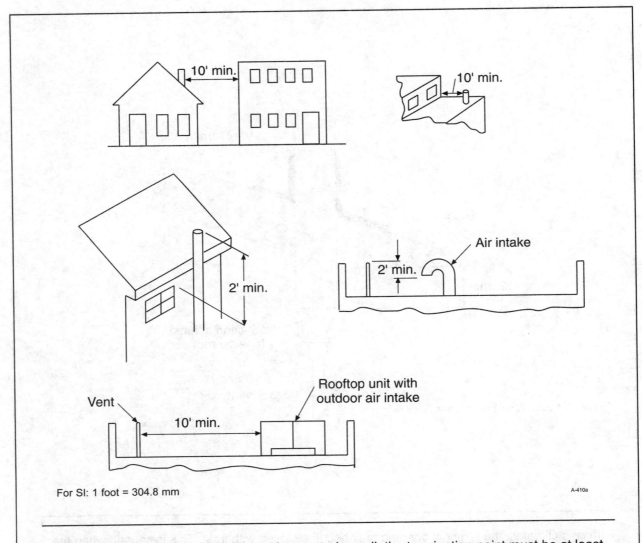

For SI: 1 foot = 304.8 mm

A-410a

Where a vent terminal extends through an exterior wall, the termination point must be at least 10 feet from any adjoining property line. Vertically, the vent terminal must extend at least 10 feet above the highest grade located within 10 feet horizontally.

2003 IRC Study Companion

Topic: Distance of Trap from Vent, Fixture Drains
Reference: IRC P3105.1, P3105.2

Category: Vents
Subject: Fixture Vents

Code Text: *Each fixture trap shall have a protecting vent located so that the slope and the developed length in the fixture drain from the trap weir to the vent fitting are within the requirements set forth in Table P3105.1.* See exception for self-siphoning fixtures. *The total fall in a fixture drain due to pipe slope shall not exceed one pipe diameter, nor shall the vent pipe connection to a fixture drain, except for water closets, be below the weir of the trap.* See provisions of Section P3105.3 for a vertical leg within a fixture drain.

Discussion and Commentary: The length of each trap arm should not exceed that specified in Table P3105.1, measured from the weir of the trap to the vent along the centerline of the pipe. The measured length must include turns and offsets.

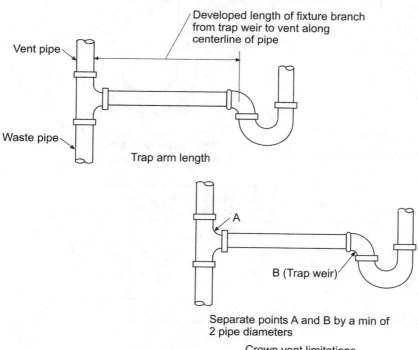

TABLE P3105.1
MAXIMUM DISTANCE OF FIXTURE TRAP FROM VENT

SIZE OF TRAP (inches)	SLOPE (inch per foot)	DISTANCE FROM TRAP (feet)
$1^1/_4$	$^1/_4$	5
$1^1/_2$	$^1/_4$	6
2	$^1/_4$	8
3	$^1/_8$	12
4	$^1/_8$	16

For SI: 1 inch = 25.4 mm, 1 foot = 304.8 mm, 1 inch per foot = 83.3 mm/m.

Vent pipe

Developed length of fixture branch
from trap weir to vent along
centerline of pipe

Waste pipe

Trap arm length

A

B (Trap weir)

Separate points A and B by a min of
2 pipe diameters

Crown vent limitations

A vent must be installed at least two pipe diameters from the trap weir. Where a lesser distance is provided, blockage is possible due to the action of the drainage flowing through the trap. The flow direction and velocity forces waste up into the connection, eventually clogging it fully.

Topic: Crown Vents
Reference: IRC P3105.4

Category: Vents
Subject: Fixture Vents

Code Text: *A vent shall not be installed within two pipe diameters of the trap weir.*

Discussion and Commentary: Crown venting is an arrangement where a vent connects at the top of the weir (crown) of a trap. The problem with this type of connection is that the vent opening can become blocked, thus closing the vent and allowing the trap to siphon. The blockage is a result of the action of the drainage flowing through the trap. Flow direction and velocity will force waste up into the vent connection. This problem can be avoided when the vent is connected at least two pipe diameters downstream from the trap weir. Therefore, the code allows such a condition where the appropriate distance is provided.

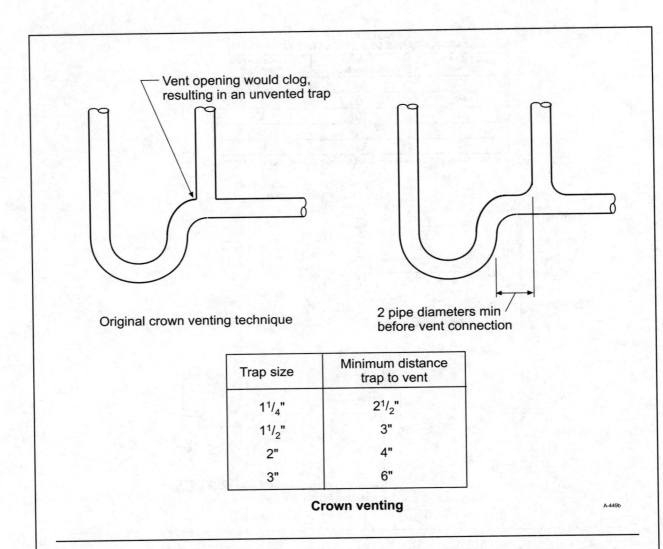

Vent opening would clog, resulting in an unvented trap

Original crown venting technique

2 pipe diameters min before vent connection

Trap size	Minimum distance trap to vent
1¹/₄"	2¹/₂"
1¹/₂"	3"
2"	4"
3"	6"

Crown venting

A-449b

In Section 3105.3, a configuration is permitted that allows the vent takeoff to be below the weir of the trap. Laboratory tests have shown that when installed following the specific requirements, this trap seal will not siphon. However, strict adherence to the requirements is necessary.

Topic: Connections
Reference: IRC P3107.1, P3107.2

Category: Vents
Subject: Common Vent

Code Text: *An individual vent is permitted to vent two traps or trapped fixtures as a common vent. The traps or trapped fixtures being common vented shall be located on the same floor level. Where the fixture drains being common vented connect at the same level, the vent connection shall be at the interconnection of the fixture drains or downstream of the interconnection.*

Discussion and Commentary: A common vent installed vertically may be used to protect two fixture traps when the drains connect at the same level on a vertical drain. A typical installation is two horizontal drains connecting together to a vertical drain to a double-pattern fitting. The extension of the vertical pipe serves as the vent. The vent may connect at the interconnection of the fixture drains or downstream along the horizontal drain.

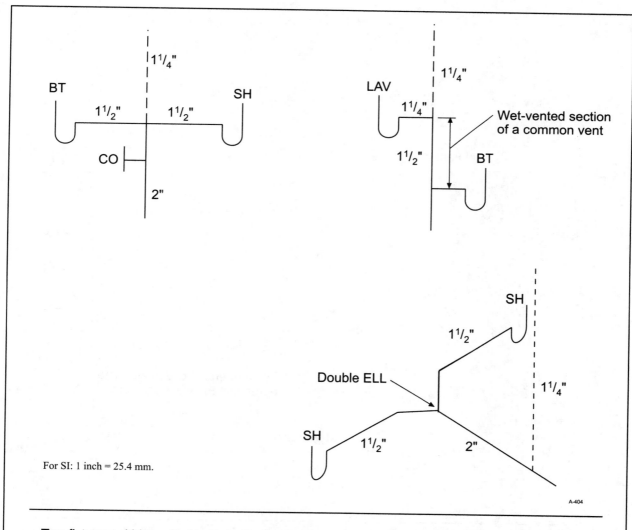

For SI: 1 inch = 25.4 mm.

Two fixtures within a single story may connect at different levels to a vertical drain and still be common-vented, provided the drain pipe between the upper and lower fixtures is oversized. Although drainage flows in the pipe between the two fixtures, the piping is not classified as a wet vent.

Topic: Wet Vent Permitted
Reference: IRC P3108.1

Code Text: *Any combination of fixtures within two bathroom groups located on the same floor level are permitted to be vented by a wet vent. The wet vent shall be considered the vent for the fixtures and shall extend from the connection of the dry vent along the direction of the flow in the drain pipe to the most downstream fixture drain connection to the horizontal branch drain.*

Discussion and Commentary: A wet vent is a vent pipe that is wet because is also conveys drainage. The wet venting concept is based on employing oversized piping to allow for the flow of air above the waste flow. The low probability of simultaneous fixture discharges and the low-flow velocity that results from fixtures on the same floor level provide the necessary degree of reliability that there will always be adequate volume within the wet vent pipe to permit required airflow.

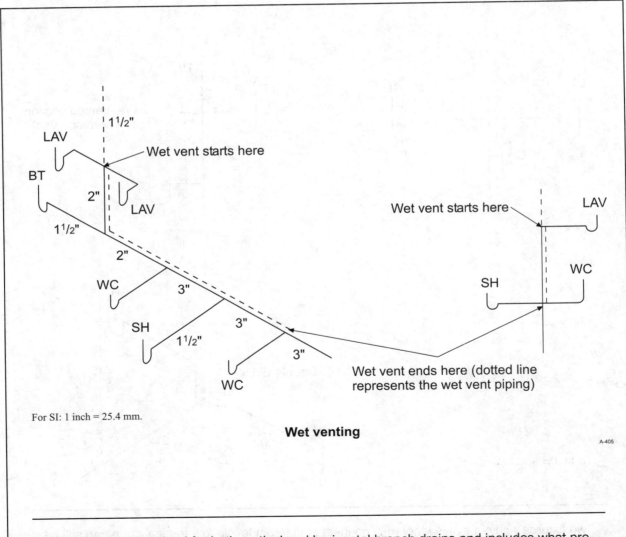

For SI: 1 inch = 25.4 mm.

Wet venting

A-405

Wet venting is permitted for both vertical and horizontal branch drains and includes what previously was referred to as stack venting. The vertical wet vent must extend from its connection to the dry vent down to the lowest fixture drain connection.

Topic: Installation

Reference: IRC P3109

Category: Vents

Subject: Waste Stack Vent

Code Text: *A waste stack shall be considered a vent for all of the fixtures discharging to the stack where installed in accordance with the requirements of Section P3109. The waste stack shall be vertical, and both horizontal and vertical offsets shall be prohibited. Every fixture drain shall connect separately to the waste stack. The stack shall not receive the discharge of water closets or urinals. A stack vent shall be provided for the waste stack. The size of the stack vent shall be equal to the size of the waste stack.*

Discussion and Commentary: A waste stack vent employs a waste stack as a vent for the fixtures. The principles of use are based on some of the original research that was done in plumbing. The system has been identified by a variety of names, including vertical wet vent, Philadelphia single-stack, and multi-floor stack venting.

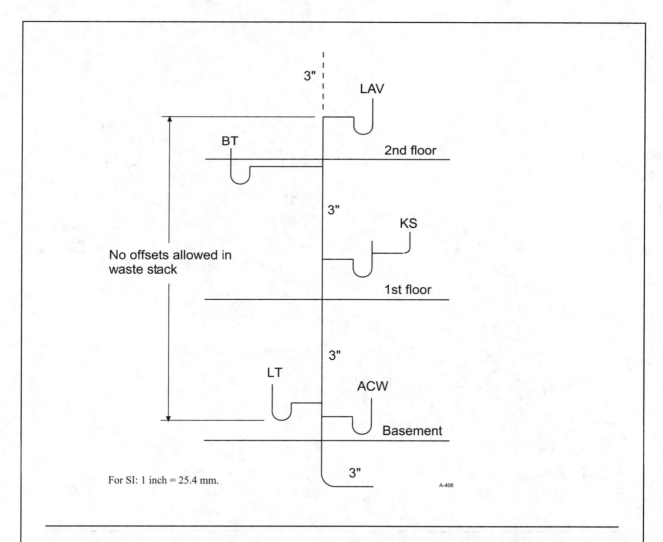

Because the drainage stack serves as the vent, there are certain limitations placed on the design of a waste stack vent. The system is specifically identified as a "waste" stack vent because it prohibits the connection of water closets and urinals. Only "waste" can discharge to the stack.

Topic: Permitted Uses
Reference: IRC P3110.1

Category: Vents
Subject: Circuit Venting

Code Text: *A maximum of eight fixtures connected to a horizontal branch drain shall be permitted to be circuit vented. Each fixture drain shall connect horizontally to the horizontal branch being circuit vented. The horizontal branch drain shall be classified as a vent from the most downstream fixture drain connection to the most upstream fixture drain connection to the horizontal branch.*

Discussion and Commentary: The principle of circuit venting is that the flow of drainage never exceeds a half-full flow condition. The air for venting the fixtures circulates in the top half of the horizontal branch drainpipe. The flow velocity in the horizontal branch is slow and nonturbulent, thereby preventing pressure differentials from affecting the connecting fixtures.

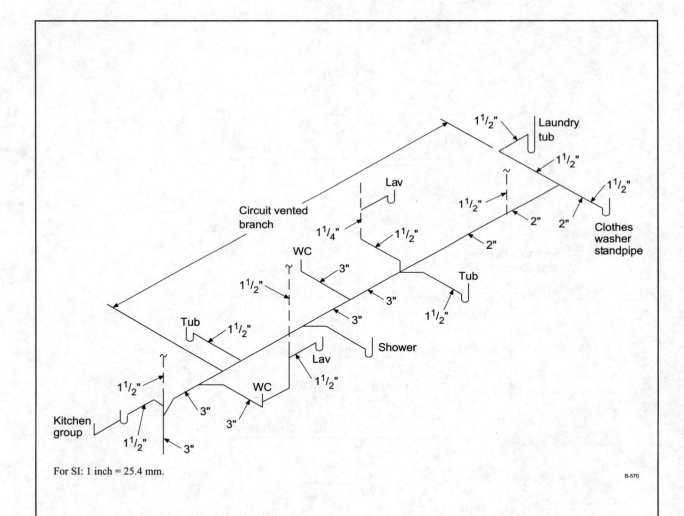

For SI: 1 inch = 25.4 mm.

B-570

The connection of the circuit vent must occur between the two most upstream fixture drains. The circuit vent pipe cannot receive the drainage of any soil or waste. In addition, the slope of the vent section of the horizontal branch drain is limited to 1:12 (8-percent slope).

Topic: Installation
Reference: IRC P3111.2

Category: Vents
Subject: Combination Waste and Vent System

Code Text: *The only vertical pipe of a combination drain and vent system shall be the connection between the fixture drain of a sink, lavatory, standpipe or drinking fountain, and the horizontal combination waste and vent pipe. The maximum vertical distance shall be 8 feet (2438 mm).*

Discussion and Commentary: A combination drain and vent system is a means of extending the distance from a trap to its vent for an unlimited distance because the vent and drain are literally one. The number of fixtures connecting to a combination drain and vent system is also unlimited, provided that the fixtures are floor drains, sinks, standpipes or lavatories. The system is useful in areas that cannot accommodate vertical vent risers, such as where island sinks are installed or where floor drains are located in large open areas.

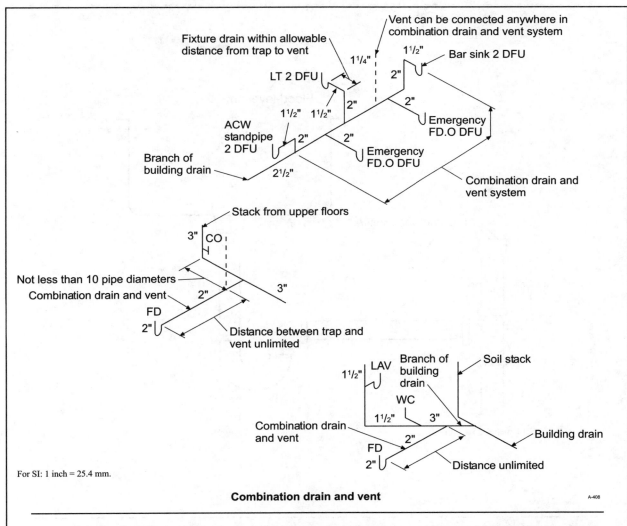

For SI: 1 inch = 25.4 mm.

Combination drain and vent

A-408

In a combination drain and vent system, the drain also serves as the vent for the fixture. The system is intended to be a horizontal piping system, with the only vertical piping (limited to 8 feet) being the connection to a sink, lavatory standpipe or drinking fountain located above the level of the vent.

Topic: Location

Reference: IRC P3114.4

Category: Vents

Subject: Air Admittance Valves

Code Text: *Individual and branch air admittance valves shall be located a minimum of 4 inches (102 mm) above the horizontal branch drain or fixture drain being vented. Stack-type air admittance valves shall be located a minimum of 6 inches (152 mm) above the flood level rim of the highest fixture being vented. The air admittance valve shall be located within the maximum developed length permitted for the vent. The air admittance valve shall be installed a minimum of 6 inches (152 mm) above insulation materials where installed in attics.*

Discussion and Commentary: Individual vents, branch vents, circuit vents and stack vents may all terminate with a connection to an air admittance valve rather than extend to the open air. Air enters the upper plumbing drainage system when negative pressures develop in the piping system. The device closes by gravity and seals the vent terminal when the internal pressure is equal to or greater than atmospheric pressure.

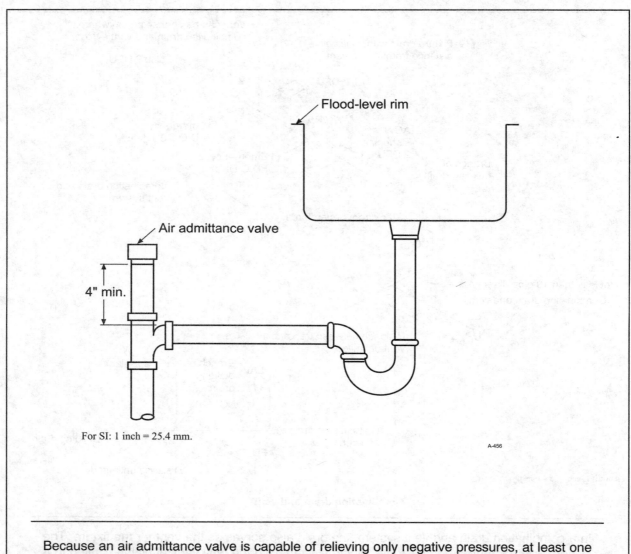

Flood-level rim

Air admittance valve

4" min.

For SI: 1 inch = 25.4 mm.

A-456

Because an air admittance valve is capable of relieving only negative pressures, at least one stack vent or vent stack must extend to the outdoors. The vent to the open air serves as the positive pressure relief for the drainage system.

Topic: Trap Seals and Trap Seal Protection
Reference: IRC P3201.2, P3201.3

Category: Traps
Subject: Fixture Traps

Code Text: *Traps shall have a liquid seal not less than 2 inches (51 mm) and not more than 4 inches (102 mm).* See exception for floor drains. *Traps shall be set level with respect to their water seals and shall be protected from freezing. Trap seals shall be protected from siphonage, aspiration or back pressure by an approved system of venting.*

Discussion and Commentary: A trap is a simple method used to keep sewer gases from emanating out of the drainage system. The water seal prevents the sewer gases and aerosol-borne bacteria from entering the building space. The configuration of a trap interferes with the flow of the drainage; however, the interference is minimal because of the construction of the trap and the relatively high inlet velocity of the waste flow.

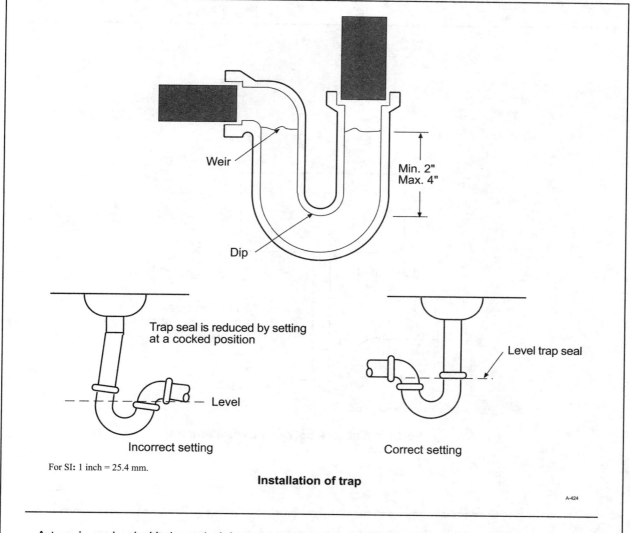

For SI: 1 inch = 25.4 mm.

Installation of trap

A-424

A trap is a simple U-shaped piping arrangement that offers minimal resistance to flow. The only type of fixture trap permitted is the "P" trap. All other types, such as bell traps, drum traps and traps with moving parts, have undesirable characteristics and are prohibited.

Topic: Number of Fixtures per Trap
Reference: IRC P3201.6

Code Text: *Each plumbing fixture shall be separately trapped by a water seal trap placed as close as possible to the fixture outlet.* See three conditions where separate trapping is not required. *The vertical distance from the fixture outlet to the trap weir shall not exceed 24 inches (610 mm). Fixtures shall not be double trapped.*

Discussion and Commentary: The 24-inch limit is based on two criteria. It is desirable to locate the trap as close as possible to the fixture to minimize the amount of drainpipe on the inlet side of the trap. Buildup on the wall of the fixture outlet pipe will breed bacteria and cause odors to develop. The vertical distance is also limited to control the velocity of the drainage flow. If the trap has too great a vertical separation from the fixture, the velocity of flow at the trap inlet may create self-siphoning of the water in the trap.

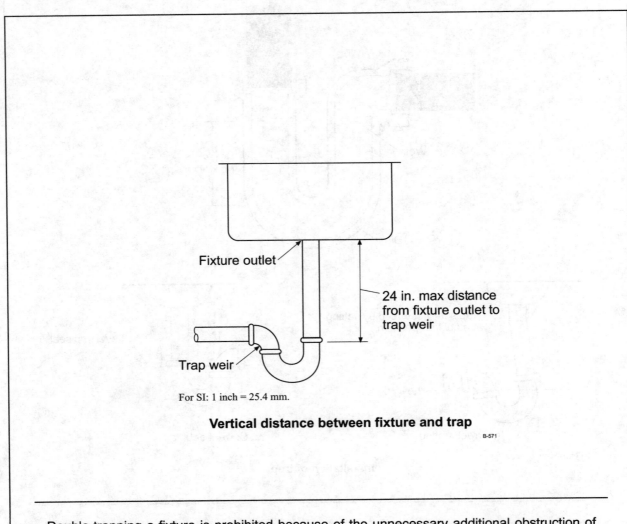

For SI: 1 inch = 25.4 mm.

Vertical distance between fixture and trap

B-571

Double-trapping a fixture is prohibited because of the unnecessary additional obstruction of flow and potential for stoppages. Double-trapping will cause a volume of air to be trapped between two trap seals, and the "air-bound" drain will impede the flow.

QUIZ

Study Session 16

1. Every building shall have a main vent that is _____.
    a. a vent stack
    b. a stack vent
    c. either a vent stack or a stack vent
    d. both a vent stack and a stack vent

2. Where the vent stack connects to the building drain, the connection shall be located within _____ of the drainage stack.
    a. 36 inches downstream
    b. 36 inches upstream
    c. 10 pipe diameters downstream
    d. 10 pipe diameters upstream

3. Every vent stack or stack vent shall extend outdoors and terminate to the open air or terminate to a stack-type _____.
    a. relief vent
    b. vent extension
    c. vent terminal
    d. air admittance valve

4. Where a roof is to be used for any purpose other than weather protection, the vent extension shall run a minimum of _____ above the roof.
    a. 2 feet
    b. 5 feet
    c. 7 feet
    d. 10 feet

5. Unless located a minimum of _____ above the top of the opening, an open vent terminal from a drainage system shall be located at least 10 feet horizontally from a door, openable window or other air intake opening of the building.
    a. 12 inches
    b. 24 inches
    c. 36 inches
    d. 60 inches

6. Every dry vent shall rise vertically a minimum of _____ above the flood level rim of the highest trap or trapped fixture being vented.
    a. 1 inch
    b. 2 inches
    c. 4 inches
    d. 6 inches

7.   A connection between a vent pipe and a vent stack or stack vent shall be made a minimum of
      _____ above the flood level rim of the highest fixture served by the vent.
      a. 1 inch
      b. 2 inches
      c. 4 inches
      d. 6 inches

8.   A maximum distance of _____ is permitted between a $1^1/_2$-inch fixture trap and the vent
      fitting.
      a. 4 feet
      b. 5 feet
      c. 6 feet
      d. 8 feet

9.   The total fall in a fixture drain due to pipe slope is limited to a maximum of _____.
      a. 1 inch
      b. one pipe diameter
      c. 2 inches
      d. two pipe diameters

10.   A vent shall be installed a minimum of _____ from the trap weir.
      a. 1 inch
      b. one pipe diameter
      c. 2 inches
      d. two pipe diameters

11.   An individual vent may vent a maximum of _____ trap(s) or trap fixture(s) as a common
      vent.
      a. one
      b. two
      c. three
      d. four

12.   Where fixture drains connect at different levels and the vent connects as a vertical extension of
      the vertical drain, the vertical drain pipe connecting the two fixture drains shall be considered
      the vent for the lower fixture drain. If the vent pipe size is 2 inches, what is the maximum
      permitted discharge from the upper fixture drain?
      a. 1 d.f.u.
      b. 4 d.f.u.
      c. 6 d.f.u.
      d. 8 d.f.u.

13. A vertical wet vent shall have a minimum pipe size of _____ where serving a total fixture unit load of 9 d.f.u.
    a. 2 inches
    b. 2$\frac{1}{2}$ inches
    c. 3 inches
    d. 4 inches

14. A 3-inch waste stack vent may be used for a maximum total discharge to the stack of _____.
    a. 8 d.f.u.
    b. 12 d.f.u.
    c. 24 d.f.u.
    d. 32 d.f.u.

15. What is the maximum number of fixtures connected to a horizontal branch drain that are permitted to be circuit vented?
    a. two
    b. four
    c. six
    d. eight

16. Which one of the following fixtures shall not be served by a combination waste and vent system?
    a. laundry tub
    b. floor drain
    c. standpipe
    d. lavatory

17. The maximum permitted slope of a horizontal combination waste and vent pipe shall be _____ unit vertical in 12 units horizontal.
    a. one-eighth
    b. one-fourth
    c. one-half
    d. one

18. A 2$\frac{1}{2}$-inch diameter pipe used as a combination waste and vent shall serve a maximum of _____ fixture units where connected to a horizontal branch.
    a. 3
    b. 6
    c. 12
    d. 26

19.   Vents having a minimum developed length of _____ shall be increased by one nominal pipe size for the entire developed length of the vent pipe.
        a. 25
        b. 40
        c. 50
        d. 60

20.   Where used in a vent system, an individual air admittance valve shall be located a minimum of _____ above the horizontal branch drain or fixture drain being vented.
        a. 1 inch
        b. 2 inches
        c. 4 inches
        d. 6 inches

21.   Traps shall have a minimum liquid seal of _____ inches and a maximum seal of _____ inches.
        a. 2, 4
        b. 2, 6
        c. 3, 4
        d. 3, 6

22.   Where a fixture is separately trapped, what is the maximum vertical distance from the fixture outlet to the trap weir?
        a. 6 inches
        b. 12 inches
        c. 18 inches
        d. 24 inches

23.   Where common trapped fixture outlets are permitted, they shall be located a maximum of _____ apart.
        a. 18 inches
        b. 24 inches
        c. 30 inches
        d. 36 inches

24.   What is the minimum required trap size for a floor drain?
        a. $1^1/_4$ inches
        b. $1^1/_2$ inches
        c. 2 inches
        d. $2^1/_2$ inches

25.   The minimum required size for a trap arm serving a clothes washer standpipe shall be
_____.
     a. $1^1/_4$ inches
     b. $1^1/_2$ inches
     c. 2 inches
     d. $2^1/_2$ inches

26.   An open vent terminal extending through an exterior wall shall terminate a minimum of _____ feet from the lot line.
     a. 4
     b. 5
     c. 6
     d. 10

27.   In circuit venting, the maximum slope of the vent section of the horizontal branch drain shall be _____ unit(s) vertical in 12 units horizontal.
     a. $^1/_4$
     b. $^1/_2$
     c. 1
     d. 2

28.   What is the maximum vertical distance between the fixture drain of a sink and a horizontal combination waste and vent pipe?
     a. 30 inches
     b. 4 feet
     c. 6 feet
     d. 8 feet

29.   The vent or branch vent for multiple island fixture vents shall extend a minimum of _____ inch(es) above the highest island fixture being vented before connecting to the outside vent terminal.
     a. 1
     b. 2
     c. 4
     d. 6

30.   Stack-type air admittance valves shall be located a minimum of _____ inches above the flood level rim of the highest fixture being vented.
     a. 4
     b. 6
     c. 12
     d. 15

NATIONAL RESIDENTIAL CODE
Study Session 17
ters 33, 34, 35 and 36—General Electrical
ents, Definitions, Services, and Branch Circuits
and Feeder Requirements

...in an understanding of the general requirements for electrical systems, ...ponents, including provisions addressing service conductors, branch circuits and ...

REFERENCE: Chapters 33, 34, 35 and 36, 2003 *International Residential Code*

KEY POINTS:
- What minimum working clearances are required at energized equipment and panelboards? How is clearance above a panelboard regulated? What is the minimum required headroom?
- What materials are permitted for use as conductors? What is the minimum permitted size? How are stranded conductors regulated? Conductors in parallel? How are conductors to be identified?
- What is the difference between "accessible" and "readily accessible"? How do "damp," "dry," and "wet" locations differ? How is "rainproof" different from "rain tight"?
- What are the requirements for "labeled" and "listed" equipment or materials? How do these terms related to that of "identified"?
- Where must the service disconnecting means be located? What is the maximum number of disconnects?
- How is the minimum load for ungrounded service conductors to be calculated?
- What minimum clearances are mandated for overhead service-drop conductors from doors, porches, decks, stairs and balconies? Above roofs? Above pedestrian areas and sidewalks? Over residential property and driveways? Over public streets and alleys or parking areas subject to truck traffic?
- How are connections at service heads to be installed?
- What is the basic method for creating a grounding electrode system? When is the use of a metal underground water pipe permitted for such a such system? A concrete-encased electrode? Ground rings?
- What is a "made" electrode? When are such electrodes permitted? What minimum sizes are required for rod and pipe electrodes? How are they to be installed? What is the minimum required resistance to ground? How may this be achieved?
- Where is bonding required to ensure electrical continuity? What methods of bonding are permitted?
- What are the rating requirements for branch circuits serving lighting units and general utilization equipment? Fixed utilization equipment? Motors? Ranges and cooking appliances? Heating loads? Air-conditioning and heat pumps?
- How many branch circuits are required to serve the kitchen and dining area? The laundry area? Bathrooms?
- How is the size of feeders to be determined? Conductors?

Topic: Drilling, Notching and Penetrations
Reference: IRC E3302

Category: Electrical Requirements
Subject: Building Structure Protection

Code Text: *Electrical installations in hollow spaces, vertical shafts, and ventilation or air-handling ducts shall be made so that the possible spread of fire or products of combustion will not be substantially increased. Electrical penetrations through fire-resistance-rated walls, partitions, floors or ceilings shall be protected by approved methods to maintain the fire-resistance rating of the element penetrated. Penetrations through fire blocking and draftstopping shall be protected in an approved manner to maintain the integrity of the element penetrated.*

Discussion and Commentary: Through penetrations and membrane penetrations of fire-resistive elements must be protected in accordance with the requirements of Section R317.3. Where fire blocking and draftstopping is mandated by Section R602.8 and Section R502.12, respectively, the penetration by electrical installations must also be done in an appropriate manner.

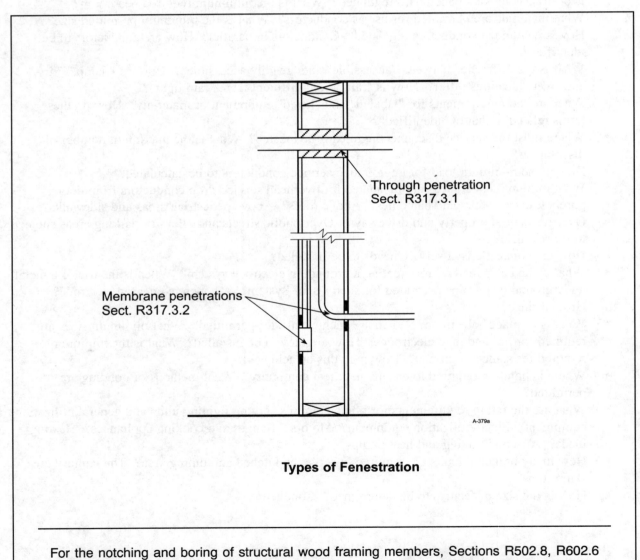

Through penetration
Sect. R317.3.1

Membrane penetrations
Sect. R317.3.2

A-379a

Types of Fenestration

For the notching and boring of structural wood framing members, Sections R502.8, R602.6 and R802.7 must be followed. These requirements are in addition to those mandated in Table E3702.1 for the protection of the electrical cables and conductors installed in such framing.

Topic: Working Clearances
Reference: IRC E3305.2

Category: Electrical Requirements
Subject: Equipment Location and Clearances

Code Text: *Except as otherwise specified in Chapters 33 through 42, the dimension of the working space in the direction of access to panelboards and live parts likely to require examination, adjustment, servicing or maintenance while energized shall be not less than 36 inches (914 mm) in depth. In addition to the 36-inch dimension (914 mm), the work space shall not be less than 30 inches (762 mm) wide in front of the electrical equipment and not less than the width of such equipment. The work space shall be clear and shall extend from the floor or platform to a height of 6.5 feet (1981 mm).*

Discussion and Commentary: A minimum depth of 3 feet in front of the access point to an electrical panelboard is needed to allow for adequate clearance for a person from any live parts. It provides enough working depth to both maintain a safe distance and allow free movement when a person is servicing or maintaining the equipment.

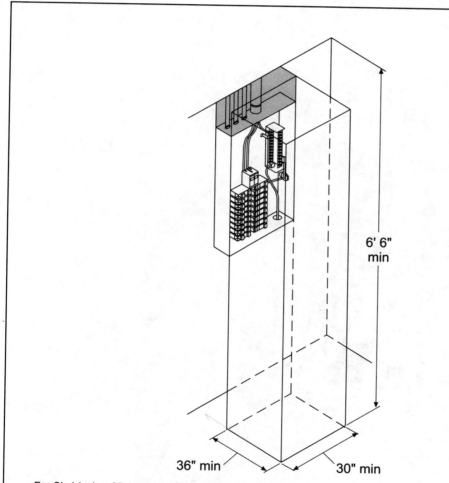

6' 6"
min

36" min

30" min

For SI: 1 inch = 25.4 mm, 1 foot = 304.8 mm.

B-572

Regardless of the size of the panelboard, a minimum clear width of 30 inches is required for working space. It need not be centered in front of the equipment but should be located for safe access. If the panelboard exceeds 30 inches in width, the clear space must extend the full width.

Topic: Clearances Over Panelboards

Reference: IRC E3305.3

Category: Electrical Requirements

Subject: Equipment Locations and Clearances

Code Text: *A dedicated space directly over a panelboard that extends from the panelboard to the structural ceiling or to a height of 6 feet (1829 mm) above the panelboard, whichever is lower, and has a width and depth equal to the equipment shall be dedicated and kept clear of equipment unrelated to the electrical equipment. Piping, ducts or equipment unrelated to the electrical equipment shall not be installed in such dedicated space.*

Discussion and Commentary: In addition to the working space that must extend vertically in front of a panelboard, the area directly above the panelboard itself is regulated. This area must be free of any equipment or building features that would interfere with the direct run of cables and/or conduits to the equipment.

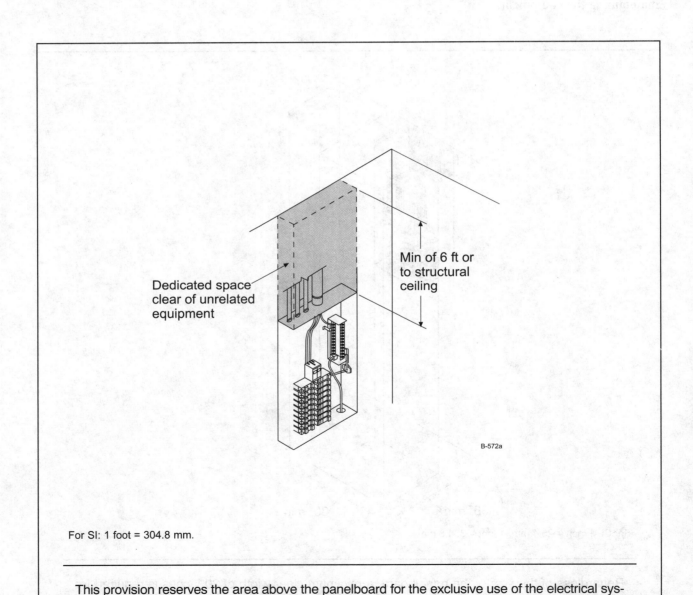

Dedicated space clear of unrelated equipment

Min of 6 ft or to structural ceiling

B-572a

For SI: 1 foot = 304.8 mm.

This provision reserves the area above the panelboard for the exclusive use of the electrical system and its components. Plumbing and HVAC piping, ducts and equipment must be located outside the designated area.

Topic: Splices
Reference: IRC E3306.10.3

Category: Electrical Requirements
Subject: Conductors and Connections

Code Text: *Where conductors are to be spliced, terminated or connected to fixtures or devices, a minimum length of 6 inches (152 mm) of free conductor shall be provided at each outlet, junction or switch point. The required length shall be measured from the point in the box where the conductor emerges from its raceway or cable sheath. Where the opening to an outlet, junction, or switch point is less than 8 inches (203 mm) in any dimension, each conductor shall be long enough to extend at least 3 inches (76 mm) outside of such opening.*

Discussion and Commentary: In order to provide adequate conductor length for connection purposes, at least 6 inches of free conductor must be available. The conductor extension also ensures that undue stress will not be placed on the conductor during or after the splice or connection.

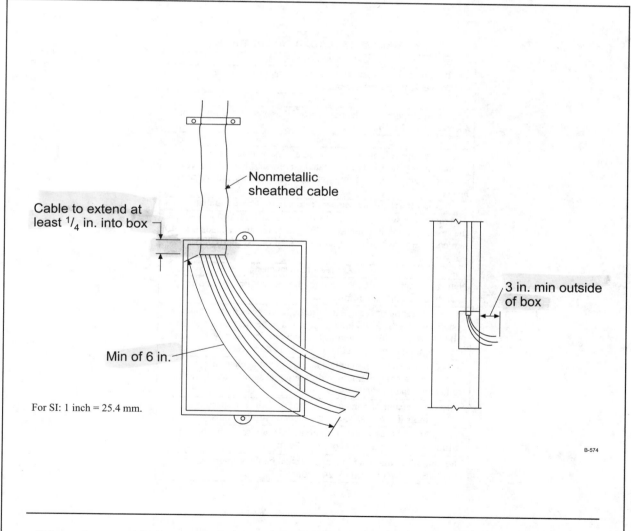

Nonmetallic sheathed cable

Cable to extend at least ¹/₄ in. into box

3 in. min outside of box

Min of 6 in.

For SI: 1 inch = 25.4 mm.

B-574

Where the opening of the outlet, switch or junction box is limited in size to the point where the connection work must be done outside of the box, a minimum conductor length of 3 inches is required beyond the front edge of the box.

Topic: Service Load

Category: Services

Reference: IRC E3502.2

Subject: Service Size and Rating

Code Text: *The minimum load for ungrounded service conductors and service devices that serve 100 percent of the dwelling unit load shall be computed in accordance with Table E3502.2. Ungrounded service conductors and service devices that serve less than 100 percent of the dwelling unit load shall be computed as required for feeders.*

Discussion and Commentary: Table E3502.2 sets forth a procedure for the calculation of the minimum service load. It is based on the square footage of the dwelling unit, the number of small appliance circuits and the rating of any fastened-in-place, permanently connected or dedicated circuit-supplied appliances such as ranges, ovens, clothes dryers and water heaters. A demand factor is applied to the calculated sum, to which is added a load for the air-conditioning and/or heating equipment.

TABLE E3502.2
MINIMUM SERVICE LOAD CALCULATION

LOADS AND PROCEDURE
3 volt-amperes per square foot of floor area for general lighting and convenience receptacle outlets.
Plus
1,500 volt-amperes × total number of 20-ampere-rated small appliance and laundry circuits.
Plus
The nameplate volt-ampere rating of all fastened-in-place, permanently connected or dedicated circuit-supplied appliances such as ranges, ovens, cooking units, clothes dryers and water heaters.
Apply the following demand factors to the above subtotal:
The minimum subtotal for the loads above shall be 100 percent of the first 10,000 volt-amperes of the sum of the above loads plus 40 percent of any portion of the sum that is in excess of 10,000 volt-amperes.
Plus the largest of the following:
Nameplate rating(s) of air-conditioning and cooling equipment including heat pump compressors.
Nameplate rating of the electric thermal storage and other heating systems where the usual load is expected to be continuous at the full nameplate value. Systems qualifying under this selection shall not be figured under any other category in this table.
Sixty-five percent of nameplate rating of central electric space-heating equipment, including integral supplemental heating for heat pump systems.
Sixty-five percent of nameplate rating(s) of electric space-heating units if less than four separately controlled units.
Forty percent of nameplate rating(s) of electric space-heating units of four or more separately controlled units.
The minimum total load in amperes shall be the volt-ampere sum calculated above divided by 240 volts.

Undergrounded service conductors must have an ampacity of no less than the load to be served. The minimum rating of the undergrounded service conductors for a single-family dwelling is 100 amperes, with an allowance of 60 amperes for other installations.

Topic: Conductor Size
Reference: IRC E3503.1

Category: Services
Subject: Service and Feeder Conductors

Code Text: *Conductors used as ungrounded service entrance conductors, service lateral conductors, and feeder conductors that serve as the main power feeder to a dwelling unit shall be those listed in Table E3503.1. Ungrounded service conductors shall have a minimum size in accordance with Table E3503.1. The grounded conductor size shall not be less than the maximum unbalance of the load and shall not be smaller than the required minimum grounding electrode conductor size specified in Table E3503.1.*

Discussion and Commentary: The minimum conductor size listed in Table E3503.1 is based on the maximum allowable load. Conductor size differs based on the allowable ampacity and the type of conductor material, either copper or aluminum. The same criteria are used for sizing the grounding electrode conductor.

TABLE E3503.1
SERVICE CONDUCTOR AND GROUNDING ELECTRODE CONDUCTOR SIZING

CONDUCTOR TYPES AND SIZES—THHW, THW, THWN, USE, XHHW (Parallel sets of 1/0 and larger conductors are permitted in either a single raceway or in separate raceways)		ALLOWABLE AMPACITY	MINIMUM GROUNDING ELECTRODE CONDUCTOR SIZE[a]	
Copper (AWG)	Aluminum and copper-clad aluminum (AWG)	Maximum load (amps)	Copper (AWG)	Aluminum (AWG)
4	2	100	8[b]	6[c]
3	1	110	8[b]	6[c]
2	1/0	125	8[b]	6[c]
1	2/0	150	6[c]	4
1/0	3/0	175	6[c]	4
2/0	4/0 or two sets of 1/0	200	4[d]	2[d]
3/0	250 kcmil or two sets of 2/0	225	4[d]	2[d]
4/0 or two sets of 1/0	300 kcmil or two sets of 3/0	250	2[d]	1/0[d]
250 kcmil or two sets of 2/0	350 kcmil or two sets of 4/0	300	2[d]	1/0[d]
350 kcmil or two sets of 3/0	500 kcmil or two sets of 250 kcmil	350	2[d]	1/0[d]
400 kcmil or two sets of 4/0	600 kcmil or two sets of 300 kcmil	400	1/0[d]	3/0[d]

For SI: 1 inch = 25.4 mm.

a. Where protected by a metal raceway, grounding electrode conductors shall be electrically bonded to the metal raceway at both ends.

b. No. 8 grounding electrode conductors shall be protected with metal conduit or nonmetallic conduit.

c. Where not protected, No. 6 grounding electrode conductors shall closely follow a structural surface for physical protection. The supports shall be spaced not more than 24 inches on center and shall be within 12 inches of any enclosure or termination.

d. Where the sole grounding electrode system is a ground rod or pipe as covered in Section E3508.2, the grounding electrode conductor shall not be required to be larger than No. 6 copper or No. 4 aluminum. Where the sole grounding electrode system is the footing steel as covered in Section E3508.1.2, the grounding electrode conductor shall not be required to be larger than No. 4 copper conductor.

An underground service conductor that serves a workshop, detached garage or similar accessory structure need only have a minimum rating of 60 amperes. The sizing of such conductors shall be in accordance with Chapter 36 for feeders.

Topic: Clearance from Building Openings
Reference: IRC E3504.1

Category: Services
Subject: Overhead Service-drop and
Service Conductor Installation

Code Text: *Open conductors and multiconductor cables without an overall outer jacket shall have a clearance of not less than 3 feet (914 mm) from the sides of doors, porches, decks, stairs, ladders, fire escapes, and balconies, and from the sides and bottom of windows that open.*

Discussion and Commentary: The clearances for overhead service drops are limited to conductors and cables that are not installed in a raceway or provided with an approved outer jacket. The requirement reduces the potential for damage to the service conductors and limits the risk of accidental contact. The clearance requirement from openable windows applies only to the sides and below the window opening. Service conductors and drip loops located just above an opening are deemed to be out of reach.

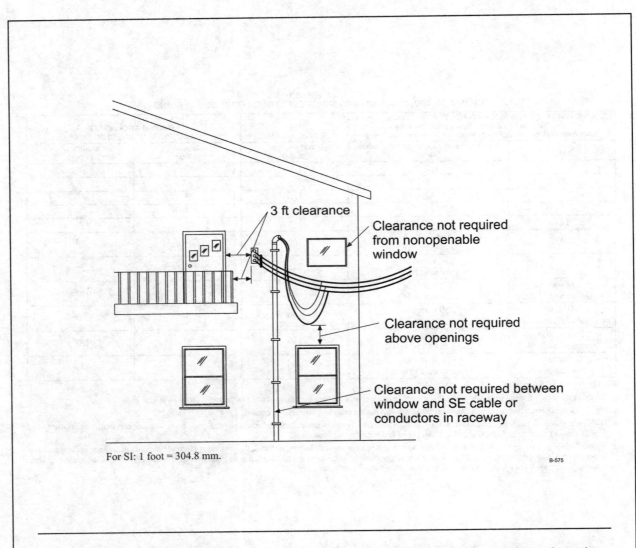

3 ft clearance

Clearance not required from nonopenable window

Clearance not required above openings

Clearance not required between window and SE cable or conductors in raceway

For SI: 1 foot = 304.8 mm.

B-575

The need for a minimum 3-foot separation is based on the potential proximity of people to the service conductors. It is expected that contact is possible from open windows, porches, stairs, decks and similar occupiable areas. No clearance is required from nonopenable windows.

Topic: Vertical Clearances
Reference: IRC E3504.2

Category: Services
Subject: Overhead Service-drop and
Service Conductor Installation

Code Text: *Conductors shall have a vertical clearance of not less than 8 feet (2438 mm) above the roof surface. The vertical clearance above the roof level shall be maintained for a distance of not less than 3 feet (914 mm) in all directions from the edge of the roof.* See four exceptions allowing for reduced clearances.

Discussion and Commentary: The general provisions mandate at least an 8-foot clearance between a service-drop conductor and a roof. This clearance must not only occur directly above the roof surface but also extend horizontally at least 3 feet, except for the point where the service drop is attached to the side of the building. However, the separation may be reduced to 3 feet where the roof slope is 4:12 or greater. It is anticipated that travel across such a steeply-sloping roof will be minimal.

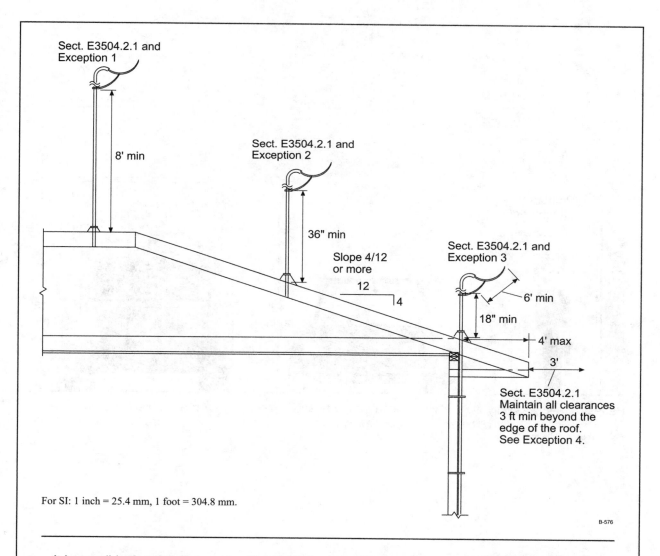

For SI: 1 inch = 25.4 mm, 1 foot = 304.8 mm.

B-576

It is possible that the clearance will need to be increased to more than 8 feet where the roof is expected to have pedestrian access. This could occur where there is a rooftop court or sundeck or where the roof is also the upper deck of a parking garage.

Topic: Vertical Clearance from Grade
Reference: IRC E3504.2.2

Code Text: *Service-drop conductors shall have the following minimum clearances from final grade: 1) for service-drop cables supported on and cabled together with a grounded bare messenger wire, the minimum vertical clearance shall be 10 feet (3048 mm) at the electric service entrance to buildings, at the lowest point of the drip loop of the building electric entrance, and above areas or sidewalks accessed by pedestrians only; 2) twelve feet (3658 mm)—over residential property and driveways; and 3) eighteen feet (5486 mm)—over public streets, alleys, roads or parking areas subject to truck traffic.*

Discussion and Commentary: The minimum clearances for service-drop conductors are based on the expected height of potential hazards. Where there is no anticipated vehicle traffic, at least 10 feet of vertical clearance is mandated. Where vehicles are present, higher clearances are necessary.

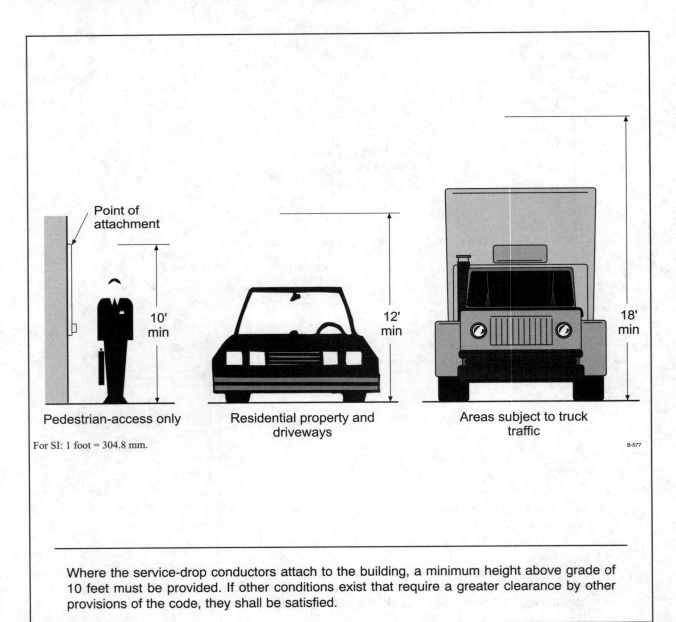

Point of
attachment

10'
min

Pedestrian-access only

For SI: 1 foot = 304.8 mm.

12'
min

Residential property and
driveways

18'
min

Areas subject to truck
traffic

B-577

Where the service-drop conductors attach to the building, a minimum height above grade of 10 feet must be provided. If other conditions exist that require a greater clearance by other provisions of the code, they shall be satisfied.

Topic: Grounding Methods
Reference: IRC E3508.1

Category: Services
Subject: Grounding Electrode System

Code Text: *Where available on the premises at each building or structure served, electrodes specified in Sections E3508.1.1, E3508.1.2 and E3508.1.3, and any made electrodes specified in Section E3508.2, shall be bonded together to form the grounding electrode system. Interior metal water piping located more than 5 feet (1524 mm) from the point of entrance to the building shall not be used as part of the grounding electrode system or as a conductor to interconnect the electrodes that are part of the grounding electrode system.*

Discussion and Commentary: One method of providing a grounding electrode is the use of a metal underground water pipe that is in direct contact with the earth for a minimum distance of 10 feet. Where this method is used, it is necessary to also provide a supplementary electrode, such as a made electrode of pipe, conduit or rod.

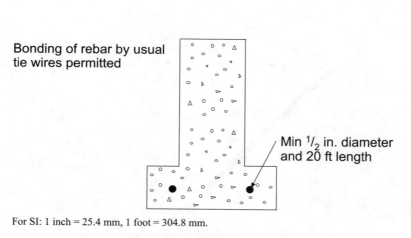

Bonding of rebar by usual tie wires permitted

Min ¹/₂ in. diameter and 20 ft length

For SI: 1 inch = 25.4 mm, 1 foot = 304.8 mm.

Concrete-encased electrode

B-578

One or more complying reinforcing bars can be used as a grounding electrode when encased in concrete. The bar must be at least ¹/₂ inch in diameter, be at least 20 feet in length, and be located within and near the bottom of the concrete foundation or footing.

Topic: Made Electrodes

Reference: IRC E3508.2

Category: Services

Subject: Grounding Electrode System

Code Text: *Where none of the electrodes specified in Section E3508.1 is available, one or more* rod or pipe electrodes *shall be used. Where more than one electrode is used, each electrode of one grounding system shall be not less than 6 feet (1829 mm) from any other electrode of another grounding system. Two or more grounding electrodes that are effectively bonded together shall be considered as a single grounding electrode system.*

Discussion and Commentary: Electrodes of rods or pipes are permitted where the other approved grounding electrode systems are not available. The minimum length of such made electrodes shall be 8 feet, embedded below the permanent moisture level if possible. Pipes or conduit must be a minimum of $3/4$-inch diameter in trade size, while iron or steel rods are to be at least $5/8$-inch diameter.

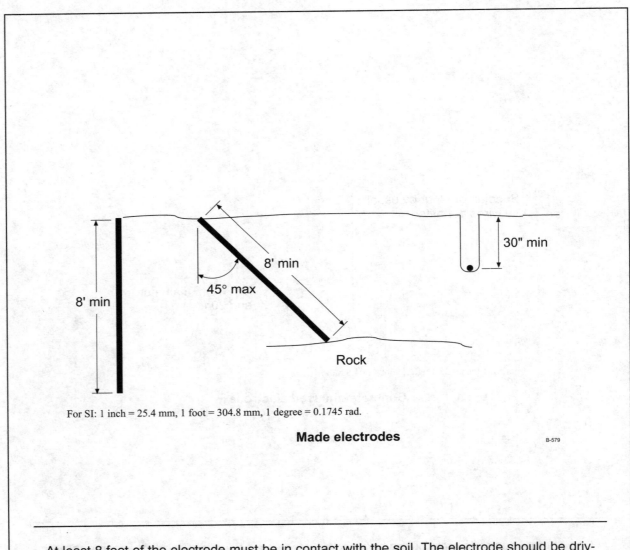

For SI: 1 inch = 25.4 mm, 1 foot = 304.8 mm, 1 degree = 0.1745 rad.

Made electrodes

B-579

At least 8 feet of the electrode must be in contact with the soil. The electrode should be driven vertically unless soil conditions are prohibitive. In such cases, the electrode can be driven at an angle of up to 45 degrees or placed in a minimum 30-inch deep trench.

Topic: Locations and Ratings

Category: Branch Circuits and Feeder Requirements

Reference: IRC E3603

Subject: Required Branch Circuits

Code Text: *Central heating equipment other than fixed electric space heating shall be supplied by an individual branch circuit. A minimum of two 20-ampere-rated branch circuits shall be provided to serve receptacles located in the kitchen, pantry, breakfast area, dining area or similar area of a dwelling. A minimum of one 20-ampere-rated branch circuit shall be provided for receptacles located in the laundry area and shall serve only receptacle outlets located in the laundry area. A minimum of one 20-ampere branch circuit shall be provided to supply the bathroom receptacle outlet(s). The minimum number of branch circuits shall be determined from the total computed load and the size or rating of the circuits used.*

Discussion and Commentary: In addition to general lighting and convenience outlet circuits, a number of specific areas of a dwelling unit must be provided with their own dedicated branch circuits.

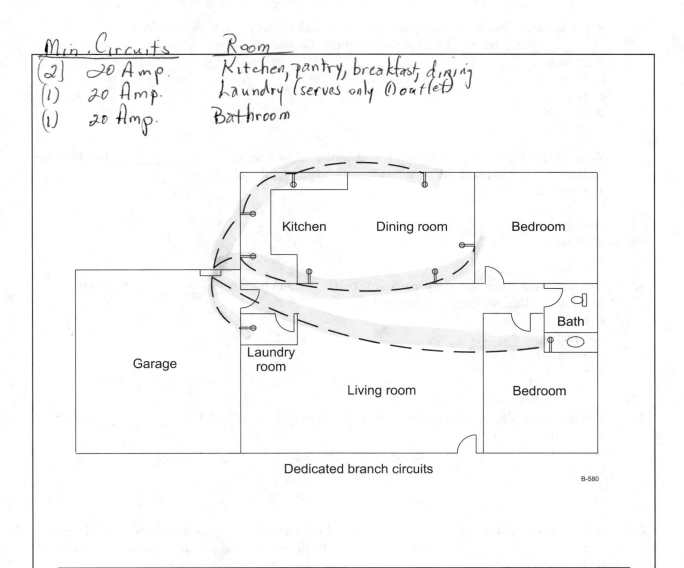

Min. Circuits — Room

(2) 20 Amp. — Kitchen, pantry, breakfast, dining
(1) 20 Amp. — Laundry (serves only (1) outlet)
(1) 20 Amp. — Bathroom

Dedicated branch circuits

B-580

The receptacles located at the kitchen countertop must be served by a minimum of two 20-ampere-rated branch circuits. One or both of the circuits may also be used to satisfy the requirements for the other outlets located in the kitchen and dining areas.

QUIZ

Study Session 17

1. Energized parts operating at a minimum of _____ shall be guarded against accidental contact by people through the use of approved enclosures.
   a. 50 volts
   b. 60 volts
   c. 90 volts
   d. 110 volts

2. A panelboard requiring access while energized shall be provided with a minimum of _____ in depth measured in the direction of access.
   a. 30 inches
   b. 36 inches
   c. 42 inches
   d. 48 inches

3. Work space in front of an electrical panelboard shall be a minimum of _____ in width, but no less than the width of the panelboard.
   a. 30 inches
   b. 36 inches
   c. 42 inches
   d. 48 inches

4. The dedicated space above a panelboard shall be a minimum of _____ high or to the structural ceiling, whichever is lower.
   a. 4 feet
   b. 5 feet
   c. 6 feet
   d. 6.5 feet

5. The minimum size of conductors for feeders and branch circuits shall be _____ copper.
   a. No. 16
   b. No. 14
   c. No. 12
   d. No. 10

6. Where conductors are to be spliced, terminated or connected to fixtures or devices, a minimum length of _____ of free conductor shall be provided at each outlet, junction or switch point.
   a. 4 inches
   b. 6 inches
   c. 8 inches
   d. 12 inches

7.   Insulated grounded conductors of sizes No. 6 and smaller may be identified by all but which one of the following methods?
     a. continuous white outer finish
     b. continuous natural gray outer finish
     c. continuous black outer finish
     d. three continuous white stripes (on other than green insulation)

8.   Equipment grounding conductors may be identified by all but which one of the following methods?
     a. continuous white color
     b. continuous green color
     c. bare
     d. continuous green color with one or more yellow stripes

9.   Which of the following continuous colors is not permitted for ungrounded conductors?
     a. gray
     b. white
     c. green
     d. red

10. What is the minimum permitted size of a grounded service conductor serving a maximum load of 150 amps?
     a. No. 1 aluminum
     b. No. 1 copper
     c. No. 1/0 aluminum
     d. No. 1/0 copper

11. An open service conductor without an overall outer jacket shall have a minimum clearance of _____ from the sides of doors, porches and openable windows.
     a. 3 feet
     b. 4 feet
     c. 6 feet
     d. 8 feet

12. An overhead service conductor shall have a minimum clearance above a 3:12 roof of _____.
     a. 3 feet
     b. 6 feet
     c. 7 feet
     d. 8 feet

13. Service-drop conductors shall have a minimum clearance of _____ over residential property and driveways.
   a. 8 feet
   b. 10 feet
   c. 12 feet
   d. 15 feet

14. In no case shall the point of attachment of service-drop conductors to a building be less than _____ above finished grade.
   a. 8 feet
   b. 10 feet
   c. 12 feet
   d. 14 feet

15. A metal underground water pipe used as part of the grounding electrode system shall be in direct contact with the earth for a minimum of _____.
   a. 8 feet
   b. 10 feet
   c. 15 feet
   d. 20 feet

16. Where more than one made electrode is used in a grounding electrode system, each electrode shall be located a minimum of _____ from any other electrode of another grounding system.
   a. 5 feet
   b. 6 feet
   c. 10 feet
   d. 20 feet

17. Where a grounding electrode is made of an iron or steel rod, the rod shall be a minimum of _____ in diameter and a minimum of _____ in length.
   a. $1/2$ inch, 8 feet
   b. $1/2$ inch, 10 feet
   c. $5/8$ inch, 8 feet
   d. $5/8$ inch, 10 feet

18. Where used outdoors, an aluminum or copper-clad aluminum grounding conductor shall be installed a minimum of _____ from the earth.
   a. 6 inches
   b. 12 inches
   c. 18 inches
   d. 30 inches

19.   The rating of any one cord- and plug-connected utilization equipment shall be a maximum of
        _____ of the branch-circuit ampere rating.
        a. 80 percent
        b. 100 percent
        c. 110 percent
        d. 125 percent

20.   A minimum of _____ branch circuit(s) shall be provided to serve receptacles located in
        the kitchen, pantry, breakfast area and dining area.
        a. one 15-ampere-rated
        b. one 20-ampere-rated
        c. two 15-ampere-rated
        d. two 20-ampere-rated

21.   A minimum of _____ branch circuit(s) shall be provided for receptacles located in the
        laundry room and a minimum of _____ branch circuit(s) shall be provided to supply the
        bathroom receptacle outlet(s).
        a. one 15-ampere-rated, one 15-ampere-rated
        b. one 20-ampere-rated, one 20-ampere-rated
        c. one 15-ampere-rated, two 15-ampere-rated
        d. one 20-ampere-rated, two 20 ampere-rated

22.   Where serving two or more three-wire branch circuits supplied by a three-wire feeder, feeder
        conductors shall have a minimum size of _____.
        a. No. 6 aluminum
        b. No. 6 copper
        c. No. 8 aluminum
        d. No. 8 copper

23.   A minimum unit load of _____ shall constitute the minimum lighting and convenience
        receptacle load for each square foot of floor area.
        a. two volt-amperes
        b. three volt-amperes
        c. five volt-amperes
        d. six volt-amperes

24.   What is the maximum overcurrent-protection-device rating permitted for No. 12 copper
        conductors?
        a. 15 amps
        b. 20 amps
        c. 25 amps
        d. 30 amps

25. Which one of the following ratings is not a standard ampere rating for fuses and inverse time circuit breakers?
   a. 25 amperes
   b. 50 amperes
   c. 75 amperes
   d. 100 amperes

26. Where metallic plugs or plates are used with nonmetallic enclosures, they shall be recessed a minimum of _____ inch from the outer surface of the enclosure.
   a. $^1/_8$
   b. $^1/_4$
   c. $^3/_8$
   d. $^1/_2$

27. Panelboards and overcurrent protection devices are prohibited in _____.
   a. clothes closets
   b. sleeping rooms
   c. garages
   d. storage rooms

28. Where the opening to an outlet, junction, or switch point is less than 8 inches in any dimension, each conductor shall be long enough to extend a minimum of _____ inches outside of such opening.
   a. 2
   b. 3
   c. 4
   d. 6

29. Service disconnecting means shall not be installed in _____.
   a. storage closets
   b. laundry rooms
   c. sleeping rooms
   d. bathrooms

30. In order to be used as a part of the grounding electrode system, interior metal water pipe shall be located a maximum of _____ feet from the entrance to the building.
   a. 3
   b. 5
   c. 6
   d. 10

INTERNATIONAL RESIDENTIAL CODE
Study Session 18
Chapters 37, 38, 39, 40 and 41—Wiring Methods, Power and Lighting Distribution, Devices and Lighting Fixtures, Appliance Installation and Swimming Pools

OBJECTIVE: To gain an understanding of the provisions regulating wiring methods, power and lighting distribution, devices and lighting fixtures, appliance installation and swimming pools.

REFERENCE: Chapters 37, 38, 39, 40 and 41, 2003 *International Residential Code*

KEY POINTS:

- How are electrical cables to be installed where they are run in attics across the top of structural members? Where they are run parallel to framing members? How must they be protected from physical damage?

- What are the minimum cover requirements for direct buried cable or raceways installed underground? How are direct buried cables to be protected from damage where emerging from the ground?

- How must backfill be placed in an excavation containing electrical cables or raceways?

- What method is used to determine the minimum spacing requirements for convenience receptacles? What wall spaces are to be considered? How are floor receptacles regulated?

- Where are small appliance receptacles regulated? How many receptacles must be provided? How must they be distributed?

- In what manner are counter receptacles to be located? How do the provisions differ for island counter spaces? For peninsular counter spaces?

- How many wall receptacles are required in a bathroom? Outdoors? In laundry areas? In basements and garages? In hallways?

- Where is ground-fault circuit-interrupter protection mandated?

- In what locations are arc-fault circuit interrupter protection mandated?

- In what rooms and spaces must wall-switch-controlled lighting outlets be installed? Under what conditions are wall-switch-controlled receptacles permitted as an alternate to lighting outlets?

- How is nonmetallic-sheathed cable to be installed in nonmetallic boxes? What is the minimum required depth of such boxes? How is box volume and fill calculated? How are boxes to be installed?

- How are luminaires to be installed in wet or damp locations? In bathtub and shower areas? In clothes closets?

- What are the limits for supporting luminaires by the screw shell of a lampholder? By an outlet box? What clearances are required between a recessed luminaire and thermal insulation? Other combustible material?

- How is track lighting to be installed? Where is track lighting prohibited?

- Where shall receptacle outlets be located in relationship to a pool, spa or hot tub? Which receptacles require ground-fault circuit-interrupter protection? How are luminaires, lighting outlets and ceiling-suspended paddle fans regulated?

- What parts of a swimming pool, spa or hot tub must be bonded? What are the acceptable bonding methods? Which equipment must be grounded? How is pool equipment regulated?

Topic: Allowable Methods

Category: Wiring Methods

Reference: IRC E3701.2

Subject: General Requirements

Code Text: *The allowable wiring methods for electrical installations shall be those listed in Table E3701.2. Single conductors shall be used only where part of one of the recognized wiring methods listed in Table E3701.2. As used in* the IRC, *abbreviations of the wiring-method types shall be as indicated in Table E3701.2.*

Discussion and Commentary: Single insulated conductors are not permitted as a wiring method without being part of a cable assembly or installed in tubing or conduit. In many older houses, a wiring method known as "knob-and-tube" was used, consisting of single insulated conductors supported in free air on porcelain insulators and, where run through a wood framing member, installed through insulating tubes. Where remodeling, retrofitting or repair is being done in such a house, the transition to new wiring should be to cable.

TABLE E3701.2
ALLOWABLE WIRING METHODS

ALLOWABLE WIRING METHOD	DESIGNATED ABBREVIATION
Armored cable	AC
Electrical metallic tubing	EMT
Electrical nonmetallic tubing	ENT
Flexible metal conduit	FMC
Intermediate metal conduit	IMC
Liquidtight flexible conduit	LFC
Metal-clad cable	MC
Nonmetallic sheathed cable	NM
Rigid nonmetallic conduit	RNC
Rigid metallic conduit	RMC
Service entrance cable	SE
Surface raceways	SR
Underground feeder cable	UF
Underground service cable	USE

In many older houses, the wiring is still sound and in good shape, but when additions or extensions are made, the possibility for arcing, bad connections, heating and overloading come into play. Thus, one of the allowable wiring methods must be utilized.

Topic: Installation and Support
Reference: IRC E3702.1

Category: Wiring Methods
Subject: Above-Ground Installation Requirements

Code Text: *Wiring methods shall be installed and supported in accordance with Table E3702.1.*

Discussion and Commentary: A number of different installation conditions are addressed in an effort to protect and support various wiring methods permitted by the code. For NM cable and other materials subject to damage, the installation procedures address the common situation of cable within stud cavities. If the cable is run parallel along the side of the stud, it must be located at least $1^{1}/_{4}$ inches from the stud edge or otherwise physically protected. The same minimum dimension is necessary where cable is run perpendicular to the vertical framing members through bored holes. Where the mandated depth is not provided, it is common to install a minimum $^{1}/_{16}$-inch-thick steel plate across the stud edge.

TABLE E3702.1
GENERAL INSTALLATION AND SUPPORT REQUIREMENTS FOR WIRING METHODS[a,b,c,d,e,f,g,h,i]

INSTALLATION REQUIREMENTS (requirement applicable only to wiring methods marked "A")	AC MC	EMT IMC RMC	ENT	FMC LFC	NM UF	RNC	SE	SR[a]	USE
Where run parallel with the framing member, the wiring shall be 1.25 inches from the edge of a framing member such as a joist, rafter or stud or shall be physically protected.	A	—	A	A	A	—	A	—	—
Bored holes in studs and vertical framing members for wiring shall be located 1.25 inches from the edge or shall be protected with a minimum 0.0625-inch steel plate or sleeve or other physical protection.	A	—	A	A	A	—	A	—	—
Where installed in grooves, to be covered by wallboard, siding, paneling, carpeting, or similar finish, wiring methods shall be protected by 0.0625-inch-thick steel plate, sleeve, or equivalent or by not less than 1.25-inch free space for the full length of the groove in which the cable or raceway is installed.	A	—	A	A	A	—	A	A	A
Bored holes in joists, rafters, beams and other horizontal framing members shall be 2 inches from the edge of the structural framing member.	A	A	A	A	A	A	A	—	—
Securely fastened bushings or grommets shall be provided to protect wiring run through openings in metal framing members.	—	—	A	—	A	—	A	—	—
The maximum number of 90-degree bends shall not exceed four between junction boxes.	—	A	A	A	—	A	—	—	—
Bushings shall be provided where entering a box, fitting or enclosure unless the box or fitting is designed to afford equivalent protection.	A	A	A	A	—	A	—	A	—
Ends of raceways shall be reamed to remove rough edges.	—	A	A	A	—	A	—	A	—
Maximum allowable on center support spacing for the wiring method in feet.	4.5[b, c]	10	3[b]	4.5[b]	4.5[i]	3[d]	2.5[e]	—	2.5[e]
Maximum support distance in inches from box or other terminations.	12[b, f]	36	36	12[b, g]	12[h, i]	36	12	—	12

For SI: 1 inch = 25.4 mm, 1 foot = 304.8 mm, 1 degree = 0.009 rad.

a. Installed in accordance with listing requirements.

b. Supports not required in accessible ceiling spaces between light fixtures where lengths do not exceed 6 feet.

c. Six feet for MC cable.

d. Five feet for trade sizes greater than 1 inch.

e. Two and one-half feet where used for service or outdoor feeder and 4.5 feet where used for branch circuit or indoor feeder.

f. Twenty-four inches where flexibility is necessary.

g. Thirty-six inches where flexibility is necessary.

h. Within 8 inches of boxes without cable clamps.

i. Flat cables shall not be stapled on edge.

A concern in accessible attics is the presence of cables that may be subject to contact and potential physical damage. The code requires protection for these cables, particularly where they are located within 6 feet of the attic access opening.

Topic: Minimum Cover, Protection from Damage
Reference: IRC E3703.1, E3703.4

Category: Wiring Methods
Subject: Underground Installation
Requirements

Code Text: *Direct buried cable or raceways shall be installed in accordance with the minimum cover requirements of Table E3703.1. Direct buried conductors and cables emerging from the ground shall be protected by enclosures or raceways extending from the minimum cover distance below grade required by Section 3703.1 to a point at least 8 feet (2438 mm) above finished grade. In no case shall the protection be required to exceed 18 inches (457 mm) below finished grade.*

Discussion and Commentary: To protect cables or raceways buried below ground level from damage, the code establishes a minimum depth. The burial depth is dependent upon two factors: the location or method of burial, and the type of wiring method. Where direct buried conductors and cables are permitted, they must be protected from below ground level to a point high enough that physical damage is improbable.

TABLE E3703.1
MINIMUM COVER REQUIREMENTS, BURIAL IN INCHES[a,b,c,d]

LOCATION OF WIRING METHOD OR CIRCUIT	TYPE OF WIRING METHOD OR CIRCUIT				
	1 Direct burial cables or conductors	2 Rigid metal conduit or intermediate metal conduit	3 Nonmetallic raceways listed for direct burial without concrete encasement or other approved raceways	4 Residential branch circuits rated 120 volts or less with GFCI protection and maximum overcurrent protection of 20 amperes	5 Circuits for control of irrigation and landscape lighting limited to not more than 30 volts and installed with type UF or in other identified cable or raceway
All locations not specified below	24	6	18	12	6
In trench below 2-inch-thick concrete or equivalent	18	6	12	6	6
Under a building	0 (In raceway only)	0	0	0 (In raceway only)	0 (In raceway only)
Under minimum of 4-inch-thick concrete exterior slab with no vehicular traffic and the slab extending not less than 6 inches beyond the underground installation	18	4	4	6 (Direct burial) 4 (In raceway)	6 (Direct burial) 4 (In raceway)
Under streets, highways, roads, alleys, driveways and parking lots	24	24	24	24	24
One- and two-family dwelling driveways and outdoor parking areas, and used only for dwelling-related purposes	18	18	18	12	18
In solid rock where covered by minimum of 2 inches concrete extending down to rock	2 (In raceway only)	2	2	2 (In raceway only)	2 (In raceway only)

For SI: 1 inch = 25.4 mm.

a. Raceways approved for burial only where encased concrete shall require concrete envelope not less than 2 inches thick.

b. Lesser depths shall be permitted where cables and conductors rise for terminations or splices or where access is otherwise required.

c. Where one of the wiring method types listed in columns 1 to 3 is combined with one of the circuit types in columns 4 and 5, the shallower depth of burial shall be permitted.

d. Where solid rock prevents compliance with the cover depths specified in this table, the wiring shall be installed in metal or nonmetallic raceway permitted for direct burial. The raceways shall be covered by a minimum of 2 inches of concrete extending down to the rock.

The backfill used to cover cables or raceways must be placed in such a manner as to avoid damage to the wiring method. It may be necessary to use boards, sleeves, granular material or other suitable methods to protect the raceway or cable from physical damage.

Topic: Convenience Receptacle Distribution
Reference: IRC E3801.2

Category: Power and Lighting Distribution
Subject: Receptacle Outlets

Code Text: *In every kitchen, family room, dining room, living room, parlor, library, den, sun room, bedroom, recreation room, or similar room or area of dwelling units, receptacle outlets shall be installed . . . so that no point along the floor line in any wall space is more than 6 feet (1829 mm), measured horizontally, from an outlet in that space. Receptacles shall, insofar as practicable, be spaced equal distances apart.*

Discussion and Commentary: The maximum spacing between convenience receptacles is basically 12 feet, allowing flexibility in the placement of electric appliances without the need of any extension cord. Those spaces that are subject to the requirement tend to be habitable spaces, the areas of a dwelling used for living, sleeping, dining or cooking.

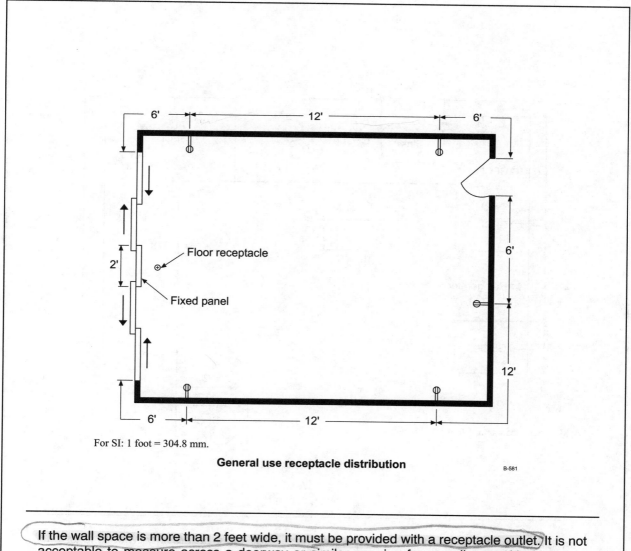

For SI: 1 foot = 304.8 mm.

General use receptacle distribution

B-581

If the wall space is more than 2 feet wide, it must be provided with a receptacle outlet. It is not acceptable to measure across a doorway or similar opening for compliance with the 6-foot rule. An extension cord across a door opening is a considerable hazard.

Topic: Countertop Receptacles
Reference: IRC E3801.4

Category: Power and Lighting Distibution
Subject: Receptacle Outlets

Code Text: *In kitchens and dining rooms of dwelling units, receptacle outlets for counter spaces shall be installed . . . at each wall counter space 12 inches (305 mm) or wider. Receptacle outlets shall be installed so that no point along the wall line is more than 24 inches (610 mm), measured horizontally from a receptacle outlet in that space. Receptacle outlets shall be located not more than 20 inches (508 mm) above the countertop. Receptacle outlets shall not be installed in a face-up position in the work surfaces or countertops.*

Discussion and Commentary: Receptacle outlets are mandated in a repetitive fashion along a kitchen counter for both safety and convenience. With a number of appliances potentially in use at the same time, and because they are likely to be moved to a number of different positions, it is important that access to multiple receptacles be provided.

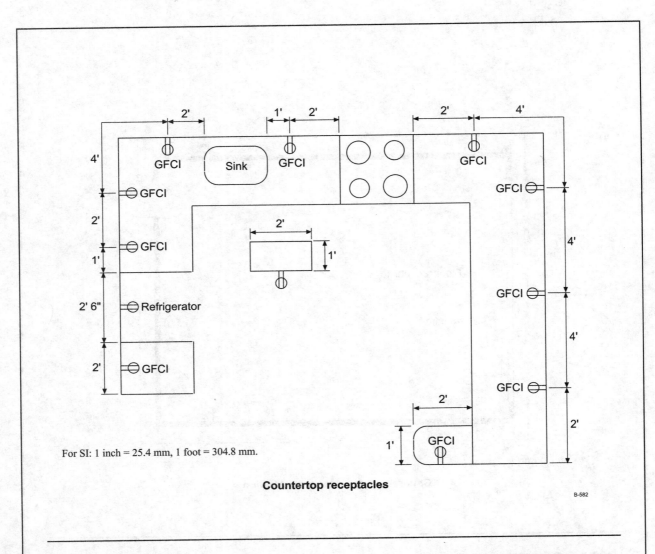

For SI: 1 inch = 25.4 mm, 1 foot = 304.8 mm.

Countertop receptacles

B-582

If an island counter space exceeds a specified size, it too must be provided with at least one receptacle outlet. The same holds true for peninsular counter spaces of similar dimensions. Appliance cords pose an obvious problem when extended through work areas or traffic paths.

Topic: Miscellaneous Receptacle Outlets
Reference: IRC E3801.6 through E3801.10

Category: Power and Lighting Distibution
Subject: Receptacle Outlets

Code Text: *At least one wall receptacle outlet shall be installed in bathrooms and such outlet shall be located within 36 inches (914 mm) of the outside edge of each lavatory basin. At least one receptacle outlet accessible at grade level . . . shall be installed outdoors at the front and back of each dwelling unit having direct access to grade. At least one receptacle outlet shall be installed to serve laundry appliances. At least one receptacle outlet, in addition to any provided for laundry equipment, shall be installed in each basement and in each attached garage. Hallways of 10 feet (3048 mm) or more in length shall have at least one receptacle outlet.*

Discussion and Commentary: In addition to habitable spaces, those spaces typically defined as not habitable also require a limited number of receptacle outlets. Although the number of outlets mandated may be fewer than required for habitable rooms, it is important that outlets be provided to serve any equipment or activity that is anticipated for the space.

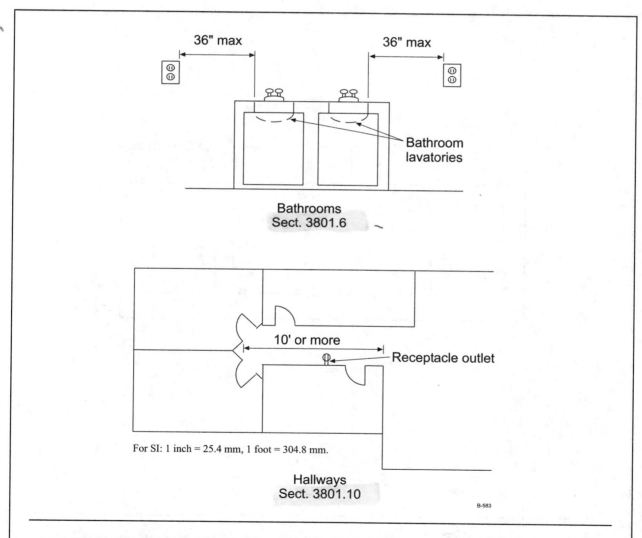

For SI: 1 inch = 25.4 mm, 1 foot = 304.8 mm.

Where heating, air-conditioning and refrigeration equipment is located in an attic or crawl space, a convenience outlet must be installed within 25 feet of the equipment. To be used during maintenance or repair activities, the outlet must be located on the same level as the equipment.

Topic: Required GFCI Protection

Reference: IRC E3802.1 - E3802.7

Category: Power and Lighting Distribution

Subject: Ground-fault Protection

GFCI

GFCI

Code Text: (Summary) All 125-volt, single-phase, 15- and 20-ampere receptacles installed in bathrooms, garages, grade-level portions of unfinished accessory buildings used for storage or work areas, outdoors, in a crawl space at or below grade, and unfinished basements shall have ground-fault circuit-interrupter protection. See Exceptions. All 125-volt, single-phase, 15- and 20-ampere receptacles that serve kitchen countertop surfaces, and all 125-volt, single-phase, 15- and 20-ampere receptacles that serve a bar sink countertop surface and are located within 6 feet (1829 mm) of the outside edge of a wet bar sink, shall have ground-fault circuit-interrupter protection.

Discussion and Commentary: Where there is an unbalance between the ungrounded "hot" conductor and the grounded "neutral" conductor, the purpose of a ground-fault circuit interrupter (GFCI) is to trip the circuit off. Where the unbalance is caused by human contact, this trip prevents serious injury or possibly death.

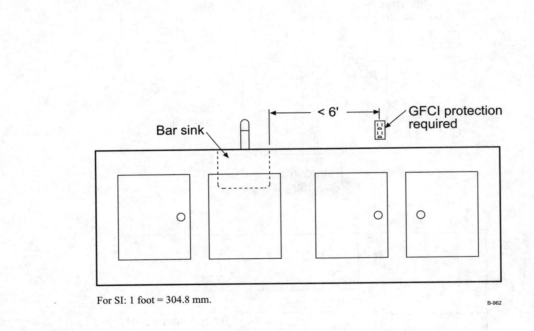

For SI: 1 foot = 304.8 mm.

B-962

The purpose of the GFCI is to recognize an unbalanced load and shut off the circuit, but it does not reduce the magnitude of the ground-fault current. A very strong, but brief, shock is still encountered by a person during the time period the GFCI device needs to recognize the unbalance and trip.

Topic: Nonmetallic Boxes
Reference: IRC E3805.3

Category: Power and Lighting Distribution
Subject: Boxes, Conduit Bodies and Fittings

Code Text: *Nonmetallic boxes shall be used only with nonmetallic-sheathed cable, cabled wiring methods, flexible cords and nonmetallic raceways. See Exceptions. Where nonmetallic-sheathed cable is used, the cable assembly, including the sheath, shall extend into the box not less than $^1/_4$ inch (6.4 mm) through a nonmetallic-sheathed cable knockout opening. Where nonmetallic-sheathed cable is used with boxes not larger than a nominal size of $2^1/_4$ inches by 4 inches (57 mm by 102 mm) mounted in walls or ceilings, and where the cable is fastened within 8 inches (203 mm) of the box measured along the sheath, and where the sheath extends through a cable knockout not less than $^1/_4$ inch (6.4 mm), securing the cable to the box shall not be required.*

Discussion and Commentary: To provide adequate insulation for the conductors up and into the box, the sheathing must extend into the box a limited distance.

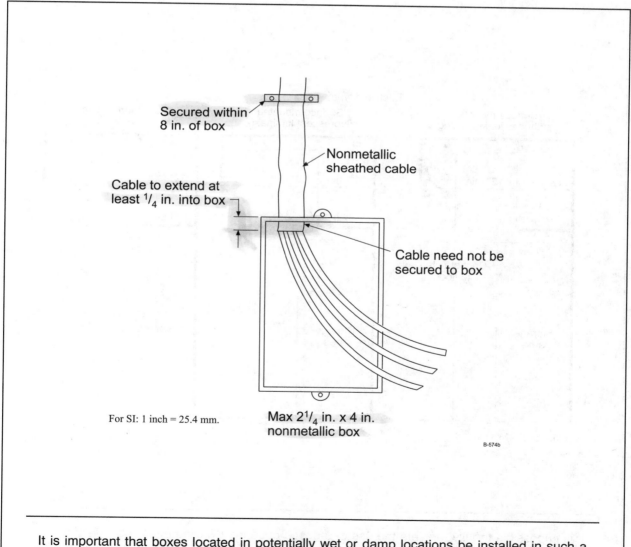

Secured within 8 in. of box

Nonmetallic sheathed cable

Cable to extend at least $^1/_4$ in. into box

Cable need not be secured to box

For SI: 1 inch = 25.4 mm.

Max $2^1/_4$ in. x 4 in. nonmetallic box

B-574b

It is important that boxes located in potentially wet or damp locations be installed in such a manner as to prevent the entry and accumulation of moisture. Where installed in such areas, the boxes must be listed for use in wet locations.

Topic: Luminaires in Clothes Closets

Reference: IRC E3903.11

Category: Devices and Lighting Fixtures

Subject: Fixtures

Code Text: *The types of luminaires installed in clothes closets shall limited to surface-mounted or recessed incandescent luminaires with completely enclosed lamps, and surface-mounted or recessed fluorescent luminaires. Surface-mounted incandescent luminaires shall be installed on the wall above the door or on the ceiling, provided there is a minimum clearance of 12 inches (305 mm) (6 inches for surface-mounted fluorescent luminaires) between the fixture and the nearest point of a storage space. Recessed incandescent luminaires with a completely enclosed lamp (and recessed fluorescent luminaires) shall be installed in the wall or on the ceiling provided there is a minimum clearance of 6 inches (152 mm) between the luminaire and the nearest point of a storage space.*

Discussion and Commentary: It is quite probable that a clothes closet is the most common place in a house where a light source will be located adjacent to combustible materials. Therefore, various clearances are set forth based on the type of luminaire installed.

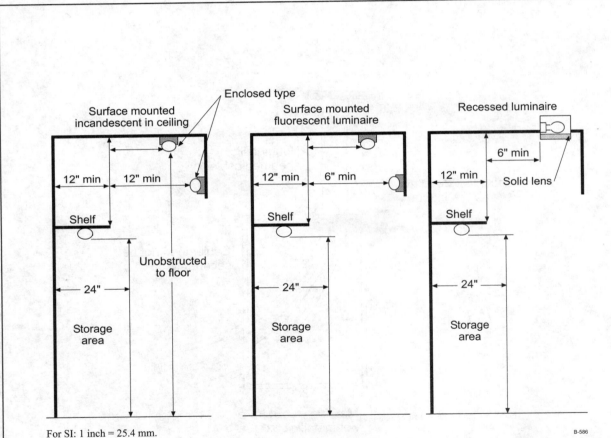

For SI: 1 inch = 25.4 mm.

B-586

It is impossible to fully regulate the use of a space once the dwelling unit is occupied; therefore, the code defines for its purposes the extent of a "storage space." It is bounded by the sides and back of the closet walls and extends to a height of 6 feet or the highest clothes-hanging rod.

2003 IRC Study Companion

Topic: Supports and Clearances
Reference: IRC E3904.4, E3904.5, E3904.8

Category: Devices and Lighting Fixtures
Subject: Fixture Installation

Code Text: *Luminaires and lampholders shall be securely supported. A luminaire that weighs more than 6 pounds (2.72 kg) or exceeds 16 inches (406 mm) in any dimension shall not be supported by the screw shell of a lampholder. Outlet boxes or fittings . . . shall be permitted to support luminaires weighing 50 pounds (22.7 kg) or less. A luminaire that weighs more than 50 pounds (22.7 kg) shall be supported independently of an outlet box except where the outlet box is listed for the weight to be supported. A recessed luminaire that is not identified for contact with insulation shall have all recessed parts spaced at least 0.5 inch (12.7 mm) from combustible materials.*

Discussion and Commentary: Lampholders and luminaires should only be carried by the lampholder screw shell when they are of limited weight. The screw shell is not designed as a supporting element beyond supporting a bulb and simple shade or diffuser.

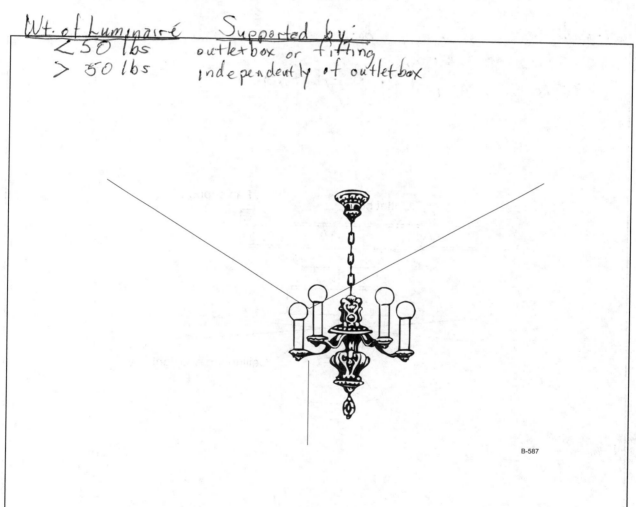

Wt. of Luminaire   Supported by:
< 50 lbs   outlet box or fitting
> 50 lbs   independently of outlet box

B-587

Where a luminaire exceeds 50 pounds, such as a large chandelier or similar decorative element, it cannot be adequately supported by an outlet box. Therefore, it is necessary to provide an independent means of support, typically addressed in the luminaire's installation instructions.

Topic: Support of Ceiling-suspended Paddle Fans
Reference: IRC E4001.6

Category: Appliance Installation
Subject: General

Code Text: *Ceiling-suspended paddle fans that do not exceed 35 pounds (15.88 kg) in weight, with or without accessories, shall be permitted to be supported by outlet boxes identified for such use and supported in accordance with Sections E3805 and E3806. Ceiling-suspended paddle fans exceeding 35 pounds (15.88 kg) in weight, with or without accessories, shall be supported independently of the outlet box.*

Discussion and Commentary: Ceiling fans can produce a potentially hazardous condition if not installed correctly. The combination of fan weight and movement of the fan creates a condition that is specifically addressed by the code. It is important to consider both the weight of the fan and any light fixture or other accessory when determining the requirements.

Weight   Supported by:
<35 lbs   outlet box
>35 lbs.   independently of outlet box

Outlet box   Fan support

Ceiling fan and light

B-588

Although most ceiling fans installed in residential applications weigh less than 35 pounds, some fan assemblies weigh up to 70 pounds. For these fans, the outlet box or support system must be listed for use with that weight of fan and accessories.

Topic: Receptacle Outlets

Category: Swimming Pools

Reference: IRC E4103.1

Subject: Equipment Location and Clearances

Code Text: *Receptacles that provide power for water-pump motors or other loads directly related to the circulation and sanitation system shall be permitted to be located between 5 feet and 10 feet (1524 mm and 3048 mm) from the inside walls of pools and outdoor spas and hot tubs. Other receptacles on the property shall be located not less than 10 feet (3048 mm) from the inside walls of pools and outdoor spas and hot tubs. At least one 125-volt 15- or 20-ampere receptacle supplied by a general-purpose branch circuit shall be located a minimum of 10 feet (3048 mm) from and not more than 20 feet (6096 mm) from the inside wall of pools and outdoor spas and hot tubs.*

Discussion and Commentary: Convenience receptacles must be located at least 10 feet from both indoor and outdoor swimming pools, as well as outdoor spas and hot tubs. At least one outlet shall be provided, but it cannot be located more than 20 feet from the pool edge.

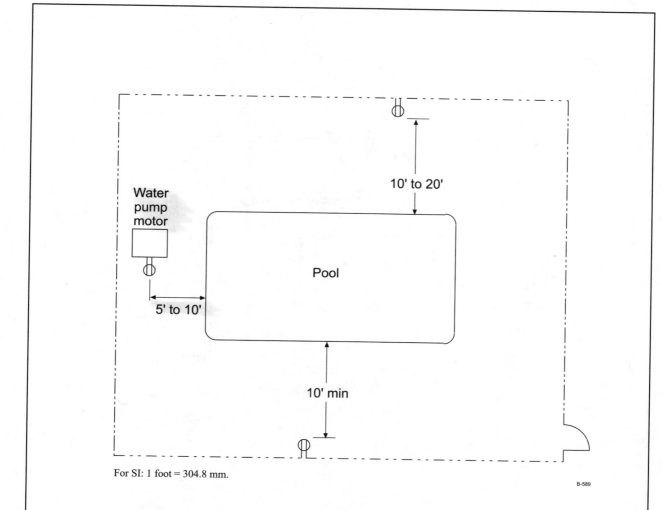

For SI: 1 foot = 304.8 mm.

B-589

An accesible disconnecting means is required in sight of and adjacent to all swimming pools, spas and hot tubs. The disconnection switch must be located no more than 5 feet from the inside wall of the pool, spa or hot tub.

Topic: Receptacle Outlets - Indoors
Reference: IRC E4103.1.4, E4103.1.5

Category: Swimming Pools
Subject: Equipment Location and Clearances

Code Text: *Receptacles shall be located not less than 5 feet (1524 mm) from the inside walls of indoor spas and hot tubs. A minimum of one 125-volt receptacle shall be located between 5 feet (1524 mm) and 10 feet (3048 mm) from the inside walls of indoor spas or hot tubs. One hundred twenty-five-volt receptacles located within 10 feet (3048 mm) of the inside walls of spas and hot tubs installed indoors shall be protected by ground-fault circuit-interrupters.*

Discussion and Commentary: The requirements applicable to receptacles adjacent to indoor hot tubs and spas are not as restrictive as those for both indoor and outdoor swimming pools and outdoor hot tubs and spas. The limited space available in most indoor spa and hot tub areas would make it impractical to apply the provisions mandated for outdoor facilities.

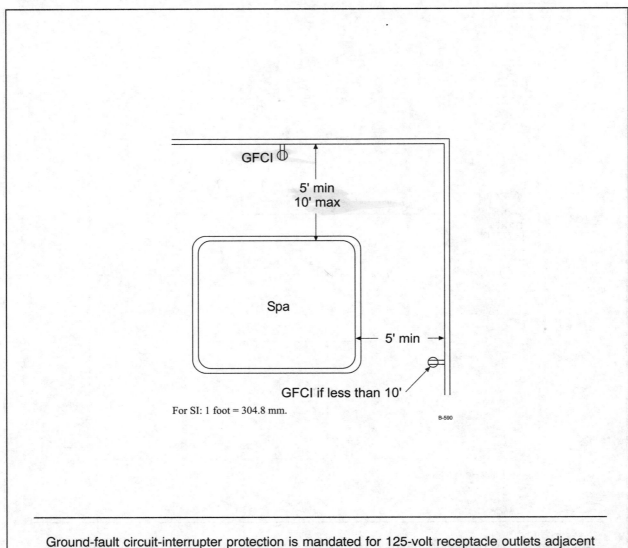

For SI: 1 foot = 304.8 mm.

B-590

Ground-fault circuit-interrupter protection is mandated for 125-volt receptacle outlets adjacent to indoor pools, spas and hydromassage bathtubs. Although any receptacle within 10 feet of a pool or spa requires a GFCI, such protection is only mandated within 5 feet of a hydromassage tub.

Topic: Luminaires and Ceiling Fans
Reference: IRC E4103.4

Category: Swimming Pools
Subject: Equipment Location and Clearances

Code Text: *In outdoor pool, outdoor spas and outdoor hot tubs areas, luminaires, lighting outlets and ceiling-suspended paddle fans shall not be installed over the pool or over the area extending 5 feet (1524 mm) horizontally from the inside walls of a pool except where no part of the luminaire or ceiling-suspended paddle fan is less than 12 feet (3658 mm) above the maximum water level.*

Discussion and Commentary: The code mandates clearances for luminaires in connection with both outdoor and indoor swimming pools. Typically the provisions for outdoor pools are consistent with those located indoors; however, a reduction in the luminaire or paddle fan height requirements to 7 feet, 6 inches is permitted. To provide this reduced clearance, the luminaires must be of the totally-enclosed type, and a GFCI must be installed in the branch circuit serving the fan or luminaire.

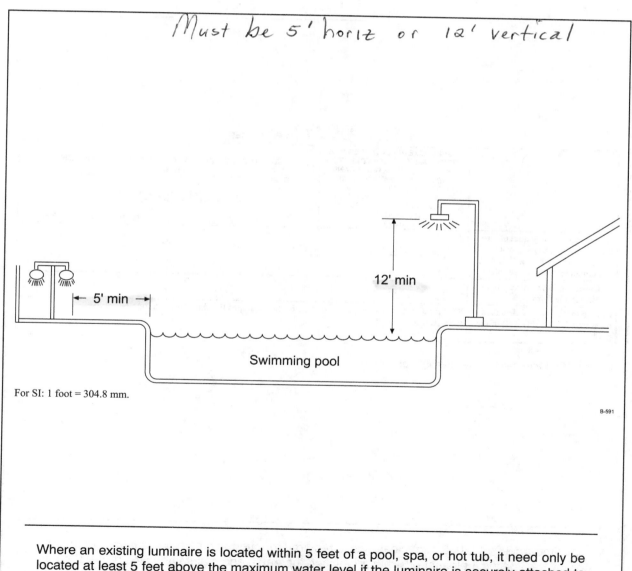

Must be 5' horiz or 12' vertical

5' min

12' min

Swimming pool

For SI: 1 foot = 304.8 mm.

B-591

Where an existing luminaire is located within 5 feet of a pool, spa, or hot tub, it need only be located at least 5 feet above the maximum water level if the luminaire is securely attached to the existing support and GFCI protection is provided on the luminaire's branch circuit.

Topic: Overhead Conductors
Reference: IRC E4103.5

Category: Swimming Pools
Subject: Conductor Clearances

Code Text: *Except where installed with the clearances specified in Table E4103.5, the following parts of pools and outdoor spas and hot tubs shall not be placed under existing service-drop conductors or any other open overhead wiring; nor shall such wiring be installed above the following: 1) pools and the areas extending 10 feet (3048 mm) horizontally from the inside of the walls of the pool, 2) diving structures, or 3) observation stands, towers, and platforms.*

Discussion and Commentary: When installing a new swimming pool, spa or hot tub, it is important to take into account the position of any overhead service-drop conductors or wiring. If the location of the pool, spa or hot tub brings it within the clearances required by the code, the conductors or wiring must be rerouted to bring the installation into compliance.

TABLE E4103.5
OVERHEAD CONDUCTOR CLEARANCES

	INSULATED SUPPLY OR SERVICE DROP CABLES, 0-750 VOLTS TO GROUND, SUPPORTED ON AND CABLED TOGETHER WITH AN EFFECTIVELY GROUNDED BARE MESSENGER OR EFFECTIVELY GROUNDED NEUTRAL CONDUCTOR (feet)	ALL OTHER SUPPLY OR SERVICE DROP CONDUCTORS (feet)	
		Voltage to ground	
		0-15 kV	Greater than 15 to 50 kV
A. Clearance in any direction to the water level, edge of water surface, base of diving platform, or permanently-anchored raft	22	25	27
B. Clearance in any direction to the diving platform	14	17	18
C. Horizontal limit of clearance measured from inside wall of the pool	This limit shall extend to the outer edge of the structures listed in Rows (A) and (B) above but not less than 10 feet.		

For SI: 1 foot = 304.8 mm.

Source: 2003 *International Residential Code*

A minimum 10-foot (3048 mm) clearance is required for overhead communications conductors, coaxial cables and their supports that are owned, operated and maintained by a utility. The separation must be maintained to pools, diving boards, observation stands and similar structures.

QUIZ
Study Session 18

1. Electrical nonmetallic tubing is an allowable wiring method in all but which one of the following applications? *(ENT)*
   a. feeders
   b. branch circuits
   c. embedded in masonry
   d. direct burial

2. Unless installed parallel to framing members, NM wiring run in an attic accessed by a portable ladder shall be protected from damage where located within _____ of the nearest edge of the attic entrance.
   a. 6 feet
   b. 7 feet
   c. 8 feet
   d. 10 feet

3. Type NM wiring shall be supported a maximum of _____ on center.
   a. 3 feet
   b. 4.5 feet
   c. 6 feet
   d. 12 feet

4. In general, direct burial cable installed below a driveway serving a dwelling unit shall have a minimum burial depth of _____.
   a. 6 inches
   b. 12 inches
   c. 18 inches
   d. 24 inches

5. Direct buried conductors and cables emerging from the ground shall be protected by enclosures or raceways extending to a minimum height of _____ above finished grade.
   a. 7 feet
   b. 8 feet
   c. 10 feet
   d. 12 feet

6. Receptacle outlets shall be installed so that all points along the floor line in any wall space are a maximum of _____ from an outlet in that space when measured horizontally.
   a. 4 feet
   b. 6 feet
   c. 8 feet
   d. 12 feet

7.   An unbroken wall space having a minimum width of _____ shall be considered for the purpose of determining receptacle outlet distribution.
       a. 12 inches
       b. 18 inches
       c. 24 inches
       d. 48 inches

8.   Floor receptacle outlets cannot be counted as part of the required number of receptacle outlets unless located a maximum of _____ from the wall.
       a. 4 inches
       b. 6 inches
       c. 12 inches
       d. 18 inches

9.   At a kitchen counter, countertop receptacle outlets shall be installed with a maximum horizontal distance of _____ from any point along the wall line to an outlet.
       a. 12 inches
       b. 18 inches
       c. 24 inches
       d. 36 inches

10.   In a bathroom, at least one receptacle outlet shall be located a maximum of _____ from the outside edge of each lavatory.
       a. 18 inches
       b. 24 inches
       c. 30 inches
       d. 36 inches

11.   A hallway a minimum of _____ in length shall be provided with at least one receptacle outlet.
       a. 6 feet
       b. 8 feet
       c. 10 feet
       d. 12 feet

12.   A 125-volt, single-phase, 20-ampere receptacle serving a countertop surface shall be provided with ground-fault circuit-interrupter protection where located a maximum of _____ of the outside edge of a wet bar sink.
       a. 2 feet
       b. 3 feet
       c. 4 feet
       d. 6 feet

13.   Where nonmetallic-sheathed cable is used, the cable assembly including the sheath shall extend
        into the box a minimum of _____ through a nonmetallic-sheathed cable knockout
        opening.
        a. $1/4$ inch
        b. $3/8$ inch
        c. $1/2$ inch
        d. $3/4$ inch

14.   Outlet boxes shall have a minimum depth of _____.
        a. $1/2$ inch
        b. $5/8$ inch
        c. $3/4$ inch
        d. 1 inch

15.   What is the maximum number of No. 12 conductors permitted for a 4-inch by $1^1/_2$-inch round
        box?
        a. 4
        b. 6
        c. 9
        d. 10

16.   In a wall constructed of wood or other combustible material, outlet boxes shall be installed so
        that the front edge of the box is set back a maximum of _____ from the finished surface.
        a. 0 inches; (the box must be flush or project)
        b. $1/8$ inch
        c. $1/4$ inch
        d. $3/8$ inch

17.   All switches shall be installed so that the center of the grip of the operating handle, in its highest
        position, shall be located a maximum of _____ above the floor or working platform.
        a. 6 feet, 0 inches
        b. 6 feet, 6 inches
        c. 6 feet, 7 inches
        d. 6 feet, 9 inches

18.   What is the required receptacle rating for a 20-ampere branch circuit supplying two or more
        receptacles or outlets?
        a. 15 amperes only
        b. 20 amperes only
        c. 30 amperes only
        d. 15 amperes or 20 amperes

19. Hanging luminaires, lighting track and pendants shall be located a minimum of _____ horizontally and _____ vertically from the top of a bathtub rim or shower stall threshold.
   a. 2 feet, 7 feet
   b. 2 feet, 8 feet
   c. 3 feet, 7 feet
   d. 3 feet, 8 feet

20. Where installed in a storage closet, a recessed fluorescent luminaire shall be located so as to provide a minimum clearance of _____ between the fixture and the nearest point of the storage space.
   a. 3 inches
   b. 6 inches
   c. 8 inches
   d. 12 inches

21. A luminaire that weighs a minimum of _____ or exceeds _____ in any dimension shall not be supported by the screw shell of a lampholder.
   a. 6 pounds, 14 inches
   b. 6 pounds, 16 inches
   c. 12 pounds, 14 inches
   d. 12 pounds, 16 inches

22. Ceiling-suspended paddle fans weighing a maximum of _____ are permitted to be supported by outlet boxes identified for such use.
   a. 18 pounds
   b. 25 pounds
   c. 32 pounds
   d. 35 pounds

23. All 125-volt receptacles located a maximum of _____ from the inside walls of a swimming pool shall be protected by a ground-fault circuit-interrupter.
   a. 10 feet
   b. 15 feet
   c. 20 feet
   d. 30 feet

24. Receptacles shall be located a minimum of _____ from the inside walls of indoor spas and hot tubs.
   a. 5 feet
   b. 6 feet
   c. 10 feet
   d. 12 feet

25. Where underground wiring is installed within 5 feet of the inside walls of a swimming pool, a minimum burial depth of _____ is required for intermediate metal conduit.
   a. 6 inches
   b. 12 inches
   c. 18 inches
   d. 30 inches

26. Where three or more conductors of minimum size _____ AWG are run at angles with joists in unfinished basements, additional protection is not required if the cable assembly is attached directly to the bottom of the joists.
   a. 6
   b. 8
   c. 10
   d. 12

27. Underground service conductors not encased in concrete and buried 18 inches or more below grade shall have their location identified by a warning ribbon placed a minimum of _____ above the underground installation.
   a. 4
   b. 6
   c. 8
   d. 12

28. All branch circuits that supply 125-volt, single-phase, 15- and 20-ampere outlets installed in dwelling unit _____ shall be protected by an arc-fault circuit interrupter.
   a. bathrooms
   b. bedrooms
   c. kitchens
   d. garages

29. A wall-mounted luminaire weighing a maximum of _____ pounds is permitted to be supported on boxes other than those specifically designed for luminaires, provided the luminaire is secured to the box with at least two No. 6 or larger screws.
   a. 6
   b. 10
   c. 14
   d. 18

30. Lighting track shall be located a minimum of _____ above the finished floor unless protected from physical damage or the track operates at less than 30 volts rms open-circuit voltage.
   a. 5 feet, 0 inches
   b. 6 feet, 0 inches
   c. 6 feet, 4 inches
   d. 6 feet, 8 inches

ANSWER KEY

Study Session 1

1.	c	Sec. R101.2
2.	b	Sec. R102.1
3.	c	Sec. R102.5
4.	a	Sec. R103.1
5.	d	Sec. R104.1
6.	b	Sec. R104.9.1
7.	b	Sec. R104.10
8.	a	Sec. R104.10.1
9.	c	Sec. R104.11.1
10.	d	Sec. R105.2
11.	b	Sec. R105.2
12.	b	Sec. R105.5
13.	a	Sec. R105.7
14.	a	Sec. R106.3.1
15.	b	Sec. R106.5
16.	d	Sec. R109.1
17.	d	Sec. R109.3
18.	d	Sec. R110.3
19.	d	Sec. R110.4
20.	b	Sec. R112.2
21.	c	Sec. R112.3
22.	b	Sec. R113.2
23.	c	Sec. R114.1
24.	d	Chapter 43
25.	a	Chapter 43
26.	d	Sec. R101.3
27.	d	Sec. R105.2
28.	d	Sec. R105.2
29.	a	Sec. R109.3
30.	b	Sec. R112.1

Study Session 2

1.	a	Figure R301.2(3)
2.	d	Figure R301.2(4)
3.	d	Figure R301.2(5), Note
4.	c	Figure R301.2(6)
5.	d	Figure R301.2(7)
6.	a	Table R301.2.1.2
7.	a	Table R301.2.1.3
8.	b	Sec. R301.2.1.4
9.	b	Table R301.2.2.1.1
10.	b	Sec. R301.2.2.3.1
11.	c	Sec. R301.2.2.2.1, #5
12.	d	Sec. R301.2.2.2.2, #4
13.	c	Sec. R301.2.3
14.	c	Table R301.5
15.	a	Table R301.4, Note b
16.	c	Table R301.4
17.	c	Table R301.5
18.	d	Table R301.6
19.	b	Table R301.6
20.	a	Sec. R302.1
21.	c	Sec. R302.1
22.	d	Sec. R302.1, R105.2
23.	d	Sec. R302.2
24.	d	Sec. R302.3, R317.3.1.2
25.	a	Sec. R302.1, Exception
26.	c	Sec. R301.2.1.4, #3
27.	b	Sec. R301.2.2, Exception
28.	b	Sec. R301.2.2.2.1
29.	a	Sec. R301.3, #2
30.	a	Table R301.5, Note f

ANSWER KEY

Study Session 3

1.	c	Sec. R303.1
2.	a	Sec. R303.1
3.	a	Sec. R303.1, Exception 1
4.	d	Sec. R303.3, Exception
5.	b	Sec. R304.1
6.	d	Sec. R304.4
7.	b	Sec. R304.3
8.	b	Sec. R305.1
9.	a	Sec. R305.1, Exception 2
10.	a	Figure R307.2
11.	a	Figure R307.2
12.	b	Sec. R307.2
13.	b	Sec. R308.4, #6
14.	c	Sec. R308.4, #9
15.	a	Sec. R308.6.1
16.	b	Sec. R308.6.8
17.	b	Sec. R309.1
18.	c	Sec. R309.1.1
19.	a	Sec. R309.2
20.	c	Sec. R310.1
21.	c	Sec. R310.1.1, Exception
22.	a	Sec. R310.1.2, R310.1.3
23.	d	Sec. R310.2
24.	d	Sec. R309.2
25.	c	Sec. R308.6.2, #1
26.	c	Sec. R303.4.1
27	a	Sec. R303.6
28.	c	Sec. R305.1, #4
29.	c	Sec. R308.4, Exception 4
30.	a	Sec. R308.4, Exception 1

Study Session 4

1.	b	Sec. R311.3
2.	c	Sec. R311.4.3
3.	b	Sec. R311.6.1
4.	c	Sec. R3116.3
5.	b	Sec. R311.6.2
6.	c	Sec. R311.5.1
7.	a	Sec. R311.5.1
8.	c	Sec. R311.5.3.1, R311.5.3.2
9.	b	Sec. R311.5.3.1
10.	c	Sec. R311.5.3.3, Exception 1
11.	b	Sec. R311.5.3.3
12.	b	Sec. R311.5.2
13.	d	Sec. R311.5.8.1
14.	c	Sec. R311.5.6.1
15.	b	Sec. R311.5.6.2
16.	a	Sec. R311.5.6.3, #1
17.	c	Sec. R312.1
18.	b	Sec. R312.2
19.	b	Sec. R313.1
20.	b	Sec. R314.1.2
21.	c	Sec. R315.1
22.	c	Sec. R317.1
23.	b	Sec. R317.2, Exception
24.	b	Sec. R319.1, #2
25.	d	Sec. R323.1.3
26.	a	Sec. R311.2.2
27.	c	Sec. R314.2.7, #2
28.	d	Sec. R311.5.6.3, #2
29.	c	Sec. R311.5.6
30.	d	Sec. R311.4.2

ANSWER KEY

Study Session 5

1.	b	Sec. R401.3
2.	b	Table R401.4.1
3.	b	Table R402.2
4.	c	Table R402.2, Note d
5.	c	Table R403.1
6.	b	Sec. R403.1.1
7.	b	Sec. R403.1.1
8.	b	Figure R403.1(2)
9.	a	Sec. R403.1.3.1
10.	d	Sec. R403.1.4
11.	b	Sec. R403.1.5
12.	c	Sec. R403.1.6
13.	a	Sec. R403.1.6
14.	d	Sec. R403.1.7.3
15.	b	Sec. R403.3
16.	c	Table R403.3
17.	d	Table R404.1.1(1)
18.	a	Table R404.1.1(2)
19.	a	Sec. R404.1.6
20.	b	Sec. R404.1.7, Exception
21.	d	Sec. R404.2.2
22.	b	Sec. R404.2.3
23.	d	Sec. R405.1
24.	c	Sec. R408.1
25.	b	Sec. R408.3
26.	c	Sec. R403.1.4.1, Exception
27.	c	Sec. R403.2
28.	a	Sec. R403.3.2
29.	d	Sec. R406.2
30.	b	Sec. R408.2

Study Session 6

1.	c	Sec. R502.3.1
2.	b	Table R502.3.1(1)
3.	a	Table R502.3.1(2)
4.	d	Table R502.5(1)
5.	b	Table R502.5(1)
6.	b	Table R502.5(2)
7.	c	Sec. R502.4
8.	b	Sec. R502.6
9.	a	Sec. R502.6.1
10.	c	Sec. R502.7.1
11.	c	Sec. R502.8.1
12.	d	Sec. R502.8.1
13.	d	Sec. R502.10
14.	c	Sec. R502.12
15.	b	Sec. R502.12.1
16.	d	Table R503.2.1.1(1), Note h
17.	b	Table R503.2.1.1(2)
18.	a	Sec. R505.1.1
19.	c	Sec. R505.2.4
20.	a	Table R505.3.1(2)
21.	c	Table R505.3.2
22.	b	Sec. R506.1
23.	c	Sec. R506.2.1
24.	a	Sec. R506.2.2, Exception
25.	c	Sec. R506.2.3
26.	a	Table R502.3.3(1)
27.	c	Table R502.3.3(2)
28.	d	Sec. R502.4
29.	b	Sec. R502.10
30.	b	Sec. R506.2.2

ANSWER KEY

Study Session 7

1.   c   Table R602.3(5)
2.   b   Table R602.3.1
3.   a   Sec. R602.3.2
4.   b   Sec. R602.3.2, Exception
5.   b   Table R602.3(3)
6.   c   Sec. R602.6
7.   c   Sec. R602.6
8.   d   Sec. R602.6, Exception 1
9.   d   Table R602.7.2
10.   c   Sec. R602.8.1
11.   a   Sec. R602.8.1.1
12.   c   Sec. R602.9
13.   b   Table R602.10.1
14.   d   Sec. R602.10.1
15.   b   Sec. R603.1.2
16.   d   Sec. R603.2
17.   b   Sec. R606.7
18.   a   Figure R606.10(1)
19.   d   Table R702.3.5
20.   b   Sec. R702.3.6
21.   a   Sec. R702.4.2
22.   d   Table R703.4
23.   a   Sec. R703.7.4.1, Exception
24.   c   Sec. R703.9
25.   c   Table R703.7.3
26.   a   Sec. R602.3.1, Exception 1
27.   d   Table R602.3(1)
28.   d   Table R602.3(2)
29.   c   Sec. R602.6.1
30.   b   Sec. R703.6.2.1

Study Session 8

1.   b   Sec. R801.3
2.   b   Sec. R802.3
3.   b   Sec. R802.3.1
4.   b   Sec. R802.3.2
5.   c   Table R802.4(1)
6.   a   Table R802.5.1(4)
7.   a   Sec. R802.5.1
8.   a   Sec. R802.6
9.   c   Sec. R802.7.1
10.   a   Sec. R802.9
11.   c   Sec. R803.2.1, Table R503.2.1.1(1)
12.   b   Tables R804.3.3(2), R804.3.3(1)
13.   c   Sec. R804.3.3.2
14.   a   Sec. R806.1
15.   c   Sec. R807.1
16.   c   Sec. R808.1
17.   c   Sec. R902.1
18.   d   Sec. R905.2.6
19.   d   Sec. R905.2.7
20.   a   Sec. R905.2.8.3
21.   c   Sec. R905.3.6
22.   d   Sec. R905.7.5
23.   b   Table R905.7.5
24.   c   Sec. R905.8.6
25.   c   Sec. R905.8.8
26.   d   Sec. R802.7.1
27.   b   Sec. R802.9
28.   b   Sec. R806.3
29.   a   Sec. R807.1
30.   c   Sec. R903.4.1

ANSWER KEY

Study Session 9

1.	d	Sec. R1001.1.1
2.	a	Sec. R1001.3
3.	d	Sec. R1001.6
4.	c	Sec. R1001.9
5.	b	Sec. R1001.10
6.	a	Sec. R1001.14
7.	c	Sec. R1001.15
8.	b	Sec. R1003.2
9.	d	Sec. R1003.3.1
10.	c	Sec. R1003.5
11.	b	Sec. R1003.5
12.	c	Sec. R1003.7
13.	d	Sec. R1003.9.1
14.	d	Sec. R1003.10
15.	b	Sec. R1003.9.2, Exception
16.	d	Sec. R1003.11
17.	d	Sec. R1003.12
18.	c	Sec. R1003.12
19.	d	Sec. R1004.1
20.	b	Sec. R1005.4
21.	b	Tables N1101.2, N1102.1
22.	c	Table N1102.1
23.	a	Sec. N1102.1.3
24.	a	Sec. N1102.1.7
25.	a	Sec. N1102.1.9
26.	b	Sec. R1001.2
27.	d	Sec. R1001.6.1
28.	c	Table R1001.17
29.	c	Sec. R1003.6
30.	b	Sec. R1006.2

Study Session 10

1.	b	Sec. M1305.1
2.	b	Sec. M1305.1.1
3.	a	Sec. M1305.1.1
4.	a	Sec. M1305.1.2
5.	a	Sec. M1305.1.3
6.	c	Sec. M1305.1.4.1
7.	c	Sec. M1305.1.4.2
8.	b	Sec. M1306.2
9.	b	Table M1306.2
10.	d	Sec. M1306.2.1, Table M1306.2
11.	b	Sec. M1307.3
12.	b	Sec. M1402.3
13.	a	Sec. M1403.1
14.	b	Sec. M1403.2
15.	c	Sec. M1406.5
16.	d	Sec. M1407.3
17.	c	Sec. M1408.3, #3
18.	a	Sec. M1408.4
19.	a	Sec. M1408.5, #3
20.	b	Sec. M1409.2, #2
21.	c	Sec. M1411.2
22.	b	Sec. M1411.3.1
23.	b	Sec. M1411.3.1, #1
24.	a	Sec. M1413.1
25.	c	Sec. M1414.21.
26.	a	Sec. M1305.1.4
27.	b	Sec. M1307.2
28.	c	Sec. M1307.4
29.	d	Sec. M1308.2
30.	b	Sec. M1410.2

ANSWER KEY

Study Session 11

1.	b	Sec. M1501.1
2.	c	Sec. M1501.3
3.	d	Sec. M1501.3
4.	d	Sec. M1502.1
5.	a	Sec. M1502.2, Exception, #4
6.	b	Sec. M1504.1
7.	a	Sec. M1504.1
8.	d	Sec. M1601.1.1, #1
9.	c	Sec. M1601.1.1, #2
10.	b	Table M1601.1.1(2)
11.	a	Sec. M1601.1.1, #5
12.	c	Sec. M1601.1.1, #6
13.	b	Sec. M1601.1.2
14.	b	Sec. M1601.1.2
15.	a	Sec. M1601.2.1, #1
16.	a	Sec. M1601.2.1, #3
17.	b	Sec. M1601.2.2
18.	c	Sec. M1601.3.2
19.	d	Sec. M1602.4.2
20.	c	Sec. M1601.4.3
21.	d	Sec. M1602.2, #1
22.	a	Sec. M1602.2, #4
23.	a	Sec. M1602.3
24.	b	Sec. M1901.1
25.	d	Sec. M1902.4
26.	b	Table M1506.3
27.	c	Table M1506.3
28.	a	Sec. M1601.4.4
29.	b	Sec. M1601.4.5
30.	c	Sec. M1902.4

Study Session 12

1.	d	Sec. M1701.1.1
2.	c	Sec. M1701.4
3.	d	Sec. M1701.5
4.	a	Sec. M1701.5
5.	d	Sec. M1702.2
6.	c	Sec. M1702.2
7.	a	Sec. M1702.2
8.	d	Sec. M1703.2.1
9.	b	Sec. M1703.2.1
10.	a	Sec. M1703.2.1
11.	b	Sec. M1703.5
12.	c	Sec. M1803.3.1
13.	a	Sec. M1804.2.6, #5
14.	c	Sec. M1703.3, #2
15.	c	Sec. M1703.4
16.	b	Table M1803.2
17.	b	Sec. M1803.3
18.	c	Sec. M1803.3.2
19.	c	Table M1803.3.4
20.	d	Sec. M1804.2.3
21.	a	Sec. M1805.3
22.	c	Sec. M2003.1.1
23.	d	Sec. M2103.3
24.	b	Sec. M2203.3
25.	c	Sec. M2301.2.9
26.	d	Sec. M2005.2
27.	a	Table M2101.9
28.	d	Sec. M2101.10
29.	b	Sec. M2201.2
30.	a	Sec. M2204.3

ANSWER KEY

Study Session 13

1.	d	Sec. G2406.2
2.	c	Sec. G2408.2
3.	b	Sec. G2408.3
4.	d	Table G2409.2
5.	b	Table G2409.2
6.	a	Sec. G2412.6
7.	a	Sec. G2413.6
8.	d	Sec. G2413.6.1
9.	c	Sec. G2414.5.2
10.	b	Table G2414.9.2
11.	d	Sec. G2415.1
12.	b	Sec. G2415.5
13.	b	Sec. G2415.9
14.	a	Sec. G2415.13
15.	c	Sec. G2415.14.3
16.	a	Sec. G2416.2, #5
17.	b	Sec. G2417.4.1
18.	a	Sec. G2417.4.2
19.	d	Sec. G2419.1
20.	c	Sec. G2420.5
21.	a	Sec. G2422.1.2
22.	c	Table G2424.1
23.	a	Sec. G2425.9
24.	a	Sec. G2440.7
25.	b	Sec. G2445.3
26.	c	Sec. G2408.4
27.	c	Sec. G2415.7
28.	d	Sec. G2417.2
29.	c	Table G2424.1
30.	a	Sec. G2450.3

Study Session 14

1.	d	Sec. P2503.4
2.	b	Sec. P2503.5.1, #2
3.	d	Sec. P2503.6
4.	d	Sec. P2603.2.1
5.	c	Sec. P2603.2.1
6.	d	Sec. P2603.5
7.	b	Sec. P2603.6
8.	c	Sec. P2603.6
9.	d	Sec. P2604.1
10.	b	Sec. P2604.3
11.	d	Table P2605.1
12.	a	Table P2605.1
13.	c	Sec. P2703.1
14.	a	Sec. P2704.1
15.	b	Sec. P2706.2
16.	b	Sec. P2706.2.1
17.	a	Sec. P2708.1
18.	d	Sec. P2709.1
19.	b	Sec. P2711.3
20.	b	Sec. P2712.4
21.	c	Sec. P2713.1
22.	b	Sec. P2714.1
23.	b	Sec. P2716.1
24.	c	Sec. P2719.1
25.	a	Sec. P2723.2
26.	b	Sec. P2503.5.2, #2.1
27.	a	Sec. P2503.8
28.	b	Sec. P2604.4
29.	d	Table P2608.4
30.	b	Sec. P2705.1, #5

ANSWER KEY

Study Session 15

1.	b	Sec. P2801.5.1
2.	d	Sec. P2801.5.2
3.	c	Sec. P2801.6
4.	d	Sec. P2803.3
5.	d	Sec. P2803.4
6.	b	Table P2902.2.1
7.	a	Sec. P2902.3.2
8.	c	Table P2903.1
9.	b	Table P2903.2
10.	d	Table P2903.2, Note a
11.	a	Sec. P2903.3, P2903.3.1
12.	c	Table P2903.6
13.	d	Table P2903.7
14.	d	Sec. P2904.2
15.	d	Sec. P2904.3
16.	b	Sec. P2904.5
17.	a	Sec. P2905.1
18.	a	Table P3004.1
19.	d	Sec. P3005.2.2
20.	b	Table P3005.2.9
21.	b	Sec. P3005.2.4
22.	a	Sec. P3005.2.5
23.	a	Sec. P3005.3
24.	a	Sec. P3005.4.1, #1
25.	d	Table P3005.4.1
26.	d	Sec. P2803.6.1
27.	c	Table P2902.2
28.	c	Sec. P2904.4
29.	a	Sec. P3005.1.1, Exception
30.	a	Sec. P3007.1, Exception

Study Session 16

1.	c	Sec. P3102.1
2.	c	Sec. P3102.2
3.	d	Sec. P3102.3
4.	c	Sec. P3103.1
5.	b	Sec. P3103.5
6.	d	Sec. P3104.4
7.	d	Sec. P3104.5
8.	c	Table P3105.1
9.	b	Sec. P3105.2
10.	d	Sec. P3105.4
11.	b	Sec. P3107.1
12.	b	Table P3107.3
13.	c	Table P3108.3
14.	c	Table P3109.4
15.	d	Sec. P3110.1
16.	a	Sec. P3111.1
17.	c	Sec. P3111.2.1
18.	b	Table P3111.3
19.	b	Sec. P3113.1
20.	c	Sec. P3114.4
21.	a	Sec. P3201.2
22.	d	Sec. P3201.6
23.	c	Sec. P3201.6, #2
24.	c	Table P3201.7
25.	c	Table P3201.7
26.	d	Sec. P3103.6
27.	c	Sec. P3110.3
28.	d	Sec. P3111.2
29.	d	Sec. P3112.2
30.	b	Sec. P3114.4

ANSWER KEY

Study Session 17

1.	a	Sec. E3304.8
2.	b	Sec. E3305.2
3.	a	Sec. E3305.2
4.	c	Sec. E3305.3
5.	b	Sec. E3306.3
6.	b	Sec. E3306.10.3
7.	c	Sec. E3307.1
8.	a	Sec. E3307.2
9.	d	Sec. E3307.3
10.	b	Table E3503.1
11.	a	Sec. E3504.1
12.	d	Sec. E3504.2.1
13.	c	Sec. E3504.2.2, #2
14.	b	Sec. E3504.3
15.	b	Sec. E3508.1.1
16.	b	Sec. E3508.2
17.	c	Sec. E3508.2.1
18.	c	Sec. E3510.1
19.	a	Sec. E3602.3
20.	d	Sec. E3603.2
21.	b	Sec. E3603.3, E3603.4
22.	c	Sec. E3604.2
23.	b	Sec. E3604.5
24.	b	Table E3605.5.3
25.	c	Sec. E3605.6
26.	b	Sec. E3304.5
27.	a	Sec. E3305.4
28.	b	Sec. E3306.10.3
29.	d	Sec. E3501.6.2
30.	b	Sec. E3508.1.1

Study Session 18

1.	d	Table E3701.4
2.	a	Sec. E3702.2.1
3.	b	Table E3702.1
4.	c	Table E3703.1
5.	b	Sec. E3703.4
6.	b	Sec. E3801.2.1
7.	c	Sec. E3801.2.2, #1
8.	d	Sec. E3801.2.3
9.	c	Sec. E3801.4.1
10.	d	Sec. E3801.6
11.	c	Sec. E3801.10
12.	d	Sec. E3802.7
13.	a	Sec. E3805.3.1
14.	a	Sec. E3805.4
15.	b	Table E3805.12.1
16.	a	Sec. E3806.5
17.	c	Sec. E3901.6
18.	d	Table E3902.1.2
19.	d	Sec. E3903.10
20.	b	Sec. E3903.11, #4
21.	b	Sec. E3904.4
22.	d	Sec. E4001.6
23.	c	Sec. E4103.1.3
24.	a	Sec. E4103.1.4
25.	a	Table E4103.6
26.	b	Sec. E3702.4
27.	d	Sec. E3703.2
28.	b	Sec. E3802.11
29.	a	Sec. E3805.6
30.	a	Sec. E3905.4, #8